T0206192

Plumbing Principles and Practice

This book provides a complete introduction to plumbing services. It explains the principles and provides practical examples of the planning, design, installation and maintenance of the plumbing technologies applicable to single-storey buildings, skyscrapers and everything in between. The book begins with an introduction to plumbing technology, the trade and its evolution. Chapters then cover:

- Pipes, fittings and accessories and their installation and testing
- Pumps and pumping systems
- Hydraulic principles
- Hot and cold water supply systems
- Fixtures and appliances
- Sanitary and storm drainage systems
- Special concerns such as seismic issues, safety, security and the state of the art.

Written and the figures drawn by a registered professional engineer and experienced teacher, this book is suitable for use on a wide range of courses from building services engineering, civil engineering, construction technology, plumbing services, environmental engineering, water engineering and architectural technology.

Syed Azizul Haq, a patent holder of the "mechanical seal trap" and author of *Harvesting Rainwater from Buildings* published by Springer International Publishing, is president of the RAiN Forum – a non-profit voluntary platform for knowledge sharing and promotion of rainwater among scientist, engineers, architects, planners, academicians, practitioners, administrators, students and others who are directly or indirectly concerned about sustainable water management – and, following 35 years with the Public Works Department in Bangladesh, also serves as a part-time faculty member at a number of universities.

Plumbing Principles and Practice

Second edition

Syed Azizul Haq

LONDON AND NEW YORK

Second edition published 2022
by Routledge
2 Park Square, Milton Park, Abingdon, Oxon, OX14 4RN

and by Routledge
605 Third Avenue, New York, NY 10158

Routledge is an imprint of the Taylor & Francis Group, an informa business

First edition published by Syeda Masuda Khatoon, 2006

British Library Cataloguing-in-Publication Data
A catalogue record for this book is available from the British Library

Library of Congress Cataloging-in-Publication Data
A catalog record for this book has been requested

ISBN: 978-1-032-00004-6 (hbk)
ISBN: 978-0-367-76301-5 (pbk)
ISBN: 978-1-003-17223-9 (ebk)

DOI: 10.1201/9781003172239

Typeset in Times New Roman
by Apex CoVantage, LLC

To
my
Father-in-law: Dr. Muhammad Habibullah
and
Mother-in-law: Begum Saleha Khatun

Contents

Foreword 1

I am honoured and delighted to have been asked by Syed Azizul Haq to write this foreword to *Plumbing Principles and Practice*. Haq's earlier book, *Harvesting Rainwater from Buildings*, was published by Springer International AG and focused on plumbing associated with the harvesting of rainwater. In addition, he is a co-author of the Bangladesh National Building Code. I have known Syed for many years; he is a committed professional and, apart from pursuing his work with passion, has contributed enormously to the learned society activities of various forums. He is currently president of the RAiN Forum in Bangladesh. He also belongs to many other national and international bodies associated with his field of interest and teaches part-time at a number of academic institutions, including the University of Asia Pacific. His interests are by no means confined to writing and participation in learned society activities. He actively undertakes research and experiments to improve things and has been granted a patent by the Government of Bangladesh for a mechanical seal trap.

Syed Azizul Haq graduated with a bachelor's degree in civil engineering from the Bangladesh University of Engineering and Technology in 1980, followed by master's degree in environmental engineering from the same university in 1989. He was registered as PEng by the Bangladesh Professional Engineers' Registration Board in 2005. He joined the Public Works Department (PWD) in 1981 and retired as from a position of additional chief engineer in 2016. This book is based on Haq's own practical experience, derived from a long and distinguished career with PWD as well as other assignments he has undertaken. It is rare in Bangladesh, as elsewhere in South Asia, for practicing engineers to write books, as this is considered to be the preserve of university teachers. The book will be of immense interest and value to both practicing engineers and academics in the developed and developing world.

As the name suggests, the book deals with plumbing principles and practice. Across 13 chapters, the author covers every conceivable aspect of plumbing design and beyond starting with a fascinating introduction that explains early water supply and drainage systems, dating back to 3000 BC. There is even a chapter on building sewer systems, which generally would be considered outside the scope of this book, but its inclusion here is helpful as it gives designer a complete picture of what happens to the sewerage and storm water generated in a building. A chapter on special concerns deals with such important matters as economy, noise, safety, durability and seismic performance, which are all essential considerations for producing efficient and sustainable solutions.

The other ten chapters are well structured and well presented, starting with engineering knowledge, followed by application of engineering principles by reference to design codes and where appropriate by illustration of worked examples. The generous provision of supporting diagrams makes it easier for readers to understand the concepts of plumbing principles. Topics are by no means limited to those that are relevant to the developing world; the book also covers the requirements of the developed world, and consequently the book should have global appeal. The book is, in effect, a complete design manual which all practicing experts will find immensely useful.

Infrastructure is one of, if not the biggest contributor of carbon emissions; it is desirable that we develop and design systems that are efficient not only at the installation stage but also throughout their life cycles. This book contains all the knowledge required to produce properly engineered, efficient and sustainable plumbing solutions for all types of buildings. The book will also be useful to policy makers, infrastructure planners, university teachers and technical institutions and students.

Finally, I congratulate Syed Azizul Haq for his painstaking work in writing this book, which I am sure will be of enormous benefit to us all.

Iftikhar A. Khan OBE,
Solihull, United Kingdom

Foreword 2

Plumbing is an essential part of our daily lives.

However, it is an invisible subject, not easily understood.

Professional plumbers are needed everywhere. Yet, the supply is inadequate when it comes to both the number of plumbers and the professionalism of plumbers.

Syed Azizul Haq is a passionate man who brings professionalism to plumbing. He is a member of the World Toilet Organization. He introduced a formal design of plumbing systems and has continued to improve plumbing systems design for various prestigious projects implemented by the Public Works Department since 1995, where he served. He has also played an active role in professional development activities in the plumbing, water supply and drainage sector in Bangladesh.

I congratulate him for writing this book, which will upgrade the standards of plumbing and hopefully increase the supply of professionally trained plumbers.

Jack Sim

Founder: BoP HUB; World Toilet Organization, Singapore

Preface

It is more than a decade since the book *Plumbing Practices* was published in April 2006. By now, many developments have taken place in the field of plumbing; new technologies have evolved and a variety of modern plumbing items have come up in the market, in both the country and other parts of the globe. Incorporation of these new developments, and introduction of new items and technologies, in a book has become a need of the times. *Plumbing Practices*, as the first attempt to write a self-published technical book undoubtedly could not satisfy the multidimensional queries of its valued readers. Furthermore, many errors, minor to major, and printing mistakes, along with other obvious issues, appeared in the book, which is not desirable at all and therefore needed to be rectified. Above all, many pertinent topics of plumbing are missing, and thus they have remained unnoticed in the book. Regardless, a good number of readers appreciated the book, and many institutions have referred to the book in support of their respective curricula.

To me, the first edition of the book was, to some extent, in high enough demand that I was encouraged to attempt a second edition. Furthermore, rectification of the errors and mistakes found in the book made me realize that updating the preceding book, as a second edition, was an obligatory task to perform. Publication of my second book, *Harvesting Rainwater from Buildings*, by Springer International Publishing boosted my confidence in this regard and motivated me to query other exotic publishers for publishing the second edition of *Plumbing Practices*. At this time, I also proposed that if the publishers didn't want to publish the proposed book as a second edition, I would agree to publish it under a new title; and so, this book is the outcome of those intentions and efforts, with a new title proposed by the publisher.

In this effort, I have taken the opportunity to provide new plumbing elements, technologies and practices that were not provided in the first one, or were not well illustrated or described. So, this book includes those elaborations for better understanding that were most demanded, as well as necessary corrections for the authenticity of previous descriptions. Since the publication of the first book, much advice and suggestions have been received from peers, well-wishers and readers, with a view to improving and updating the book; this feedback was highly valued, and I have tried to incorporate these into this book.

As this book is the outcome of readers experiencing *the previous one*, the chapters are kept almost the same, with the same titles and in chronology as in the previous one, focusing on respective subjects of discussion; the exception is Chapter 11, which has purposefully been thematically changed. In the previous book, the last chapter was "Rainwater Management", in which basic concepts of rainwater harvesting and rainwater drainage were described. But in this book, this chapter focuses exclusively on rainwater drainage, and so the chapter is named "Rainwater Drainage". The reason for is that readers interested in rainwater harvesting can now go through the writer's second book on rainwater harvesting, as mentioned earlier, in which they will find almost all the information related to rainwater harvesting in any type of building elaborately detailed. In Chapter 12, building sewer systems is discussed separately and exclusively. In addition, a new chapter, on special but very pertinent plumbing concerns, is included to give readers a holistic understanding of plumbing principles and practices. The first chapter has been rewritten, after accepting the title of the book as *Plumbing Principles and Practice*, especially to focus on principles of plumbing. Last, but not the least, all the artworks are drawn by me to bring a sort of uniqueness and harmony in the figures.

The book is bound together by 13 chapters, as introduced below.

Chapter 1 – ***Introduction.*** An exclusive introduction of plumbing technology is given first, and its history of development is considered. The scope of plumbing along with its objectives also discussed. Most importantly, the principles of plumbing are thoughtfully identified. The academic qualifications, experience and skill of plumbing practitioners, including professionals, plumbers, trade groups and related business, are discussed and their respective roles in this field identified.

Chapter 2 – ***Hydraulics in Plumbing.*** Fluid flow is subject to hydraulic principles and plumbing is subject to fluid flow. So, in this chapter related aspects of hydraulics, very important and related to plumbing, are discussed. Pressure, friction loss etc., as related to the fluid flow through a pipe, are the main subjects of discussion in this chapter.

Chapter 3 – ***Pipe and Fittings.*** Pipes are the main and major element of plumbing, and fittings are their important items. So, the properties and characteristics of all sorts of piping and fittings, mostly used in plumbing, are discussed elaborately in this chapter with respect to material, strength etc.

Chapter 4 – ***Piping Installation and Testing.*** Jointing of pipes and fittings and the installation of piping systems are the basic and major jobs of plumbing. So, in this chapter, all the practices of joining pipes and fittings, their installation and testing of jobs, are elaborately illustrated and described as well.

Chapter 5 – ***Pump and Pumping System.*** Pumps, and especially centrifugal pumps, are major and very important machines used in plumbing, particularly in water supply systems. So, various types of centrifugal pumps used in water supply and drainage and their characteristics, usefulness, capacity sizing, installation, maintenance etc. are discussed in this chapter.

Chapter 6 – ***Plumbing Accessories and Appurtenances.*** Various accessories and appurtenances are indispensable items in plumbing, to some extent, used for handling controlled flow of fluids and developing better functioning systems. So, common and widely used accessories and appurtenances, such as valves, faucets, meters and water tanks, are illustrated and discussed in this chapter.

Chapter 7 – ***Water Supply Systems.*** Water supply is the primary and major purpose of plumbing. So, this chapter illustrates and discusses the planning, design, installation and maintenance aspects of water supply systems. Special considerations around developing water supply systems for skyscrapers are also discussed.

Chapter 8 – ***Hot Water Supply Systems.*** Hot water supply is another important branch of water supply systems in buildings and has some special requirements other than required in normal water supply. This chapter is exclusively dedicated to illustrating and describing various aspects of hot water supply systems in buildings of all sizes, including skyscrapers.

Chapter 9 – *Plumbing Fixtures and Appliances.* Plumbing fixtures are very important elements in plumbing, particularly for sanitary drainage systems. Plumbing fixtures are available in different shapes, styles, materials and sizes, each with different implication in usage. All the pros and cons of various natures and different types of fixtures and their installation, along with trims, are illustrated and discussed in this chapter. Some special fixtures and appliances, such as emergency fixtures, various interceptors etc., are also considered.

Chapter 10 – *Sanitary Drainage Systems.* Sanitary drainage is another major job of plumbing. So, in this chapter the planning, design, installation and maintenance aspects of all types of sanitary drainage along with venting systems are illustrated and described. Special considerations for developing sanitary drainage systems for skyscrapers are also discussed.

Chapter 11 – *Rainwater Drainage Systems.* Rainwater drainage is also another vital job of plumbing. So, in this chapter all of the planning, design, installation and maintenance aspects of rainwater drainage systems are illustrated and described. Special rainwater drainage designed for buildings with very large footprints are also discussed.

Chapter 12 – *Building Sewer Systems.* Some books on plumbing do not address building sewer systems as being a subject of the outside of the building. This chapter is dedicated to the subjects describing and illustrating all aspects of building sewer systems, including development of surface drainage systems and pumping drainage. Details of public sewer connections are also delineated.

Chapter 13 – *Special Concerns.* In all chapters, the very common and most desired plumbing concerns, such as safety, economy and durability, are discussed where needed. In addition to these concerns, there are various other concerns, such as sound, seismicity, security etc., that are equally important but poorly addressed. This chapter discusses all of these special plumbing concerns to help develop state-of-art plumbing.

Finally, I would like to say that I have tried my level best in providing the book considering global readers' demand for the subject. So, this book, in my belief, will benefit undergraduate and postgraduate students of civil and mechanical engineering and architecture, as well as professionals, practitioners and policy makers involved in the activities concerning plumbing. Due to my inability to include everything, I sincerely confess that this book might not fulfil the ever-increasing thirst and quest for knowledge in this concern. So, I humbly request that my readers not hesitate to put their valuable comments or suggestions in this regard to the writer.

Syed Azizul Haq, PEng

Acknowledgements

At the onset, I wholeheartedly acknowledge the blessing of Almighty Allah (SWT), the most gracious and the most merciful, who blessed me with all sorts of knowledge and other needs to accomplish this book.

Next, I humbly acknowledge and remember the continuous encouragement of my mother, who used to pray to Almighty Allah (SWT), for granting me the ability to accomplish benevolent deeds, and my father, who gave all-out efforts for my education.

My special thanks and gratitude to Mr. Iftikhar A. Khan, OBE, and Mr. Jack Sim, who enlightened the book and put me in their debt by providing their invaluable forewords.

It is my great pleasure to thank my peers, students and trainees for their continuous encouragement to pursue writing a new, updated version of the book. Readers of the book are, of course, very welcome to communicate with the author by e-mail to azizulhaqsyed@yahoo.com, so that the underlined corrections, comments and suggestions can be incorporated into a future edition.

Last, but not least, I apologize to my family members and relatives to whom I could not give due time and attention, being engaged in this project, and express my inability to mention the names of all concerned, who encouraged me with complimentary words and helped me in various ways in the long journey towards publishing the book.

Syed Azizul Haq, PEng

Abbreviations

°C	Degree centigrade
AAV	Air admittance valve
ABS	Acrylonitrile Butadiene Styrene
AC	Alternating current
AD	Anno Domini
BC	Before Christ
BHP	Brake horse power
BWG	Birmingham Wire Gauge
CI	Cast iron
cm	Centimetre
CPVC	Chlorinated polyvinyl chloride
cSt	Centistokes
cum	Cubic meter
DC	Direct current
DOL	Direct on-line
DU	Dwelling unit
FHR	First hour rating
FLA	Full load amps
FS	Float switch
GI	Galvanized iron
HOA	Hands-off auto
HP	Horse power
hr	Hour
J/kg	Joule per kilogram
K	Kelvin
kg	Kilogram
kJ	Kilo Joule
Kj/h	Kilo Joule per hour
kJ/s	Kilo Joule per second
kN	Kilo Newton
kPa	Kilo Pascal
kW	Kilo watt
lpcd	Litre per capita per day
lpf	Litre per flush
LPG	Liquefied petroleum gas

lph	Litre per hour
lpm	Litre per minute
Lps	Litre per second
m	Meter
mm	Millimetre
mPa	Mega Pascal
mps	Metres per second
N	Newton
NPSH	Net positive suction head
NPSHA	Net positive suction head available
NPSHR	Net positive suction head required
NTU	Nephelometric Turbidity Units
OH	Overhead
Pa	Pascale
PAPA	Positive air pressure attenuator
PB	Polybutene
PEng	Professional engineer
pH	Potential of hydrogen
ppb	Parts per billion
ppm	Parts per million
PVC	Polyvinyl chloride
PVD	Physical vapor deposition
RCC	Reinforced cement concrete
RDP	Rainwater down pipe
RPM	Revolutions per minute
s	Second
SDFU	Sanitary drainage fixture unit
SF	Service factor
Sqcm	Square centimetre
Sqmm	Square millimetre
St	Stokes
SWT	Subhanu wa ta-ala
TPDT	Triple Pole Double Throw
UG	Underground
uPVC	Unplasticized polyvinyl chloride
U/L	Underwriters label
UV	Ultra Violet
VAC	Volts alternating current
VWLP	Vertical water lifting pipe
W	Watt
W/m°C	Watt per meter per degree Centigrade
W/mK	Watt per meter per Kelvin
WSFU	Water supply fixture units
wt	Weight

Figures

Tables

1 Introduction

1.1 Introduction

Plumbing as a technology falls under the purview of building services, which helps an occupant to use their building for the purposes for which it has been developed. Plumbing may be defined as art, business or work involved in the design, installation and maintenance of pipes, fixtures, equipment and accessories that convey fluids like water, or gas and waste carried by water, within a building and around its premises. A plumber is a person engaged in the art and work of plumbing. The words "plumbing" and "plumber" are derived from the Latin word "Plumbum", or lead, denoted by the symbol "Pb" [1].

1.2 History of plumbing

The Indus Valley civilization extended from North-East Afghanistan to North-West India, and archaeological sites in Mohenjo-daro and Harappa reveal that these sites once held buildings of two or more storeys in which rainwater from the roof and wastewater from upper storey bathrooms and toilets were carried downward through enclosed terracotta pipes or open chutes, and finally disposed of in nearby street drains [2]. Even as early as 3000 BC, there are cities that show evidence of a well-integrated public water supply system and well-planned drainage system; this is thought to be the world's first urban sanitation system, in which individual homes or groups of homes obtained water from wells and wastewater was directed to covered drains in major streets [3]. History also reveals that in Athens, Rome, there were water carriage drainage systems. In Rome, there was a large drain known as "Cloaca Maxima"; measuring about 1.6 km, it was designed as a paved and vaulted tunnel for conveying the city's sewage, wastewater and rainwater into the river Tiber [4]. The drain is still running for limited use.

During the time of Roman civilization (circa 40 BC), ceramic pipes were used, and water was distributed by lead pipes [5, 6]. Since lead dissolves in water, continuous consumption of lead-contaminated water resulted in sickness and death. The Indus civilization in Mohenjo-daro and Harappa was more advanced. At that time, copper and iron pipes were commonly used [7]. In addition to executing water supply and drainage works, plumbers also worked on storm drainage and installed flue pipes, used for removing gases from kitchens and fireplaces. According to the Xinhua news agency, the world's earliest water closet is in China; quite like what is used now, this 2000-year-old toilet consisted of a stone seat and a comfortable armrest, complete with running water [8].

In the mid-14th century, developing society and builders realized the importance of the role of plumbers and demanded better quality of work at competitive prices. Plumbers also

DOI: 10.1201/9781003172239-1

felt strongly about their commitment to their jobs. This ultimately led to the formation of a guild of plumbers in England in 1365 AD, known as the “Worshipful Company of Plumbers”. The company was a trade association of plumbers and protected their professional rights and trade interests. It provided technical education to the plumbers as well.

Until the middle of the 17th century, pipes made of wood, clay and lead were used. At that time, no high pressure with standing pipes was available. Later, with the development of pumps driven by steam, cast iron pipes were developed.

In the late 19th century, the invention of the flush toilet is widely attributed to the plumber Thomas Crapper of England, who also installed toilets for Queen Victoria and patented a U-bend siphoning system for flushing the pan [8].

1.3 Plumbing practice: global perspective

Plumbing has not evolved all over the world at the same pace. In developed countries, plumbing is practiced in a well-established manner, but in developing countries this practice is yet to be developed in many respects. Plumbing practice scenarios in different parts of the globe are reflected below.

Australia

In Australia, the construction sector is booming, with a correspondingly increase in demand for plumbing professionals and plumbers. There is a national governing body for plumbing regulation, the Australian Building Codes Board. They are responsible for the creation of the National Construction Code (NCC), which includes the Plumbing Code of Australia. State governments have their own authority and regulations in place for licensing plumbers. They also administer and enforce plumbing regulations as outlined in the NCC. The qualifications required to get a license as a plumber are the following:

1 Studying blueprints, drawings and specifications to determine the layout of plumbing systems and estimation of materials, cost, time etc. required.
2 Installing hot and cold water systems and associated equipment.
3 Installing water-based fire protection systems, including fire hydrants, hose reels and sprinkler systems.
4 Installing below-ground drainage systems and associated ground support systems.
5 Installing gas appliances, flues and pressure regulating devices.

United Kingdom (UK)

In the United Kingdom, the majority of plumbing rules and practices are based on the regulations and bye-laws of Water Supply (water fittings) Regulations and Scottish Water Bye-Laws, which play an important role in protecting public health, safeguarding water supplies and promoting the efficient use of water within customers’ premises, across the UK. These regulation and by-laws set legal requirements for the design, installation, operation and maintenance of plumbing systems.

To become a plumber in the United Kingdom, an individual must have an academic qualification below level 3 (A level), with little or no work experience, and be aged between 16 and 24. After successful completion of traineeship, typically 6 months, one is on the path to becoming a plumber. Some colleges across the UK offer one-to-two-year courses designed to get a certificate to be a plumber and achieve a City and Guilds Level 3, Diploma in Plumbing and Heating (QCF). Some courses are designed for people with no plumbing experience, while others are designed for people with some experience. After achieving required trainings plumbers need to be registered at the Chartered Institute of Plumbing and Heating Engineering (CIPHE), which is a professional body. Fellows of this institution are the highest category of individual membership; those are registered with the Engineering Council as EngTech, IEng or CEng, etc.

United States of America (USA)

In the United States of America, plumbing codes and licensing of plumbers are generally administered by the state and local governments. At the national level, the Environmental Protection Agency sets guidelines about lead-free plumbing fittings and pipes, in order to comply with the Safe Drinking Water Act.

An apprenticeship completing 246 hours of technical education, in mathematics, applied physics and chemistry, including training on Occupational Safety and Health Administration, and up to 2,000 hours of practical training working with an experienced plumber is the way to become a plumber in the USA. To work independently, a plumber is required to achieve a license for the job. In most states, the prerequisite for getting a license is experiencing two to five years of practical knowledge. There is also an examination to test technical know-how and understanding of plumbing codes.

India

In India, the plumbing profession is yet to be developed at a highly satisfactory level. Due to a lack of plumbing infrastructure and a trained, certified workforce, there are considerable fatal casualties each year, primarily attributed directly to water and sanitation issues, as reported by the World Plumbing Council (WPC). According to the president of the Indian Plumbing Association (IPA), no architecture or engineering course of study in India currently offers a diploma or degree in plumbing. An overwhelming percentage of the plumbers are "casual labourers" who have learned the trade through personal experience or working with experienced plumbers [9]. To practice plumbing there is a Uniform Plumbing Code–India, an extensive code of plumbing practice. Plumbing Education to Employment Program (PEEP) India offers structured courses of study to produce plumbing design engineers, plumbing construction managers or supervisors and plumbers. There are 17 plumber colleges in India, offering 17 courses, and one training institution offering the Plumbing Education to Employment Program in India, which is accredited by the International Association of Plumbing and Mechanical Officials–India (IAPMO) and the Indian Plumbing Association (IPA).

Bangladesh

In Bangladesh, plumbing practice was good until the mid-1980s; the presence of very uncommon fittings in old buildings, as shown in Figure 1.1, endorses the fact. After this, it started to decline, mostly due to the degradation of plumbing products and skilled manpower. After the mid-1990s, the situation started to improve, though comparatively slowly, alongside the development of the building sector. Compared to development in other building services, plumbing practice lags considerably behind. Though there are opportunities to become a certified plumber, through two years of vocational and technical training courses in plumbing, a

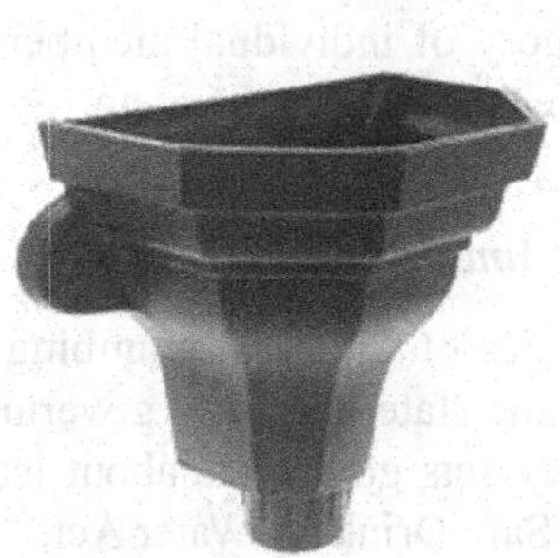

Figure 1.1a Upright-Wye in the vent pipe system and b. CI Hopper.

majority of practicing plumbers are found to be non-certified. According to the Bangladesh National Building Code, plumbers must have a license to practice, but there are not really any office or institute to give such a license, or any regulatory authority to regulate plumbing practices.

In Bangladesh, no undergraduate architecture or engineering course of study offers a diploma or degree in plumbing. In undergraduate architecture courses, there is a 1- and 2 credit-hour full course on plumbing, which makes practicing architects more knowledgeable than engineers in plumbing development.

1.4 Scope of plumbing systems

It is important to understand the scope of work of a plumbing professional and the type of service one has to deliver. Plumbing professionals shall be capable of dealing with all these services, and they will advise architects or other concerned professionals regarding the space, load requirements for equipment, pipe etc. to be used in plumbing works. Plumbing services include one or multiple systems of the following.

1 Water supply related various systems

- a Cold water supply
- b Hot water supply
- c Chilled water supply
- d Water treatment for building occupants
- e Pump selection, installation and maintenance
- f Steam and hot water boiler selection and related works
- g Fire suppression system
- h Garden hydrant system
- i Fountain, cascade and water fall
- j Swimming pool water supply, treatment and conditioning and
- k Interior-scape irrigation: manual and automatic.

2 Waste disposal-related various systems

   a Plumbing fixture, pipe equipment selection, installation and maintenance
   b Soil, waste and vent piping
   c Building sewer
   d Basement drainage and pumping
   e Sewage pump selection and installation
   f Swimming pool, fountain, water cascade, water body and
   g Wastewater recycling and reuse.

3 Storm water management-related systems

   a Rainwater harvesting: rainwater collection, storage, filtration, disinfection and distribution
   b Storm water drainage and
   c Groundwater recharging.

4 Gas supply related systems

   a Fuel gas supply in building
   b Special gas like oxygen, nitrous oxide etc. for medical applications and
   c Steam supply for heating.

1.5 Importance of plumbing

Plumbing is the most vital and basic building service to be developed in a building to be used by any living beings, even it is for a short while. Plumbing addresses not only various natural needs for living in a building but also the environment and safety of the occupants in the building. So, plumbing in a building shall have to be developed with utmost care so that it doesn't pose any negative impact on the environment and health of the occupants. Plumbing has many other implications for which its importance is well considered in building development. Following are the important reasons that plumbing is getting recognized for its importance.

1 Plumbing is related to environment, particularly water environment
2 Plumbing may pose health hazards and endanger life
3 Plumbing is a concern for energy and resource saving
4 Plumbing can cause loss of property
5 Plumbing is indicator of development and
6 Plumbing development is subject to regulatory systems.

1.5.1 Environmental aspects of plumbing

The basic objective of plumbing development is developing water supply and sanitary drainage systems in a building and its premises. The source of water is either surface water or, in some cases, underground water. Due to faulty or non-engineered plumbing development and poor maintenance, excess consumption and waste of water might happen. This will ultimately result in over-extraction of water from these sources, causing various negative impacts on the water environment. On the other hand, over-consumption of water means excess wastewater generation, which might cause an overload in treatment, resulting in poor quality effluent and degrading water environment in the disposal area. Faulty plumbing, particularly drainage systems, primarily pollutes the indoor environment through ingress of foul sewer gas into toilets or kitchens and after to the respective building floor spaces, thus polluting the indoor environment. So, good plumbing development helps to reduce water consumption and wastewater generation, thereby safeguarding both the inside and outside environment.

1.5.2 Health aspects of plumbing

Plumbing is primarily of concern for health, sanitation and agronomy aspects of any given building's occupants as well as for the plumbers. Plumbing codes provide strict guidelines for plumbing installation in the interest of public health. The majority of the guidelines are governed with respect to the environment and safety issues, as these are also concerns of public health. Plumbing addresses all sorts of microbial, chemical and physical risks in its planning, design, installation and maintenance. Even in choosing plumbing material, health risks are considered. The surface finish and features of some fixtures and appurtenances are categorically guided considering the public health benefits.

1.5.3 Energy aspects of plumbing

In plumbing, various equipment and appliances are used which need energy to operate. More and bigger sizes of equipment or appliances used mean more energy will be consumed. So, avoiding those elements or, where unavoidable, using smaller sizes would be the good engineering practice to save energy and cost. Pumps, water heaters, pressure reducing valves etc. are the important elements of plumbing from the perspective of energy consumption. So, in selecting the types and sizes of these elements, special attention is to be given.

1.5.4 Plumbing damaging property

Plumbing elements, predominantly pipes and some appurtenances, are generally installed behind the wall, under the floor or suspended from the ceiling of a building. In some cases, pipes are installed inside the masonry walls, passing through walls, beams or floors. In special cases, openings into floors, walls or beams are needed to pass through a good number of pipes. Punching holes in masonry walls, beams or floors after construction might cause damage to these structural elements, which weakens the strength and stability of the structure. In water-retaining structures, particularly in water reservoirs, installing pipe by making holes and then sealing the gap around can hardly ensure a watertight joint, resulting in leaking water accumulating on supporting floors and ultimately damaging the affected masonry elements. The most notorious occurrence in plumbing are micro leakages in pipes, in very remote, not quickly visible and hard-to-reach locations, which might cause substantial damage to a property and incur additional costs for rectification. Poor maintenance or carelessness in repairing pipe leakages intensifies the damage of a building structure, as shown in Figure 1.2.

1.5.5 Plumbing as a development indicator

Plumbing can be considered as an indicator of development. In developed countries, plumbing in all buildings is highly valued as an important building service for safety, sanitation

Figure 1.2 Damaging wall of a building by leaking pipes.

and comfort. In underdeveloped countries, plumbing is comparatively poorly valued, mostly from a sanitation perspective. Health, environmental and energy aspects of plumbing are generally well addressed in the plumbing of developed countries, where these works are highly regulated along with other building service works. In developed countries, plumbers are given a license subject to acquiring job competency by going through practical training and certificate courses, and thus becoming a comparatively high-paid, skilled worker. In developing or underdeveloped countries, many plumbers are found working without a valid license, and so, plumbing there is poorly developed in many cases.

1.5.6 Regulatory aspects of plumbing

Where there is a building to be occupied by any living beings, plumbing is there. Areas under man-made development need to be built where plumbing is unavoidable. So, all countries need some sort of guidelines to develop plumbing. In developing areas, much of the plumbing work is mostly regulated by the government or semi-government agencies due to its direct impact on the public's health, safety, welfare, environment etc. To protect the occupants of the buildings and to ensure quality and safety of their property, plumbing development must be done according to the plumbing and building codes. Plumbing works need permits to be obtained by the plumbing contractors and the plumbers and are typically secured from the relevant building regulatory, water supply and sewerage authorities on behalf of building owners.

1.6 Principles of plumbing

Plumbing is installed in a building for providing convenience, comfort and safety as well as sanitation and good environment for the occupants. It also helps to maintain a good environment inside and around the building.

In a building, water and gas are mostly supplied in adequate quantity and with desired quality to the points of use; however, human excreta, ablutionary water, kitchen wastewater and excess rain or storm water are drained off of and from buildings, through developing a plumbing and drainage system. The basic objective of plumbing installations should be such that the system provides quick supply of wholesome, sufficient water, and at the same time disposes of waste quickly and sanitarily. Furthermore, it should be kept in mind that the plumbing system shall be safe for users, particularly where high pressure, temperature and electricity are concerned. Plumbing installation should be done in such a way that it does not impair the strength of building structures or deface the building aesthetic.

To accomplish the job of plumbing, energy is consumed, mostly in the water supply system, for which dependency on power – either electricity or fuel (diesel) – increases. Loss of energy means loss of money and resources. So, energy conservation is another major concern of plumbing.

From these facts, it can be well understood that the plumbing system must be developed based on some principles. The basic principles of plumbing development are as follows.

1 The supply of any resource (water and gas) shall be made safe and sound, ensuring adequacy in quantity and acceptability in quality, collecting from a potential source and carrying to the points with demand
2 Drainage of wastewater shall be done quickly and in a sanitary way, maintaining good indoor and outdoor environments and ensuring safety as well
3 Drainage of storm and rainwater shall be done after its maximum harvesting, ensuring safety of user and environment
4 Plumbing shall be developed without impairing structural integrity and aesthetics.

In order to follow these principles, the following sub-principles shall be fulfilled.

1 All the premises and buildings made for human use or habitation for a considerable time shall be provided with the supply of pure water, having no chance of any contamination.
2 The supply of water or gas shall be made at sufficient flow rate and maintain adequate pressure for all plumbing items to function satisfactorily without creating any unwanted situation.
3 Plumbing system shall be designed and adjusted to use the minimum quantity of water or gas consistent with proper performance of devices and appliances and efficient cleaning of wastewater.
4 Plumbing elements susceptible to potential hazard shall be installed, maintained and operated properly and carefully.
5 Every premise and building located near a public sewer shall have its drainage connection to the sewer system properly.
6 Each family dwelling unit shall have a minimum of one water closet, one basin or sink, a shower to fulfil the basic requirements of sanitation and personal hygiene.
7 Plumbing fixtures should be made of non-absorbent material with smooth finished, and shall be free from sharp edges and fouling spaces, and shall be housed in ventilated enclosures.
8 The drainage system shall be designed, installed and maintained to prevent the fouling, deposit of solids, clogging, supported by adequate clean-outs so located to facilitate easy maintenance.
9 To ensure satisfactory service, all planning and design shall be done by certified professionals, elements shall be made of durable approved materials and installation shall be free from defective workmanship, done by licensed plumbers.

10 Every plumbing fixture with no integral trap built-in, connected directly to the drainage system, shall be supported by a water-seal trap.
11 The drainage pipe size shall be so chosen that it can provide adequate circulation of free air to avoid siphonage action.
12 Vent terminals shall extend to the outdoor air and be terminated at a level of general man-height, and where there is no possibility of entering foul gas from the vent to any nearby dwelling unit.
13 Piping system shall be subjected to all recommended pressure tests to effectively disclose all leaks and defects if any, in the workmanship.
14 The waste materials, which may interfere unduly with the sewage flow, clog or choke the pipes and cause potential hazard or damage the pipe or the joints, shall not be allowed to enter the drainage system.
15 Strong protection shall be taken to avoid spoilage of food, water, sterile goods etc. by backflow of wastewater. When necessary, the concern fixtures, devices or appliances shall be connected indirectly to the drainage system.
16 In any compartment or room, which is not properly ventilated or lighted, water closet and urinal should not be installed.
17 On-site primary treatment systems, like septic tanks, Imhoff tanks or Baffled reactors, must be incorporated for building sewage management where there is no provision for disposal of sewage in public sewers or other approved disposal systems.
18 Where a plumbing system may be subject to backflow of wastewater, effective measures shall be taken to prevent the backflow in the system.
19 Plumbing and drainage systems shall be installed and maintained in serviceable condition by licensed skilled plumbers.
20 All plumbing fixtures shall be installed at ideal height, properly spaced, to be accessible for their intended use.
21 Plumbing systems shall be installed with due regard to the integrity of the strength and form of the structural members and without defacing their finished surfaces.
22 Wastewater from plumbing systems, not suitable for discharging directly in to the surface or subsurface waters, shall not be discharged into the ground or into any waterway without properly treating.
23 Excess rain or storm water shall be drained out after storing it for its maximum use and recharging as much as possible.
24 Combined drain or sewer system shall have to be avoided.
25 The minimum number of faucets, fixtures and appliances are to be installed for users.

As a whole, the plumbing system should be developed in such a way, based on the principles and sub-principles of plumbing, that it achieves the following objectives.

1 Safety
2 Sanitation
3 Good environment
4 Convenience
5 Comfort
6 Economy
7 Durability
8 Maintainability and
9 Legality.

1.6.1 Achieving the objectives

When a plumbing professional designs plumbing services, he must always be aware of the objective of the job. To meet the objectives, the design approach should reflect the following achievements.

1 Safety

  a Provide safe drinking water
  b Safe disposal of wastes
  c Piping and appliances are safely installed
  d Equipment is fitted with proper safety devices and
  e Structural safety has not been impaired by plumbing installation.

2 Sanitation

  a Ensure non-polluting system
  b Every plumbing fixture shall be equipped with a water-sealed trap and
  c Waste disposal in public sewer systems or in specified disposal area.

3 Good environment

  a No hazards to the environment that might be created by improper waste disposal and
  b Surroundings shall be kept free from bad smells.

4 Convenience

  a Providing appropriate type and good quality, proper sized fixtures and fittings and
  b Installing optimum number of fixtures and fittings.

5 Comfort

  a Installed fixtures are properly spaced at appropriate positions and comfortable heights considering users' desire and disability and
  b Minimize noise and sound developed in the plumbing system.

6 Economy

  a Provide optimum sized piping, fittings and equipment and
  b Instal fixtures, fittings and appurtenances of economy price.

7 Durability

  a Provide long-lasting pipes, fitting, fixtures and accessories
  b Joint the fixtures and pipe fittings properly
  c Provide protective measures for all the elements in the plumbing system
  d Test the plumbing system to check for leakage and defective workmanship and
  e Check installation of the equipment for proper functioning before the system is put into operation.

8 Maintainability

  a Provide clear access to the plumbing items
  b Provide sufficient space and tolerable environment for working
  c Provide facilities for maintaining plumbing items and

d Provide adequate clean-outs, so arranged that it helps clean deposit of solids, clogging objects etc.

9 Legality

a Design the plumbing system in accordance with local codes of practice or by-laws.

1.7 Basic plumbing requirements

The basic plumbing requirements in any dwelling house for human beings are as follows.

1 Every building intended for human occupancy or for any living beings, should be provided with an adequate, safe, and potable water supply and sanitary drainage for generated wastewater.
2 To fulfil the basic needs of sanitation and personal hygiene, each dwelling connected to a private on-site wastewater treatment system (POWTS) or public sewer should be provided with at least the following plumbing fixtures:

   a One water closet
   b One wash basin
   c One kitchen sink and
   d One bathtub or shower.

3 Hot or tempered water should be supplied to plumbing fixtures that normally require hot or tempered water for proper use and function.
4 Where plumbing fixtures exist in a building that is not discharging in to a public sewer system, suitable provision should be made for treating or recycling or dispersing or holding the wastewater.
5 There shall be provision of rain or storm water drainage facilitating ground recharging as much as possible.

1.8 Plumbing practice

The development of plumbing systems follows a set of activities performed by different groups of practitioners related to this field. The set of activities involving plumbing development are as follows.

1 Planning
2 Designing
3 Installation and
4 Maintaining.

1.8.1 Planning of plumbing

Elements of plumbing systems serve as the arteries of a building. Planning is key for all success, and so it is unavoidable in order to achieve efficiency in all man-made systems, including the plumbing systems of a building. Unplanned plumbing systems make systems inefficient and costly.

Architects play a vital role in efficiently planning plumbing systems, as they are responsible for positioning the major elements of plumbing, such as the fixtures, pump house reservoir etc. In critical cases, architects should consult with the relevant engineers in selecting positions and allotting sufficient space for major elements of the plumbing system.

1.8.2 Designing of plumbing

Plumbing engineers are primarily responsible for the sizing of the pipes, accessories and equipment for the system. Planning the routing of the piping is another major job of plumbing engineers, for which they might need to consult with the architects and other service engineers as required. The job of design-engineers and architects, the concerns of plumbing, is presented in drawings developed by draftsman who have been specially trained for and practice plumbing drawings. Design engineers accordingly approve the plumbing drawings and send them to the field for execution.

1.8.3 Installation of plumbing

Plumbers are the trade group that installs the plumbing items in accordance with the drawings furnished by the engineers. Plumbers also carry out all of the tests for good workmanship of the system. Plumbers also carry out the job of disinfecting the water supply system.

1.8.4 Maintaining plumbing

Plumbing items are subject to regular and routine maintenance, in addition to repair works, as and when required. Both repair and maintenance works are primarily accomplished by the plumbers. In critical cases, problems are mainly identified by the engineers, but the repair works are done by the plumbers or the technician experts in the particular job.

1.9 Plumbing practitioners and traders

In a plumbing job, there are four categories of traders that are mainly involved. These are as follows.

1 The plumbing professionals
2 The plumbing contractors
3 The plumbers and
4 The manufacturers.

1.9.1 Plumbing professionals

The engineer who intends to practice plumbing studies all aspects of engineering, including the planning, design, installation and maintenance of fluid (water and gas) supply, waste disposal and storm drainage systems, together with other fields of plumbing systems developed for buildings.

This knowledge is applied in planning, designing, installing and maintaining the plumbing system. Among these important aspects of plumbing, planning for the plumbing services in a building needs additional aptitude on the part of the engineer, which is his sense of forms in art. The sense of art in plumbing develops in the minds of engineers with their experience of practicing in the field, and seldom through reading plumbing books. In this case, plumbing engineers must consult with the architects concerned so that no objections are raised after installation of the plumbing items. Not only the architects but also the professionals of other building services must be consulted so that no hindrance or conflict occurs between the plumbing service elements and other building service elements.

Plumbing professionals shall have updated knowledge regarding new developments in the code of practice, new innovations in the research and new items available in the market, in order to develop state-of-the-art plumbing for any prestigious projects.

1.9.2 Plumbing contractors

The contractor plays an important role in implementing the project properly and in a timely manner. The plumbing contractor should have a thorough idea of all the elements of his job. He should be well acquainted with the price and quality of materials or equipment in particular and should also possess the ability to select skilled manpower. Knowledge regarding prices will help in successfully bidding for a job and making a profit from it. However, engaging inexperienced and unskilled manpower may bungle the job, with disastrous consequences.

The contractor should be capable of arranging materials, equipment and sufficient labour within the time scheduled, coordinating with the consultant or the owner of the project at regular intervals to ensure smooth progress of work. The contractor should have the capability to anticipate problems that may arise during the progress of work and be ready to solve these to the satisfaction of all concerned.

Plumbing contractors shall take necessary measures to store plumbing items at a site, ensuring their safety and management. Pipes in particular shall be staggered in such a way that they do not cause any accident during handling. Small and only occasionally used pipe fittings shall be stored in such a manner that finding these when needed does not take more time and delay the job.

Contractors shall arrange the necessary equipment for performing all tests and get approval of their executed work per the rules and regulations defined by the concerned authorities. All testing data shall be well recorded and reported for submission to the concerned authorities. Contractors must also ensure all necessary measures are enacted to protect workers from any risk of casualties or hazards.

1.9.3 Plumbers

Plumbers play the vital role in plumbing work as they are the main artisans and craftsmen who joint and install pipes for water and gas supply and waste disposal, install plumbing fixtures, fit accessories and, in the long run, repair these items when necessary. So, they must acquire skill in these jobs. Plumbers must have a sufficient academic background to read and understand plumbing drawings such that they are able to install plumbing systems correctly in the field. A plumber should be acquainted with all sorts of equipment involved in plumbing works and able to use them efficiently. They must be familiar with new products and items for plumbing that are available in the market, as well as their use and installation processes. Nowadays, a plumber has to know how to weld, braze out pipes and make threads in pipes. They should also know how to install electrical or gas-burnt geysers, pumps etc. Plumbers shall have mental and physical fitness and the ability to work in cramped conditions and filthy environments, together with the stamina to continue hard work with patience for the long period of a day. Women have been found to be less interested in this career. In the United Kingdom, less than 1 per cent of women work as plumbers [10].

1.9.4 Manufacturers and traders

Manufacturers producing plumbing items like pipes, fitting and fixtures should be sincere and ethical in manufacturing their commodities. They should manufacture items that maintain good quality and are a standard size and shape. Deviation from good quality, size or shape results in losing durability and invites trouble from poor jointing and improper functioning. As a result, not only will there be squandering of money, but these will also invite miseries of the users and the owners without having any prior notice. Manufacturers must not look only for maximizing their profit from manufacturing substandard items, taking the privilege of callousness or weakness of the regulatory authorities, though the products might be sold profusely due to its high demand and the ignorance or ill motive of the buyers. Traders of plumbing items must also uphold similar business ethics to the manufactures. In addition, traders should make available all plumbing items, including the least usable items, in their trade lists.

References

[1] wikipedia.org (1942) "What Is the Origin of the Word 'Plumbing'?" *Pittsburgh Post-Gazette*, May 12, Retrieved from https://en.wikipedia.org/wiki/Plumbing, on 27 December 2013.

[2] Rodda, J.C. and Ubertini, L. (2004) "The Basis of Civilization – Water Science?" *International Association of Hydrological Sciences*, p. 161, Retrieved from https://en.wikipedia.org/wiki/Sanitation_of_the_Indus_Valley_Civilisation on 25 December 2020.

[3] wikipedia.org "Sanitation of the Indus Valley Civilisation" Retrieved from https://en.wikipedia.org/wiki/Sanitation_of_the_Indus_Valley_Civilisation.

[4] Aldrete, Gregory S. (2004) *Daily Life in the Roman City: Rome, Pompeii and Ostia*. Greenwood Publishing Group, pp. 34–35. ISBN 978-0-313-33174-9, Retrieved from https://en.wikipedia.org/wiki/Cloaca_Maxima, on 25 December 2020.

[5] Vitruvius. On Architecture; in two volumes; translated into English by Frank Granger, The Loeb Classical Library, Retrieved from www.iwapublishing.com/news/brief-history-water-and-health-ancient-civilizations-modern-times.

[6] Frontinus Sex, "Iulius. De aquaeductu urbis Romae" *The Stratagems and the Aqueducts of Rome; with an English* Translation by Charles E. Bennett; edited and prepared for the press by Mary B. McElwain, The Loeb Classical Library.; Retrieved from www.iwapublishing.com/news/brief-history-water-and-health-ancient-civilizations-modern-times.

[7] Deolalikar, S.G. (1994) *Plumbing: Design and Practice*, Tata McGraw-Hill Publishing Company Ltd, New Delhi.

[8] Patrentbureau, "History of Plumbing" *Web Magazine*, Retrieved from http://patent.net.ua/intellectus/facts/populus/86/ua.html on 25 December 2020.

[9] ITP Media Group (2009) "India Introduces Plumber Training" *Construction Week*, Retrieved from www.constructionweekonline.com/article-5383-india-introduces-plumber-training on 25 December 2020.

[10] Carol, Canavian (2018) *A Carrier in Plumbing*, 2nd Ed, CreateSpace Independent Publishing Platform, p. 2.

2 Hydraulics in plumbing

2.1 Introduction

The plumbing system involves the flow of fluids like water, gas, steam, wastewater etc. under pressure and the flow of liquid at or near atmospheric pressure under gravity, generally in water supply and drainage systems. These flows are dependent on several properties of liquids, such as density, viscosity, surface tension etc., as well as on other properties, such as temperature, pressure, condition of flowing media etc. So, knowledge regarding these parts of hydraulics is very important in designing an effective plumbing system. This chapter therefore deals with various aspects of basic hydraulics related to plumbing system design.

2.2 Hydraulics

Hydraulics is the study in the branch of science which deals with fluids in motion or at rest. Hydraulics in plumbing basically deals with the flow of fluids in pipes and channels and their confinement in tanks. French scientist-philosopher Blaise Pascal and Swiss physicist Daniel Bernoulli formulated two fundamental laws of fluids, on which modern hydraulics is based.

Pascal's law, also known as Pascal's principle, states that a pressure changes at any point in a confined incompressible fluid is transmitted throughout the fluid such that the same change occurs everywhere. Simply, it can be said that pressure in a liquid is transmitted equally in all directions; i.e. when fluid is made to fill a closed container, the application of pressure at any point will be transmitted to all sides of the container. The Pascal (Pa) is the derived unit of pressure in SI systems, used to quantify internal pressure, stress, Young's modulus and ultimate tensile strength. One Pascal is the pressure exerted by a force of magnitude of one Newton perpendicularly upon an area of one square meter.

Bernoulli's law formulated that energy in a fluid is due to its elevation, motion and pressure, and if there is no loss due to friction and no work is done, the sum of the energies remains constant. Thus, energy due to velocity, generated from motion, can be partly converted to pressure energy by enlarging the cross-section of a pipe. According to the principle of conservation of energy, in a steady flow, the sum of all forms of energy in a fluid along a streamline is the same at all points on that streamline, i.e. the sum of kinetic energy, potential energy and internal energy remains constant.

In plumbing, flow of and pressure exerted by fluid, water and wastewater are the major concerns of hydraulic performance. There are some important factors which greatly influence the hydraulic performance of water. Some of those are discussed briefly in the following.

DOI: 10.1201/9781003172239-2

2.2.1 Density

Density is the mass of material conditioned in a unit volume and can be expressed as:

$$\text{Density } \rho = \frac{m}{v} \qquad 2.1$$

Where
m = mass (kg) and
v = volume (cum).

Density of any fluid varies with temperature. As the volume of a fluid increases with the rise of temperature, its density decreases. At 4°C, the density of water is almost 1 kg/cum, and at 30°C, density is 9.95 gm/cum, as can be found in Table 2.1. Again, water expands at temperatures below 4°C. This characteristic of water may cause the bursting of a pipe in cold climates where water freezes in the pipe.

2.2.2 Viscosity

Viscosity of a fluid is a property of its resistance to deformation due to shear stress or tensile stress caused by any external force. All fluids are more or less viscous. Viscosity occurs due to cohesion between fluid layers. So, when fluid flows, shearing stress develops between moving layers and indicates relative resistance to flow. Fluids of high viscosity are generally thick in nature, and the flow is relatively slower. Thinner fluids have low viscosity and flow comparatively faster. Temperature has great effect on viscosity; as the temperature increases, the viscosity decreases.

Viscosity can be related to the concept of shear force effect between different layers of the fluid exerting shearing force on each other, or on other surfaces as they move against each other. When fluid flows, the relative velocity of adjacent layers varies continually in the direction normal to the motion, so that the stress varies likewise. Formally, viscosity

Table 2.1 Density and weight of water at various temperatures, at standard sea-level atmospheric pressure [1].

Temperature (°C)	*Density (gram/cm³)*
0	0.99987
4.0	1.00000
4.4	0.99999
10	0.99975
15.6	0.99907
21	0.99802
26.7	0.99669
32.2	0.99510
37.8	0.99318
48.9	0.98870
60	0.98338
71.1	0.97729
82.2	0.97056
93.3	0.96333
100	0.95865

(represented by the symbol η "eta") is the ratio of the shearing stress (F/A) to the velocity gradient in a fluid.

$$\tau = \mu \frac{\delta u}{\delta y} \qquad 2.2$$

Where
τ = shear stress
$\frac{\delta u}{\delta y}$ = velocity gradient and
μ = coefficient of viscosity and its values are used as a measure of fluid viscosity.

There are actually two quantities that are called viscosity. The quantity defined above is sometimes called dynamic viscosity, absolute viscosity or shear viscosity to distinguish it from the other quantity, but is usually just called viscosity. In SI systems, it is measured in Pascal-seconds or poises.

The other quantity, called kinematic viscosity (represented by the Greek letter ν "nu"), is the ratio of the viscosity of a fluid to its density.

$$\nu = \frac{\eta}{\rho} \qquad 2.3$$

Where
η (eta) = fluid viscosity and
ρ = fluid density.

Kinematic viscosity (also called "momentum diffusivity") is a measure of the rate at which momentum is transferred through a fluid: the ratio of the dynamic viscosity ν to the density of the fluid ρ. Kinematic viscosity is a measure of the resistive flow of a fluid under the influence of gravity. In SI systems, kinematic viscosity is measured in stokes (St).

2.2.3 Surface tension

The molecules of all liquids are held together by intermolecular attraction, which enables the liquid to withstand a small tensile stress. So, some tension effect occurs on the surface of liquids when the liquid surface is in contact with another liquid of different density, gas or solid. The surface appears to act as an elastic skin, which is in tension in both directions. This tension is called the surface tension. This surface tension is responsible for capillary rise of liquid in narrow spaces, the mechanics of bubble formation, the breakup of liquid jets, the formation of drops etc.

The surface tension is defined as the force in the liquid surface normal to a line of unit length down in the surface. The surface tension is dependent upon the nature of the contact surface and also on temperature. With the increase of temperature, the magnitude of surface tension will decrease.

2.2.4 Vapour pressure

All liquid possesses the tendency to vaporize i.e. to change from liquid state to gaseous state. Such vaporization occurs because molecules are continually projected through the free liquid surface and lost from the liquid. The ejected molecules, being gaseous, exert their own partial pressure, which is termed the vapour pressure of the liquid. Vapour pressure is expected to increase with the rise of temperature.

2.2.5 Temperature

All fluids, including water, are greatly affected by temperature. Density of any fluid varies with temperature. As the volume of a fluid increases with the rise of temperature, its density decreases.

The temperature rise of a fluid causes a rise in the pressure exerted by it. For a compressible fluid in a container, there would be an increase in a certain percentage of pressure for that percentage increase of temperature. For an incompressible fluid in a closed container, a minor rise in temperature would cause tremendous rise in pressure. For water, as a rule of thumb, for every 256° K rise in temperature, there would result in increase of pressure by 690 kPa.

Temperature has great effect on viscosity of fluid; when the temperature rises, the viscosity decreases. With the increase of temperature, the magnitude of surface tension will decrease. Vapor pressure is also expected to increase with the rise of temperature.

2.2.6 Other factors

In addition to the factors discussed so far that influence the hydraulic performance of water, there are other factors that might be significant in some special cases but are rarely nor widely used in general plumbing practice. So, instead of discussing these properties or factors in detail, some major concerns of properties of water are mentioned in Table 2.2.

2.3 Pressure

Pressure is the force exerted by a body per unit area divided by its mass. All fluids exert pressure on the surfaces with which they are in contact.

By intensity of pressure, we denote the pressure per unit area.

$$\text{Intensity of pressure } p = \frac{P}{A} \qquad 2.4$$

Table 2.2 Selected physical properties of water in SI metric unit [2].

Sl. No	*Properties*	*Value*
1	Molar mass	18.0151 grams per mole
2	Melting point	0.00°C
3	Boiling point	100.00°C
4	Maximum density (at 3.98°C)	1.0000 grams per cubic centimetre
5	Density (25°C)	0.99701 grams per cubic centimetre
6	Vapour pressure (25°C)	23.75 torr
7	Heat of fusion (0°C)	6.010 kilojoules per mole
8	Heat of vaporization (100°C)	40.65 kilojoules per mole
9	Heat of formation (25°C)	285.85 kilojoules per mole
10	Entropy of vaporization (25°C)	118.8 joules per °C mole
11	Viscosity	0.8903 centipoise
12	Surface tension (25°C)	71.97 dynes per centimetre

Where
P is the total pressure on area A.

2.3.1 Atmospheric pressure

Since the atmosphere is a fluid, it exerts normal pressure on all surfaces with which it is in contact. This pressure reduces with altitude until it becomes zero (vacuum) in outer space. This intensity of atmospheric pressure has been experimentally found to be 1 kg/sqcm at mean sea level, under normal barometric conditions.

2.3.2 Static water pressure

At any point within water, it exerts equal intensity of pressure in all directions. The intensity of pressure at that point depends on the depth of submergence of the point below the free surface and on the unit weight of water.

The intensity of pressure on any surface under water is expressed in kilograms per square centimetre i.e. kg/sqcm or in meters of water above atmospheric pressure at the point where the pressure is to be measured.

The pressure is a function of water depth and density of water i.e.

$$P = hd \qquad 2.5$$

Where
h = depth of water in meters (m) and
d = density of water.

For 1 m of water column, the pressure exerted is

$$\begin{aligned} P &= h\,d \\ &= h \times 9.81 \end{aligned}$$

This means that the pressure exerted on the bottom of a 1 cum. container filled with water is equal to 9.81 kPa, as shown in Figure 2.1.

So, p = 9.81 h kPa
Where
h = depth of water column in meter.
So, 1.0 kPa pressure is developed for 1/9.81 m of water column.
= 101.97 mm of water column.
Say, 102 mm of water column
1 bar (100 kPa) is developed for 100/9.81 = 10.2 m of water column.

h=10p meter of water, where pressure p in bar   2.6

2.3.3 Absolute pressure and gauge pressure

The value of p in Expression 2.6 is known as gauge pressure because it is the pressure that would be found in a pressure gauge. Absolute pressure is the gauge pressure plus 101 kPa,

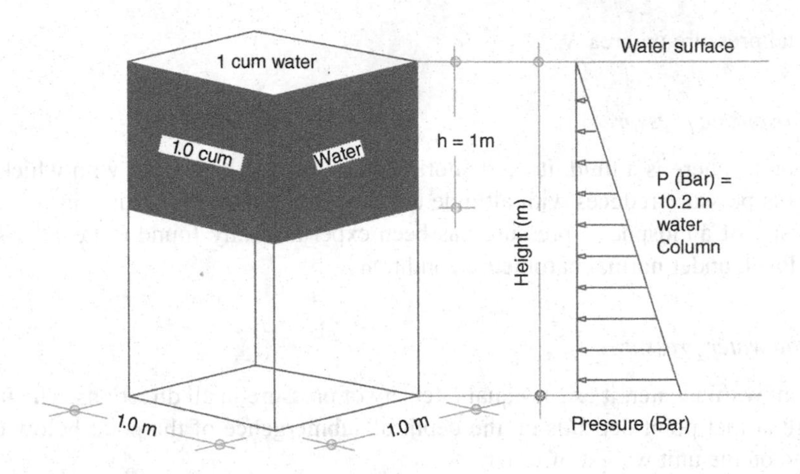

Figure 2.1 Basic relationships between water pressure in bar and head of water in m.

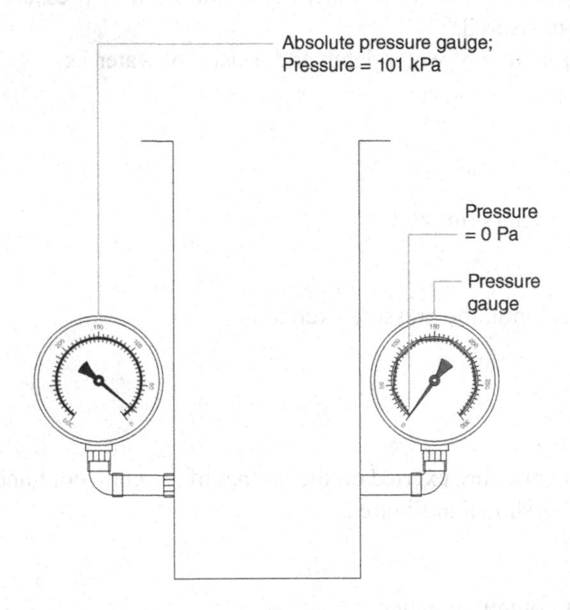

Figure 2.2 Relation between absolute pressure and gauge pressure.

because at mean sea level, the absolute atmospheric pressure is 101 kPa. But the gauge reads zero pressure there. In Figure 2.2, a reading of absolute pressure corresponding to the gauge pressure is shown. The details of pressure gauge are discussed in a later chapter.

2.3.4 Velocity and Pressure

The velocity of flow of a fluid is an important concern for plumbing design. In a fluid flow, velocity is to be controlled and kept within a specific limit. Again, velocity of flow is directly related to the pressure developed in the system. This relationship between flow velocity and pressure can be mathematically derived as below.

We know that velocity

$$v = \sqrt{2gh} \quad 2.7$$

Where
g = acceleration due to gravity = 9.81 m/sec^2 and
h = elevation of water column in m = 10 p.

So, velocity $v = \sqrt{(2 \times 9.81 \times 10)\,p} = 14\sqrt{p}$ m/sec,

Where pressure p in bar.

$$v = 14\sqrt{p} \text{ m/sec} \quad 2.8$$

2.4 Flow of water

When water remains at an elevated position, it gains potential energy. From this elevated position, water flows downward under gravitational force. The difference of height between the elevation of highest and lowest level of water is termed the static head. Water can be pushed upward if additional energy can be applied by any means. Generally, a pump is used to create energy to cause water flow through a pipe to an elevated position. The energy applied is in dynamic form. So, the energy needed for water to flow is termed the dynamic head.

The flow of water in a pipe occurs in various ways. The type of flow of water depends upon the manner in which the particles unite and move. Though there are many types of flow, the following are important in plumbing.

Uniform flow: In uniform flow, the water particles at all sections of a pipe or channel have the same velocities.

Non-uniform flow: In non-uniform flow, the water particles at different sections of a pipe or channel have different velocities.

Turbulent flow: In turbulent flow, each water particle does not have a definite path and the paths of individual particles also cross each other.

Steady flow: In steady flow, the quantity of water flowing per second is constant. A steady flow may also be uniform or non-uniform.

Unsteady flow: In unsteady flow, the quantity of water flowing per second is not constant.

Incompressible flow: In incompressible flow, the volume of a fluid and its density does not change during the flow. The flow of water is considered to have incompressible flow. On the contrary, the gases are considered to have compressible flow.

2.4.1 Flow under gravitational force

Flow of water under gravitational force may be in pipes or in an open channel or on an open surface. Flow of water from a rooftop water tank to the faucets in every floor below occurs through pipes due to gravitational force. On a flat roof, surface rainwater flow towards the inlets of rainwater down pipe due to gravitational force developed by making sloped surface, made inclined towards the inlets.

Gravitational water flow in pipes: Water flows in a pipe from a higher level to a lower level due to the gravitational force when both the ends of the pipe are open to atmospheric pressure. The velocity of flow is computed by Manning's formula.

$$V = \frac{1}{n} R^{2/3} S^{1/2} \qquad 2.9$$

Where
V = velocity of flow (m/sec)
n = Manning's coefficient of roughness of a pipe's internal surface (see Table 2.3 for value of n)
R = hydraulic mean depth = area/wetted perimeter (m)
For circular pipe $R = D/4$ where D = diameter of pipe (m)
S = slope of pipe (m/m)

The discharge capacity of a pipe is expressed by the formula

$$Q = AV \text{ (cum/sec)} \qquad 2.10$$

Where
A = cross-sectional area of pipe (sqm) and
V = velocity of flow (m/sec).

Flow in open channel: Open channel is the passage through which water flow has its free surface in contact with the atmosphere. In an open channel flow, the pressure is atmospheric pressure, which may be neglected. The hydraulic gradient is considered to be equal to the slope of the channel when the latter is uniform. Based on the above fact, Chezy developed a formula for flow in an open channel, as below.

$$V = C\sqrt{RS} \qquad 2.11$$

Table 2.3 Value of Manning's n for closed conduits of particular finish, flowing partly full [3].

Sl. No	*Type of conduit and description*	*n; Normal*
1	Cast iron: uncoated	0.014
2	Wrought iron: galvanized	0.016
3	Cement: neat surface	0.011
	Mortar	0.013
4	Concrete: culvert, straight and free of debris	0.011
	Culvert with bends, connections and some debris	0.013
	Finished	0.012
	Sewer with manholes, inlet etc. straight	0.015
	Unfinished, steel form	0.013
	Unfinished, smooth wood form	0.014
	Unfinished, rough wood form	0.017
5	Clay: common drainage tile	0.013
	Vitrified sewer	0.014
	Vitrified sewer with manholes, inlet, etc.	0.015
	Vitrified sub-drain with open joint	0.016

Where
V = velocity of flow in an open channel (m/sec)
R = hydraulic mean depth; $R = A / P$ 2.12
Where
A = a cross-sectional area of a channel (sqm)
P = wetted perimeter
S = slope of channel bed and
C = a constant depending on the shape and surface condition of the channel. The value of C is determined experimentally.

Bazin deduced the following formula for the value of C.

$$C = \frac{87}{1 + \frac{K}{\sqrt{S}}} \tag{2.13}$$

Where K is roughness factor, a constant depending on the surface of a channel. K value ranges are $0.109 < K < 3.17$ [4]. The values of K for different channel surfaces are shown in Table 2.4.

2.4.2 Flow under pressure

Water flows through a confined section under pressure when water is forced to flow through, generally by pumping. In a pipe, full flow of water occurs under pressure which is streamlined at low velocities but turns chaotic as the velocity is increased above a critical value.

To determine the flow of water in a pipe under pressure, the most common formula is the Hazen-Williams equation, used in the design and evaluation of a water distribution system. In a condition of flowing full through a pipe, the formula for velocity is expressed as follows.

$$V = 1.318CR^{0.63}S^{0.54} \tag{2.14}$$

Where
V = velocity of flow (m/sec)
C = Hazen-Williams coefficient that depends on the material and age of pipe (see Table 2.5 for value of coefficient C for different pipe materials)
R = hydraulic radius and
S = slope of the hydraulic gradient.

Table 2.4 Values of K for different channel surface [5].

Sl. No	*Channel side*	*K*
1	Very smooth cement of planed wood	0.11
2	Unplaned wood, concrete or brick	0.21
3	Ashlar, rubble masonry or poor brickwork	0.83
4	Earth channels in perfect condition	1.54
5	Earth channels in ordinary condition	2.36
6	Earth channels in rough condition	3.17

Table 2.5 Hazen-Williams coefficient "C" value for different materials of pipe [6].

Sl. No	*Material of pipe*	*Hazen-Williams coefficient: "C"*
1	Brick sewer	90–100
2	Cast iron: new unlined (CIP)	130
	40 years old	64–83
	Asphalt coated	100
	Bituminous, cement lined	140
3	Concrete	100–140
4	Copper	130–140
5	Ductile iron pipe (DIP)	140
	Cement line DIP	120
6	Fibreglass pipe – FRP	150
7	Galvanized iron	120
8	Plastic:	130–150
	Polyethylene, PE, PEH	140
	Polyvinyl chloride, PVC, CPVC	150
	ABS – acrylonite butadiene styrene	130
9	Steel: new unlined	140–150
10	Vitrified clay	110
11	Wrought iron: plain	100

We have seen from Equation 2.10 that flow in a pipe is Q = A V.

Which becomes $Q = 0.281CD^{2.63}S^{0.54}$ 2.15

Where
Q = quantity of flow in pipe in gallon per minute (cum/m)
C = coefficient (see Table 2.5 for value of C)
D = diameter of pipe (m) and
S = hydraulic gradient.

Pressure–flow relation: Pressure is a measure of the force exerted by the water column when no water is flowing i.e. static pressure. The starting water flow under such pressure can be related by the following equation.

$$q = 0.66d^2\sqrt{p} \quad 2.16$$

Where
q = rate of flow at the pipe outlet (lpm)
d = inside diameter of pipe outlet (mm) and
p = static pressure (bar)

Example: For a faucet with a 12 mm supply and a flow pressure of 1 bar, then the flow in the pipe would be as follows.

$$q = 0.66 \times 12^2 \times 1$$
$$= 95.04 \text{ lpm}$$

2.5 Nomograph charts

A nomograph, also known as nomograph, is a two-dimensional graphical diagram, so designed to help make the approximate graphical computation of a mathematical function generally consisting of three variables represented by three graphical scales (not necessarily straight), arranged in such a manner that any straight line, called an index line, cuts the scales in values of the variables satisfying the equation. For equations having auxiliary variables with more than three variables constructing additional auxiliary scales, the equations may also be represented by nomographs. By the nomographs, the user gets the opportunity for quick graphical calculations of complicated formulas, to a practical precision. In water system design, two nomographs are mostly used. These are as below.

1 Nomograph of Manning's formula and
2 Nomograph of Hazen-Williams formula.

2.5.1 Nomograph of Manning's formula

Manning's equation, as given in Equation 2.9, is for calculating the velocity (v) of flow through a circular pipe running full (but not under pressure) or a non-circular cross-section of pipe or channel. The nomograph of Manning's equation is a diagram showing the relationship between the velocity and discharge of flow in a pipe or a channel geometry, slope and a friction coefficient expressed as a Manning's "n", as shown in Figure 2.3.

From Figure 2.4, it is found that a pipe with a roughness coefficient of 0.014 and a slope of 0.0025 has a flow velocity of 0.083 m/sec for a pipe diameter of 900 mm, in order to produce a discharge of 0.9 cum/sec.

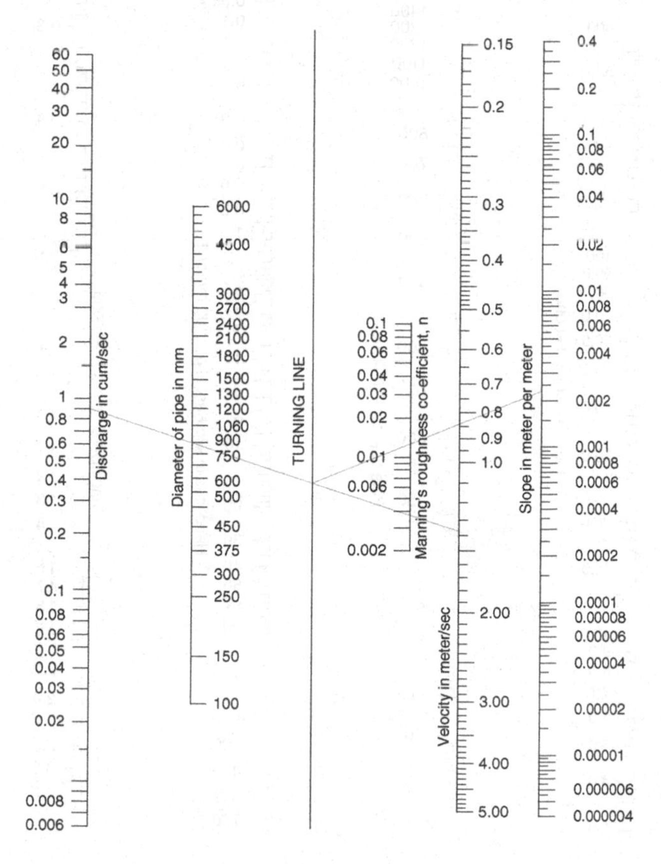

Figure 2.3 Nomograph for Manning's formula; where velocity in m/sec, loss of head in meter/meter; diameter of pipe in mm, and discharge in cum/sec.

2.5.2 Nomograph of Hazen-Williams formula

The nomograph of the Hazen–Williams equation, as represented by Equation 2.14, is a graphical representation of an empirical relationship which relates to the flow of water in a pipe with the physical properties of the pipe and the pressure drop caused by friction. It is used for a quick determination of pipe diameter for water supply piping systems.

Given any two of the parameters (discharge, diameter of pipes, loss of head or velocity), the remaining two can be determined from the intersections along a straight line drawn across the nomograph for the Hazen-Williams formula, as shown in Figure 2.4.

For example, for a flow of 60 lps in a 250 mm diameter pipe, the velocity of flow is 1.2 m/sec with a head loss of 9 kPa per 100 meter. Head losses in pipes with coefficient values other than 100 can be determined by using the correction factor given in Table 2.6. For example, if the head loss for $c = 100$ is 9 kPa/100 m, then the head loss for $c = 130$ would be $= 0.62 \times 9.0 = 5.58$ kPa/100 m.

2.6 Friction loss

While water is flowing in a pipe, friction loss or loss of pressure or "head" occurs due to the effect of the viscosity of water near the surface of the pipe. In mechanics, the term refers to the power lost in overcoming the friction between flowing water and the pipe surfaces.

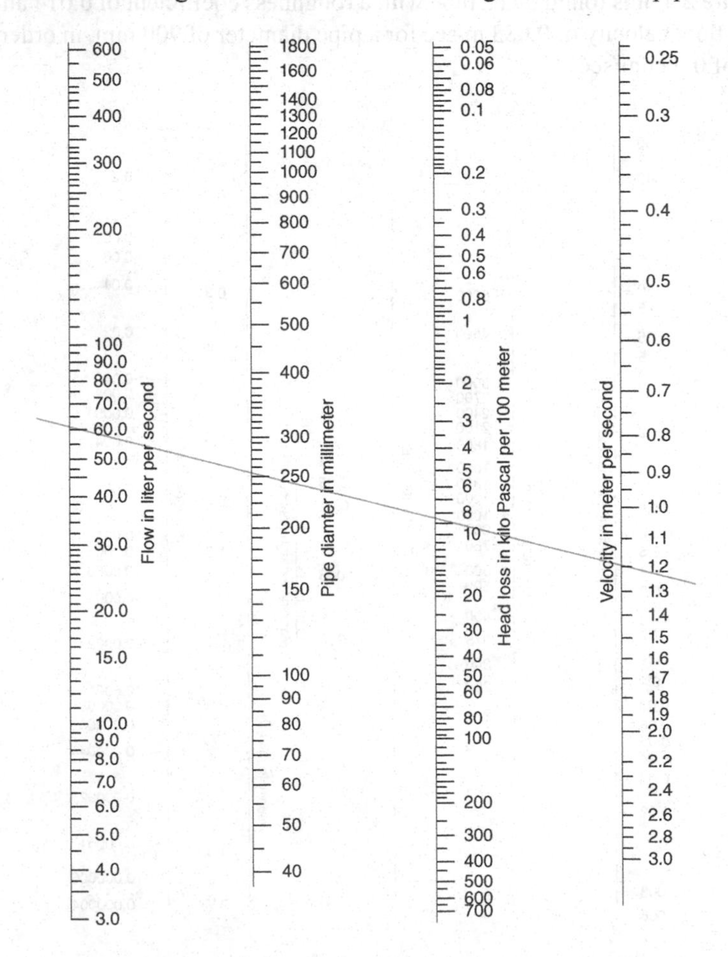

Figure 2.4 Nomograph for Hazen-Williams formula, based on C = 100; where velocity in m/sec, loss of head in kPa/100 m, diameter of pipe in mm and discharge in lps.

Table 2.6 Correction factor to determine head losses at values of C other than C = 100.

Corrected $h_l = K \times h_l$ at C = 100.					
Sl. No	*C*	*K*	*Sl. No*	*C*	*K*
1	80	1.51	4	120	0.71
2	100	1.00	5	130	0.62
3	110	0.84	6	140	0.54

Corrected head loss $h_l = K \times h_l$ when C = 100. K = 1.00; when C is 80 K = 1.51.

Friction loss basically occurs due to the shear stress between the pipe surface and the flowing water, which depends on the conditions of flow and the physical properties of the system. These conditions can be represented by a dimensionless number Re, called Reynolds number. Considering kinematic viscosity only, the number can be expressed by the following formula.

$$Re = \frac{1}{\nu} VD \qquad 2.17$$

Where
V = mean fluid velocity
D = diameter of the pipe and
ν = kinematic viscosity in sqm/sec = $\frac{\mu}{\rho}$, in which μ is viscosity of the fluid (kg/m/sec) and
ρ = density of the fluid (kg/m^3)

When the flow in a pipe is of comparatively low velocity, all liquid particles move in ordered tracks parallel to the pipe axis and the flow is called Laminar flow at Reynolds numbers of Re $\leq$ 2320. In this form of flow, the head losses occurring are determined only by the internal friction and can be expressed by the following formula.

$$H = f \frac{l}{d} \frac{V^2}{2g} \qquad 2.18$$

Where
f = a non-dimensional characteristic of the pipe friction = $\frac{64}{Re}$
Re = Reynolds number
l = length of pipe (m)
d = diameter of pipe (m)
V = velocity of flow (m/sec) and
g = acceleration due to gravity.

Reynolds number (Re): This is a dimensionless value used for identifying the flow type in a conveying media when there is a substantial velocity gradient (i.e. shear). In another way, it expresses the relative significance of the viscous effect compared to the inertia effect. The Reynolds number is proportional to the ratio of inertial force and viscous force. The types of flow, according to various ranges of Reynold number, are as follows.

1 Laminar flow; when Re < 2300
2 Transient flow; when 2300 < Re < 4000
3 Turbulent flow; when 4000 < Re

Table 2.7 Reynolds number for flowing one litre of water per minute, through pipes of different sizes [7].

Sl. No	*Pipe size (mm)*	*Reynolds number with 1 litre/min*
1	25	835
2	40	550
3	50	420
4	75	280
5	100	210
6	150	140
7	200	105
8	250	85
9	300	70
10	450	46

Table 2.7 gives the Reynolds number for flow of one litre of water per minute through pipes of different sizes.

References

[1] U.S. Department of the Interior, Bureau of Reclamation (1977) "Ground Water Manual, from the Water Encyclopedia" *Hydrologic Data and Internet Resources*, 3rd Ed, Edited by Pedro Fierro, Jr. and Evan K. Nyler, 2007, Retrieved from www.usgs.gov/special-topic/water-science-school/science/water-density?qt-science_center_objects=0#qt-science_center_objects on 25 December 2020.

[2] Universalium (2010) "Selected Physical Properties of Water" Retrieved from https://universalium.academic.ru/295229/Selected_physical_properties_of_water on 26 December 2017.

[3] Chow, V.T. (1959) *Open-Channel Hydraulics*, Tata McGraw-Hill Publishing Company Ltd, New York.

[4] Limerinos, J.T. (1970) "Determination of the Manning Coefficient from Measured Bed Roughness in Natural Channels" *Studies of Flow in Alluvial Channels*, Geological Survey Water-Supply Paper 1898-b, Washington.

[5] Ghani, N., and Adilah, A.A. (2017) "Uniform Flow in Open Channel" Chapter 2, Retrieved from http://ocw.ump.edu.my/pluginfile.php/17249/mod_resource/content/3/OCW%28UMP%29_Hydraulic_Topic%202.1.pdf on 25 December 2020.

[6] Engineering ToolBox (2004) "Hazen-Williams Coefficients" Retrieved from www.engineeringtoolbox.com/hazen-williams-coefficients-d_798.html on 25 December 2020.

[7] Engineering ToolBox (2003) "Water Flow in Tubes–Reynolds Number" Retrieved from www.engineeringtoolbox.com/reynold-number-water-flow-pipes-d_574.html on 25 December 2020.

3 Pipes and fittings

3.1 Introduction

Pipes are the most vital and major element of a plumbing system. Various kinds of pipes of different materials are available for plumbing installation. Different types and kinds of pipes are manufactured to meet the varied conditions of plumbing services. The choice of pipe depends upon cost, durability, quality and availability. Generally, the same type of pipe should be used in a plumbing system, but sometimes a combination of two or more type of pipe can be used to suit the purpose. To use the most suitable type of pipes and fitting, detailed knowledge regarding the merits and demerits of various pipes and their fittings is essential. This chapter describes various common types of pipes and fittings used in plumbing works.

3.2 Pipes

A pipe is a cylindrical or tubular hollow conduit, usually but not necessarily of circular cross-section, used mainly to convey substances which can flow. Pipes are the principal element of plumbing. Pipes are available in different sizes, lengths and strengths and made of various materials.

3.2.1 Pipe classification

Pipes may be classified in various ways as follows.

1 Material of pipe
2 Process of manufacturing
3 Method of pipe jointing and
4 Strength of operation.

A According to the material, some widely used pipes in plumbing are:

  1 Ferrous metal pipes, such as:

    a Iron pipe
    b Steel pipe and
    c Cast iron pipe

  2 Copper
  3 Brass
  4 Lead and
  5 Plastic.

DOI: 10.1201/9781003172239-3

B According to the process of manufacturing:

1 Brazed
2 Butt-welded
3 Lap welded
4 Riveted and
5 Soldered.

C According to the methods used in pipe jointing:

1 Threaded
2 Flanged
3 Spigot and
4 Slip.

D According to the strength of operation:

1 Standard
2 Medium strong or heavy and
3 Extra strong or extra heavy.

3.2.2 Pipe sizes

Globally, there are two standards of common pipe sizes: 1. the American (ANSI/ASME/API) standard, and 2. the European (DIN) system. In the American system, the pipe diameter is known as "nominal pipe size" (NPS) or "nominal bore" (NB), which is expressed in imperial units. Nominal pipe size (NPS) is the number that defines the size of the pipe i.e. a 100 mm pipe means that the nominal size of that pipe is 100 mm. In the European system, this is known as the "nominal diameter" (DN), which is expressed in metric units, generally in millimetres (mm). There are various common sizes of pipes available, along with some uncommon sizes, as shown in Table 3.1.

Table 3.1 Nominal pipe size, nominal diameter and outside diameter for sizes of pipes [1].

Sl. No	*Nominal diameter DN [mm]*	*Outside diameter OD [mm]*
1	15	21.3
2	20	26.7
3	25	33.4
4	32	42.16
5	40	48.26
6	50	60.3
7	65	73.03
8	80	88.9
9	100	114.3
10	125	141.3
11	150	168.28
12	200	219.08

Sl. No	*Nominal diameter DN [mm]*	*Outside diameter OD [mm]*
13	250	273.05
14	300	323.85
15	350	355.6
16	400	406.4
17	450	457.2
18	500	508
19	600	609.6

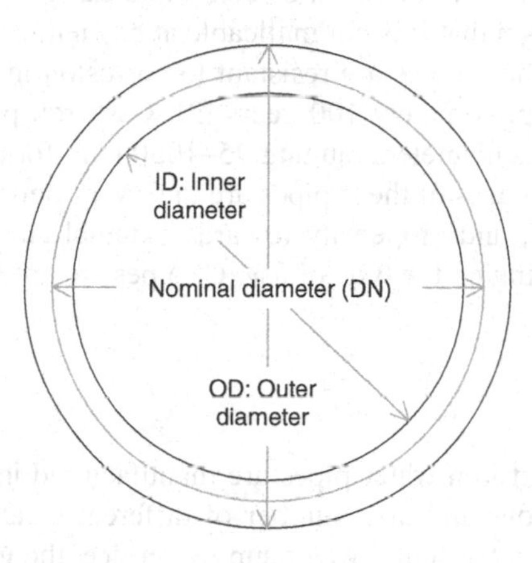

Figure 3.1 Pipe size indicators.

The pipe schedule represents pipe wall thickness. The range of manufactured pipe schedules is very wide. The specified pipe schedules are: 5, 10 (light); 20, 30, 40 or STD (standard); 60, 80 or XS (extra strong); 100, 120, 140 and 160 or XXS (double extra strong). The higher the number, the thicker the pipe. Commonly used are 40 and 80 schedule pipes. Various expressions of a cross-section of pipe are shown in Figure 3.1.

Lengths of plastic pipes are normally manufactured in overall lengths of 6 m. Pipes can also be made in overall lengths of 5.8 m to fit inside containers. Some smaller diameter pipes with plain ends are manufactured in overall lengths of 4 m.

3.3 Pipes of different materials

3.3.1 Iron pipe

Iron pipe is extremely durable and may be expected to have a service life in excess of 100 years [2]. It is, however, subject to corrosion; this may produce a phenomenon called tuberculation, in which scales of rust coat the inside of the pipe, reducing its diameter and increasing its relative roughness.

Galvanized iron (GI) pipe: Galvanized iron pipe or GI pipes are widely used in plumbing, mainly for conveyance of water. Galvanizing consists of applying zinc coating to the metal. Galvanizing is done by dipping clean-surfaced iron pipes into pure molten zinc. Galvanized iron pipes are used mainly for preventing pipes from corrosion. But if the galvanizing is done poorly and the coating does not adhere properly to the iron surface, corrosion takes place under the zinc coating. Another disadvantage is that GI pipes easily get ruptured by freezing. Furthermore, the corrosive action of the liquid conveyed and the soil in which it is buried affect galvanized iron pipes. In aggressive corrosive conditions, galvanized iron pipes should not be used.

GI pipes are manufactured in three classes: class A (light), class B (medium) and class C (heavy). The allowable working pressure for a class B GI pipe is 20 kg/sqcm and for a class C pipe is 30 kg/sqcm [3]. GI pipes are connected mostly by threaded joints. Welding of galvanized pipe should be avoided since it results in emission of a toxic zinc oxide gas from the zinc coating.

Cast iron (CI) pipe: Cast iron is defined as iron alloys having 3 to 4 percent carbon and small amount of sulfur, phosphorous, silicon and manganese. The carbon may be present as graphite as in gray cast iron or in the form of combined carbon as in white cast iron. Cast iron contains so much carbon that it is not malleable at any temperature. Cast iron pipes are abbreviated as CI pipes. These pipes are resistant to corrosion in favourable environments and have a long life of approximately 100 years [2]. Cast iron pipes of socket and spigot type are available in various diameters ranging 75–1050 mm. for an effective length of 3 m to 6 m. The major disadvantages of these pipes are their very poor or non-elastic behaviour, lower mechanical strength, and propensity towards external and internal corrosion under aggressive conditions. Basing on the type of use, CI pipes are grouped into two categories:

1 CI water pipe and
2 CI soil pipe.

Cast iron water pipe: Cast iron water pipes are manufactured in accordance with a number of standard specifications and in a number of different weight and lengths. Cast iron water pipes, selected for any particular water supply service, the internal pressure and other stresses to which it may be exposed should be compared with the characteristics given to the different classes of standard CI water pipe and fittings.

Although cast iron is resistant to corrosion, aggressive water may cause pitting of the interior or tuberculation in the interior. So, the outside is protected by providing a bitumastic tar coating and the inside is covered by a layer 3 mm thick of cement mortar.

Three types of ends are standard for cast iron water pipes and fittings. These are as follows.

a Bell and spigot
b Flanged and
c Screwed.

Cast iron soil pipe: Cast iron soil pipes are lighter than cast iron water pipes and are primarily for use as drainage pipes. Three standards of CI soil pipes are manufactured:

a Standard
b Medium and
c Heavy.

According to the manufacturing procedure, cast iron pipes are grouped in the following ways.

1 Pit cast iron pipe: These are pipes cast vertically in sand mould.
2 Spun cast iron pipe: These pipes are manufactured by pouring molten cast iron into a horizontal water-cooled metal mould, which is rotated at high speed about its longitudinal

axis. A maximum of 1200 mm in diameter and a length up to 6 m can be manufactured. Pipe manufactured in this process poses uniform thickness throughout.

Spun cast iron pipes are grouped into two categories.

1 Grey cast iron spun pipe and
2 Ductile cast iron spun pipes.

Ductile cast iron is produced by introducing a carefully controlled amount of magnesium into a molten iron of low sulphur and phosphorous content. So, pipes of this material are stronger, tougher and elastic (less brittle) than grey CI spun pipes, and can withstand more deformation and possess longer life. Both ductile and grey cast iron pipes are noted for long life, toughness, imperviousness and ease of tapping as well as the ability to withstand internal pressure and external loads.

Cast iron soil pipes are manufactured in four varieties, as shown in Figure 3.2. These are as follows.

1 Standard pipe
2 Single hub (bell)
3 Double hub and
4 Hub-less.

The end of the pipe when chamfered out to form a socket is a bell or hub. When the other end is cast with a raised bead to fit snugly into the bored-out bell, it is a spigot end. Cast iron pipes with bell or hub at one end and spigot at the other end is a standard CI pipe. The pipe is termed a single hub pipe when there is a bell at one end but no spigot at the other. For special purposes, if a bell is made at both ends, then the pipe is a double hub pipe. Hub-less pipes have neither a hub nor a spigot at the ends.

To protect CI soil pipes from corrosion, a cement lining is provided on the inner surface of the pipe. The cement lining consists of a 1:2 Portland cement mortar, which is applied centrifugally and is tapered slightly at the ends. The lining may be cured by coating it with bituminous seal or by storing the lined pipe in a moist environment. To prevent external

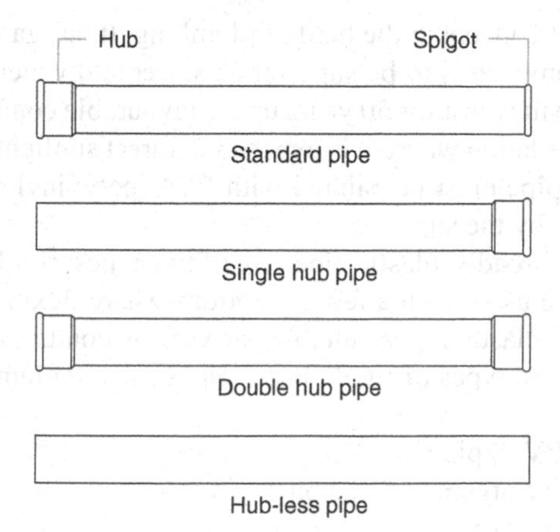

Figure 3.2 Cast iron pipes of different features.

corrosion of pipes in unfavourable soil conditions, the pipe may be encased in polyethylene tube as construction proceeds.

3.3.2 Steel pipe

In iron, rusting is greatly reduced by adding alloy, generally about 10 percent chromium, and then it is made as stainless steel. Steel pipe may be used for water lines, particularly in circumstances where larger diameter and high-pressure resistance is needed. In fire-fighting systems with water, steel pipe might be preferred. Steel pipe ensures the highest quality of water by not adding any chemicals from the pipe material. Steel pipe, weight for weight, is cheaper than iron pipe. Since its walls are relatively thin, it buckles readily.

Steel pipes are characterized as having high strength, the ability to yield without breaking and great resistance to shock. It has a smooth interior surface and can withstand pressures up to 1724 kPa [4]. So, in a water supply system where about 690 kPa or more pressure is needed, steel pipe is preferred in that cases.

Steel pipe is subjected to deposits of salt and lime, which gradually accumulates in the pipe and eventually choke the flow. Furthermore, these pipes are corroded by alkaline and acidic water. So, careful protection against corrosion is sometimes needed. Common outside protective coatings include galvanizing, paint and wrapping. Protective measures should be taken according to the intensity of the corrosive environment of the pipe and placement i.e. above ground or underground.

Steel pipes of the larger sizes are made from steel plate bent to a circular form; the edges of the plate are either lap-welded or butt-welded. The smaller sizes range from 32 mm to 450 mm diameter pipes and are made from billet bars or ingots of hot steel, which are pierced and then rolled into a cylinder of the right dimension. Steel pipes are usually connected by threaded joints and grooved joints.

3.3.3 Plastic pipe

Plastics are materials made by the chemical reactions of natural products or from synthetic organic compounds. The main components of organic compounds are carbon (C) and hydrogen (H). The basic raw materials used to make plastics are natural compounds, such as coal, petroleum, natural gas and cellulose (from wood fibres). These raw materials are made into plastic resins.

Plastic pipe is the latest concept in the field of plumbing. It has gained widespread acceptance after it proved in some cases to be superior as sewer and water supplying pipes. Tests indicated that plastic pipe may last for 50 years under favourable condition [5]. Plastic pipe is not good for outdoor installation where it is exposed to direct sunlight. If outdoor installation is unavoidable, then the pipe must be painted with PVA (polyvinyl acetate) paint to protect the pipe from being dried by the sun.

Type of plastic pipes: Broadly, plastic pipes are of two types: rigid and flexible. In plumbing, mostly rigid pipes are used, with a few exceptions where flexible pipes are used. There are various types of rigid plastic pipes suitable for various conditions and purposes of use. Following are very common types of rigid plastic pipes used in plumbing systems.

1 Polyvinyl chloride (PVC) pipe
2 Acrylonitrile butadiene styrene (ABS) pipe
3 Polyethylene (PE) pipe and
4 Polypropylene (PP) pipe.

Polyvinyl chloride (PVC) pipes: This is a rigid thermoplastic pipe possessing high tensile strength and good modulus of elasticity. PVC pipes have a good chemical resistance capacity to a wide range and strength of corrosive fluids, but may be damaged by ketones, aromatics and some chlorinated hydrocarbons. This pipe can withstand temperatures of about 51°C for type 1 (normal impact) and 40°C for type II (high impact). PVC pipes are jointed generally by using cement solvents. Jointing by threading is also done in some cases.

Unplasticized polyvinyl chloride (uPVC) pipe: These pipes are made more rigid than PVC pipes so these are also called rigid PVC pipes. uPVC pipes are manufactured by plastic resins extruded in special machines under closed temperature control and manufacturing process. uPVC pipe may deform under superimposed load. They are suitable for temperature of up to 60°C. This pipe should not be used for discharging extreme hot water or wastewater. These pipes should not be installed where exposed directly to sunlight due to being highly susceptible to loss of strength due to heat and its high coefficient of thermal expansion (70×10^{-6} mm/m/°C). So, rigidly fixed pipes may deform due to high thermal expansion. As a result, joints may fail resulting leakage. To avoid deformation provision for pipe expansion should be kept in the piping system. A variety of injection moulded uPVC fittings are manufactured to fit with the uPVC piping system and provide flexibility of design. uPVC pipe and fittings can be joined by using solvent cement or by pushing the spigot end into the socket fitted with a rubber O-ring. This type of joint permits some expansion within the socket.

Plasticized PVC Pipe: PVC pipes are plasticized by adding some kind of rubberized material with PVC to make the pipe flexible in nature. Plasticized pipe has comparatively lower strength and lower temperature withstanding capacity than uPVC pipe.

Chlorinated polyvinyl chloride (CPVC) pipe: CPVC pipes are useful for handling high temperature, corrosive fluids, having a maximum service temperature of 99°C. It has excellent chemical resistance capacity like PVC pipes. These pipes are primarily recommended for hot water supply systems and process piping for hot, corrosive liquids. It may also be used in normal water supply systems.

Polyvinyl dichloride (PVDC) pipe: These pipes are suitable for handling hot corrosive liquids at temperatures 40°C to 60°C above the limits of other vinyl plastic pipes. These pipes can be used for hot and cold water supply. These pipes are joined by solvent welding and threading.

Polyvinylidene fluoride (Kynar) pipe: These are fluorine-containing thermoplastic pipes suitable for handing chlorine-containing fluids at temperatures of up to 107°C.

Acrylonitrile butadiene styrene (ABS) pipe: ABS is a copolymer of Styrene and Acrylonitrile grafted to polybutadiene. These pipes are intended to be used in any conventional sanitary drainage or storm drainage, conveying water for irrigation, gas transmission and venting. In drainage systems, it offers good abrasion resistance and inhibits scale formation.

ABS pipes are popular for its high impact strength, and so are very tough. These pipes can be used at wide range of temperature, from −40°C to 70°C. At sub-zero temperatures, it possesses good impact strength. ABS pipes have a lower chemical resistance capacity and design strength than PVC pipes. The coefficient of linear expansion for ABS plastic pipe is 63×10^{-6} m/m/°C. According to Australian standards, a maximum static working pressure for ABS fittings ranges from 896 kPa to 1517 kPa at 20°C. Solvent cement welding and threading are the most common recommended processes for jointing these pipes. ABS plastic pipe should not be installed in areas near any high heat sources.

Polyethylene (PE) pipe: Polyethylene falls under polyolefin, a semi-crystalline thermoplastic group of plastics. The chemical formula is written as $(CH_2\text{-}CH_2)_n$. It is considered an environmentally friendly hydrocarbon product. PE pipes can be separated by its density into three groups according to crystalline structure percent, as mentioned below.

1 Low density polyethylene (LDPE)
2 Medium density polyethylene (MDPE)
3 High density polyethylene (HDPE).

The cost of PE pipe is relatively high, but it is popular due to its fast, easy, safe and reliable jointing system. The jointing of PE pipe to PE pipes and fittings etc. is done by an electro-fusion method. The melting temperature and viscosity of PE material helps in developing a stronger, welding-type joint. PE electro-fusion connection systems also make the repair work of PE pipeline systems easy.

PE pipe is popular for drinking water and natural gas transportation systems. High concentrations of chlorine in water or high residual chlorine for inappropriate disinfection procedures may harm or reduce the durability of PE piping systems.

Polypropylene (PP) pipe: This is the lightest thermoplastic piping material, but it offers higher strength and better chemical resistance than polyethylene pipes. PP pipes are excellent for laboratory and industrial drainage where acid, bases and solvents are involved. This type of pipe is widely used in the petroleum industry.

Considering the polymer structure of the PP materials, the following three categories of PP can be found:

1 PP-H (homopolymer PP)
2 PP-B (polypropylene block copolymer) and
3 PP-R (polypropylene random copolymer).

Of these three, PP-R pipe is generally used in plumbing systems. PP-R pipes are preferred in the plumbing of hot and cold water supply systems. PP-R is joined by heating both the socket and the pipe end to be inserted. When the heated parts achieve contact, both parts are fused together to become one. This is called fusion welding.

PP-R is widely used for pressurized water supply systems, mostly in hot water supply systems where use of CPVC would not be effective. PP-R pipe jointing needs special tools, namely fusion-welding tools, as well as trained professionals to ensure proper jointing, good workmanship and safety.

3.3.4 Copper pipe

Copper pipe is expensive but is highly regarded for its corrosion resistance to most water, flexibility, ease of installation, relatively light weight for a metal pipe and low resistance to flow. In plumbing, copper pipe is generally used in hot water supply systems when metal pipe is opted for the purpose. In case of fire, the copper piping may remain intact and could possibly be reused. Copper pipe is comparatively expensive and involves a high cost for installation. Copper pipes may corrode if they handle water that is too acidic. Copper pipes can slightly change the taste of water to a metallic taste.

Copper pipes are available in two principal types.

1 Flexible copper tubing and
2 Rigid copper piping.

There are three grades of copper pipes, designated as K, L, and M, are commonly used now. Type K possesses the thickest walls, preferred for water distribution and underground services and general plumbing but is not recommended for use in natural gas applications, as it can damage the joints. Type L is thinner than type K and the most common for general plumbing; and type M possesses the thinnest walls, suitable for soldered and flare fittings only.

There is another type, known as DWV copper pipe, which is only approved for use as drain and vent pipes, not for water systems. DWV copper pipes are found in larger diameters and have comparatively thinner walls than the other types of copper pipe. These pipes are now rarely found in use.

Flexible copper tubing: Flexible copper tubing is available in coils ranging from 5 m to 80 m in length. It is also manufactured in two grades. The standard grade type, L, is for indoor plumbing where internal pressure is moderate. Type K is for outdoor and underground use where pressure is considerably high.

Two type of fittings, either flare or solder type, can be used for flexible copper tubing. Flare fitting is more expensive than solder type, but the advantage is that there is no cost for heating needed as is required in a solder joint. Flared joint is recommended for repair or in old works to avoid using a torch for heating, and soldered joint is recommended for a new job.

Rigid copper tubing: Rigid copper tubing is extensively used in new construction. A full line of solder fittings and valves are available for installing with this type of piping. Two types of rigid copper tubing are manufactured: type L for indoor use and type M for outdoor and underground installations. In most cases, the direction of the pipe is changed by using appropriate fittings, but sometimes slight bending using a power bender is done where necessary.

3.3.5 Concrete pipe

Concrete pipe is supposed to be the most durable, long-lasting and cost-effective pipe among all pipes recommended for underground use. The life of concrete pipe might be as long as 150 years [6]. Considering strength, concrete pipes can be classified as (1) non-pressure pipe, (2) pressure pipe and (3) special-purpose pipe. Reinforced concrete pipe is a composite structure and is specially designed using reinforcement in cement concrete. Concrete pipe may be reinforced or non-reinforced. To make reinforced cement concrete (RCC) pipe, iron wire or steel bars are used as reinforcement. With these reinforcing wires or bars, cylindrical-shaped cages are made, which are finally embedded in cement concrete, to shape up as pipe. Although concrete pipes can be manufactured in various shapes, mostly circular pipes are used for their good self-centring and efficient hydraulic characteristics. The diameter of non-reinforced concrete pipe as generally manufactured ranges from 100 to 900 mm; reinforced concrete pipes are manufactured ranging in diameter from 150 to 3600 mm. The crushing strength of concrete pipe is one of the most important factors to be considered in selection of this pipe. The crushing strength for concrete pipes of different diameter is given in a later chapter.

Concrete pipe ends are generally constructed in three forms for different types of jointing, as shown in Figure 3.3.

1 Bell-spigot type
2 Rebated joint (tongue-and-groove) type and
3 Butt joint (plain-end) type.

Concrete pipes are manufactured in effective lengths of 1.2, 2 and 2.4 m. For special cases, 750 mm long pipes are also manufactured. The main advantage of concrete pipes is that these

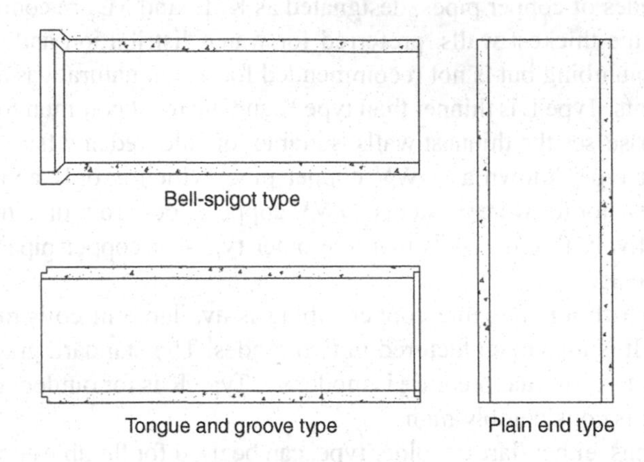

Figure 3.3 Concrete pipes of various jointing system.

are excellent for corrosion resistance. The disadvantages are that these pipes are bulky and heavy and thus require careful transportation and handling. The layout process of these pipes is costlier than that of plastic or iron pipes.

3.4 Fittings

Pipe fittings are those devices attached to pipes for the following purposes.

1. To alter the direction and diameter of pipe
2. To make branching in pipe
3. To connect two pipes, pipe with fixtures, accessories or equipment and
4. To close an end.

These are various pipe fittings classified according to different respect, such as the following.

1. With respect to material
    a. Galvanized iron
    b. Steel
    c. Cast iron
    d. Plastic
    e. Copper etc.
2. With respect to the method of connecting
    a. Screwed
    b. Flanged and
    c. Bell and spigot etc.
3. With respect to strength
    a. Standard
    b. Medium strong (heavy) and
    c. Double strong (extra heavy).

3.4.1 Uses of various fittings

There is a great multiplicity of fittings shown in Figure 3.4, divided into several groups and classified with respect to their use. The groups of fittings and the respective fittings falling in the group are as follows.

1 Fittings for pipe extension and joining.

Nipple: By definition, a nipple is a piece of pipe of length less than 300 mm, with thread on both ends. With respect to length, nipples may be classed as close, short and long, as shown in Figure 3.4a.

There is another type of nipple, which has a right-hand thread on one end and a left-hand thread on the other, with a hexagon nut in the middle forming part of the nipple. This type of nipple is used instead of unions.

Coupling: Couplings are used for joining pipes of full length. These are generally made with right-handed female (inside) threads. Another form of coupling, called an extension piece, has male (outside) thread at one end and female thread at the other end. Coupling and extension piece are shown in Figure 3.4b.

Union: Union is a fitting used for joining pipes. The additional purpose of using union is to facilitate the opening of a pipe joint in the future, for maintenance or other works. Pipe joints are opened by opening the union on the pipe initially. Union consists of four parts, as shown in Figure 3.4c. Use of union requires the appropriate alignment of pipes to be jointed to secure a tight joint. To avoid this difficulty, a ground joint union has been devised.

2 Fittings for reducing or enlarging the diameter of pipes.

Reducer: These fittings are used to reduce the diameter of larger size pipes to smaller size and at the same time to enlarge from lower size pipe to higher size pipe. But in that case it is not termed as enlarger because it is a custom. Reducers have female thread at both ends, as shown in Figure 3.4d.

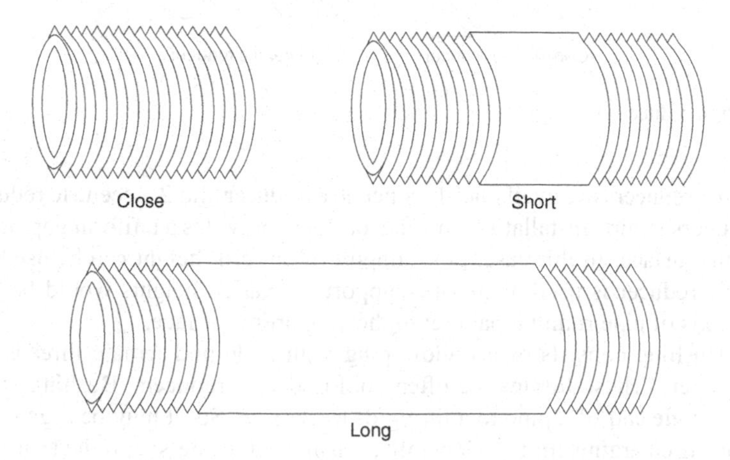

Figure 3.4a Various pipe-nipples.

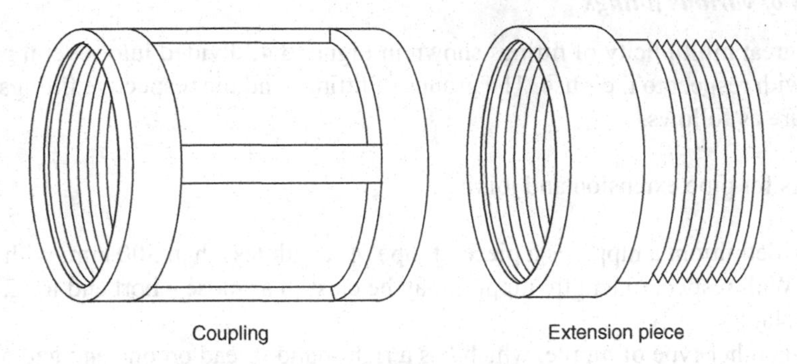

Figure 3.4b Various pipe-couplings.

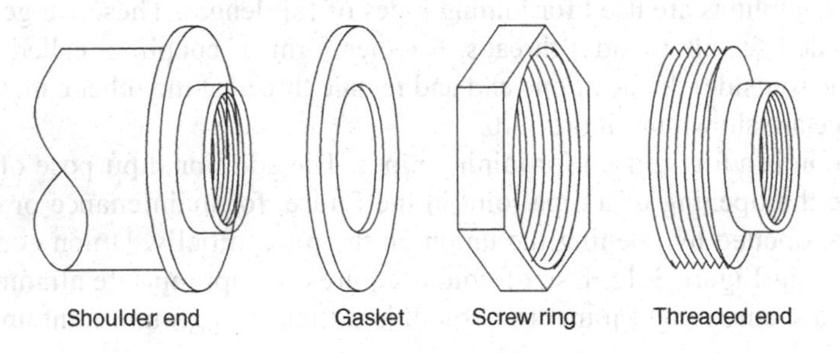

Figure 3.4c Parts of a union.

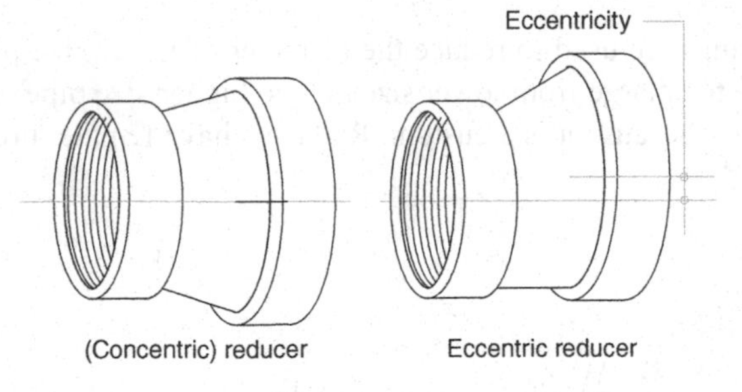

Figure 3.4d Pipe reducers.

Two types of reducers are available: 1. concentric reducer and 2. eccentric reducer. Use of eccentric reducers in pipe installation on walls or floors provides a uniform gap between pipe and supporting surface. In this case, pipe support of uniform height can be used. However, if a concentric reducer is used, then pipe-supports of varied heights would be required to maintain the axis of pipe running parallel to the supporting surface.

Bushing: Bushing consists of a hollow plug with male and female threads to suit the different diameters. These fittings are often confused with reducers. Bushing is often used to connect the male end of a pipe to a fitting of larger size. So, it may be regarded as either a reducing or an enlarging fitting. Generally, bushing of a one size reduction is used, but reducing two or more sizes is also available. Bushing should not be used for the purpose of reducing pipe sizes. Bushings are listed by the pipe size of the male thread. Thus, a 20 mm bushing joins a 20 mm fitting to a 12 mm pipe. Figure 3.4e illustrates various types of bushings.

3 Fittings used for changing the direction of pipe run.

Elbow: Elbows are the pipe-fittings for changing the direction of the pipe in any of several standard and specified angles. For water, gas and steam supply pipes, standard angles are 90° and 45° and special angles are 60° and 22½°. The angle of elbow is the angle between the axis of one arm and the projected axis of the other arm, as illustrated in Figure 3.4f.

Offset: When a pipe run is to be made in a position parallel to, but not in alignment with, the balance of the pipe run and the distance between the axes of the two pipes run is of standard dimension, then the shifting of the pipe run is done conveniently by using a fitting called an offset. Offsets shown in Figure 3.4g are commonly used in drainage pipes.

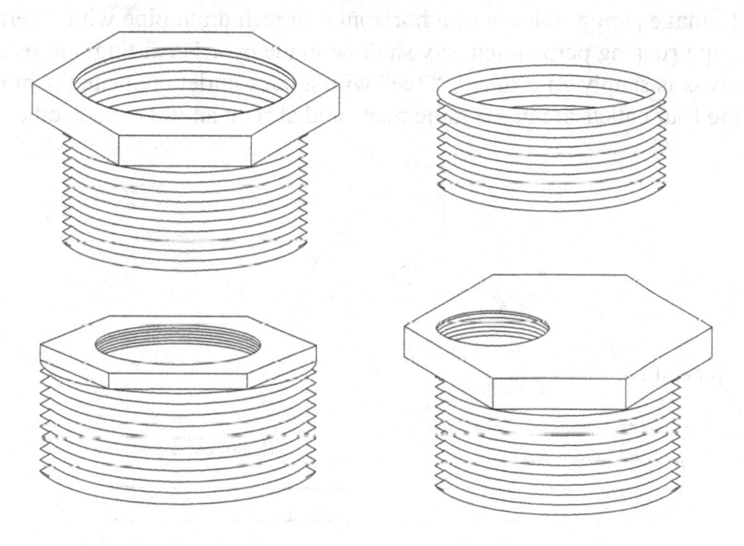

Figure 3.4e Various types of pipe-bushing.

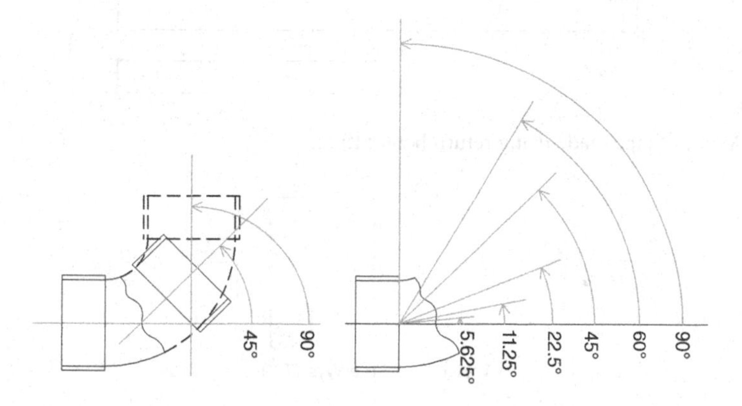

Figure 3.4f Various elbows of standard angles.

Return bends: Return bends are shown in Figure 3.4h, consisting of U-shaped fittings with a female thread at both ends. These fittings are not widely used except in steam heating and water boilers.

4 Fittings used for branching.

Tee (T): These fittings, also called sanitary T-fitting, as shown in Figure. 3.4i (i), are used for making a branch of 90° to the main pipe and always have the branch at right angles. Tees are made in a multiplicity of sizes and patterns.

Wye (Y): In Y-branches, the side outlet is set at an angle of 45° or 60°, as shown in Figure. 3.4i (ii). When a side outlet is provided in two opposite sides, then it is termed a double Y-branch.

Tee-Wye (T-Y): Tee-wye fittings are more or less similar to Tee-fittings, except for the angling of the branch line, which makes smaller, short-sweep radius bends, as shown in Figure 3.4i (iii), to reduce friction, turbulence and deposition of any entrained solids at the junction. So, in drainage piping, joining of a horizontal branch drain pipe with a vertical stacks or any drain pipe running perpendicularly shall be made by a T-Y fitting, not by a T-fitting.

Cross: A cross is simply an ordinary "Tee" with a back outlet opposite the branch outlet. The axes of the four outlets are in the same plane and at right angles to each other. In a cross,

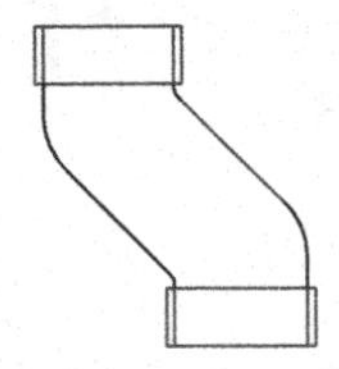

Figure 3.4g A pipe offset.

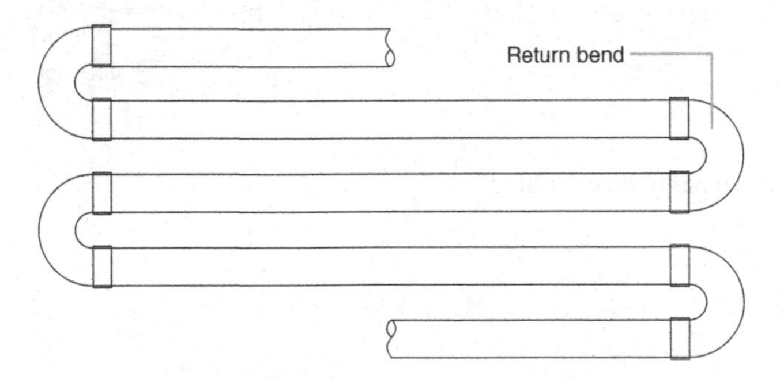

Figure 3.4h A coil of pipe made using return bend fitting.

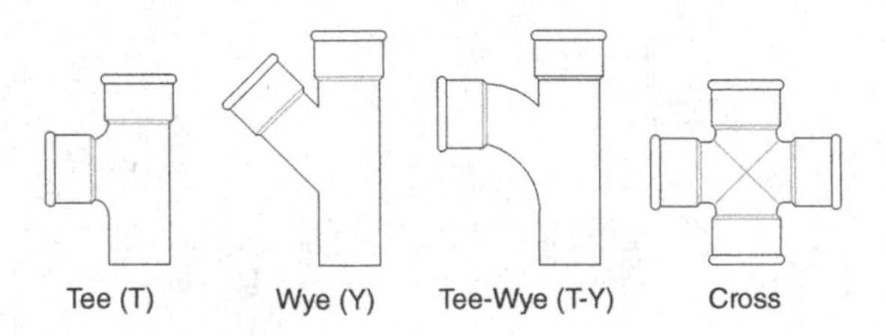

Figure 3.4i Various fittings for branching of pipe.

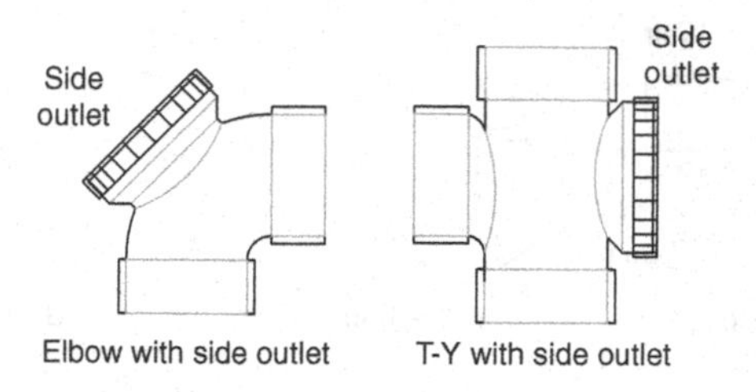

Figure 3.4j Fittings with side outlet.

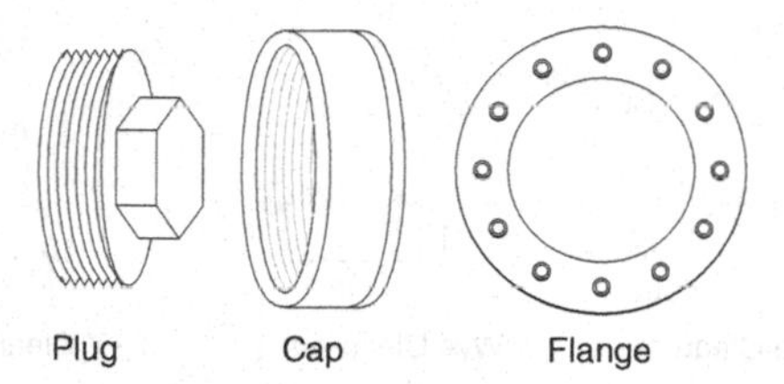

Figure 3.4k Various fittings for closing a pipe end.

the back outlet is always made the same size as the opposite side outlet of the "Tee" part. The other two sides are made in various combinations of sizes. Figure 3.4i (iv) illustrates a typical cross fitting used in a plumbing system.

Side outlet elbows/return: In this type of fitting, there is a third opening whose axis of run is at 90° to the plane of the elbow, Tee, Tee-Wye or return bend, as shown in Figure 3.4j. This opening always remains closed with a removable cap. These fittings are of much demand in drainage piping, due to the facility of inspecting the flow condition and doing corrective operations by opening the cap when required.

5 Fittings used for shut off or closing a pipe.

Plug: Plug is the fitting used only for closing the end of a pipe or a fitting which possesses a male thread outside. Figure 3.4k (i) shows various patterns of plug. Usually, a projected square head or a four-sided counter sunk head in small sizes and a hexagon head for larger sizes is provided for ease of plugging and unplugging a pipe.

Cap: Cap is the fitting used for closing the end of pipes or fittings which possess a female thread inside. Figure 3.4k (ii) shows various types of caps.

Flange: Flanges (also called blind flanges) are simply discs for closing pipes or fittings. Figure 3.4k (iii) illustrates a flange.

3.4.2 Fittings for special purposes

Trap: A trap, used in the plumbing system, is a device or fitting so constructed as to prevent the passage of sewer gas through it and yet not affect the fixture discharge to an appreciable extent. Figure 3.5 illustrates various traps: 1. P-trap 2. Q-trap, and 3. S-trap. P-trap is constructed in the form of the letter "P"; Q-trap is constructed in the form of the letter "Q"; and S-trap is shaped like the letter "S".

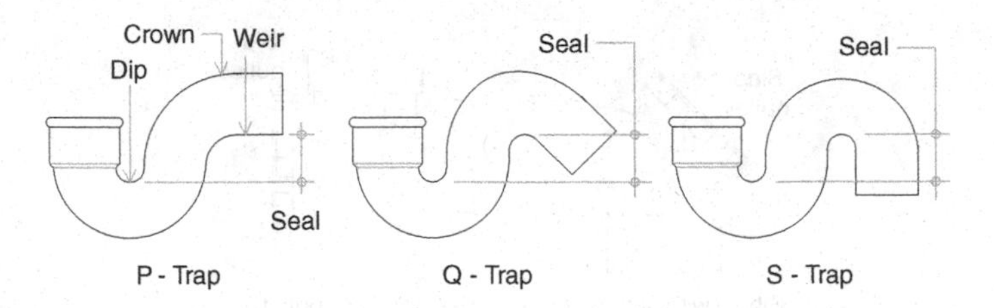

Figure 3.5 Various water seal traps.

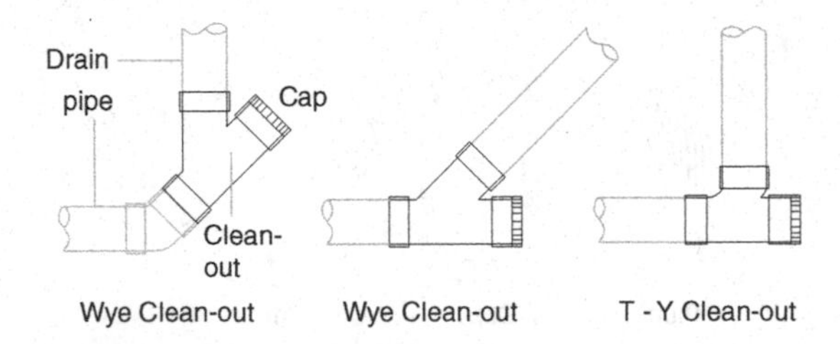

Figure 3.6 Various clean-outs.

The barrier of water column retained between the overflow and the dip of a trap is called a "water seal". There are two forms of water-sealed traps. These are known as i. common seal and ii. deep seal. Common seal has a depth of barrier of 50 mm as used in plumbing fixtures subjected to normal conditions. But for abnormal situations, such as extreme heat, and circumstances where complete ventilation cannot be obtained, a deep seal trap should be used, which has a seal depth of 75 mm to 100 mm.

The discharge pipe of all plumbing fixtures shall be supported by a water-seal trap, except as otherwise permitted by any code. It is good to install a trap just at the outlet of a fixture, but in special cases it can be installed at a vertical distance not exceeding 600 mm, and the horizontal distance not exceeding 750 mm, measured from the centreline of the fixture outlet to the centreline of the inlet of the trap.

Clean-out: Clean-outs are the fittings used to make the drainage pipe accessible without breaking it up. Clean-outs are provided by installing "Tee" or "Wye" fittings in the drainage pipe and plugging the unused opening with a removable cap, as shown in Figure 3.6. Clean-outs must be of the same diameter as the branch drainage pipe.

In drainage piping, clean-outs should be installed at some special and critical locations where there is likelihood of repeated blockage. The special locations are as follows.

1 Where changes of drainage branch pipe direction are made and located at the farthest end of the branch away from the stack, as shown in Figure 3.7
2 In very long and straight drainage pipes, clean-outs shall be provided at 15 m intervals
3 As close as practicable, to the point where the building drains leaves the building
4 At the base of every soil or wastewater stack and
5 At every 90° change of direction in sink and basin drain pipes.

Flexible jointer: Flexible joint is a connection between pipes and pipes or fittings that allows minor angular deflection or axial movement or a combination of both in service

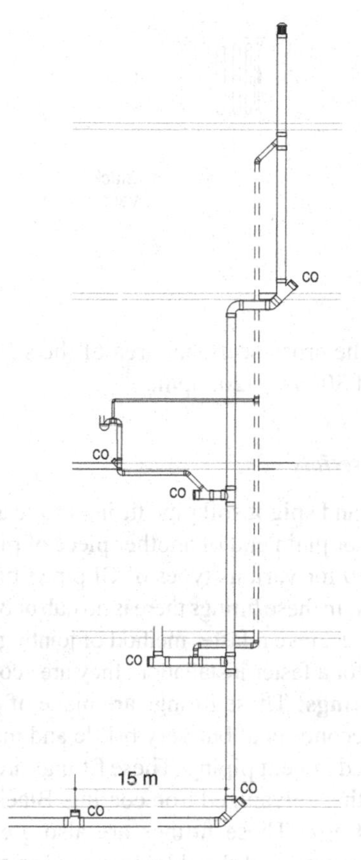

Figure 3.7 Preferred locations of clean-out installation.

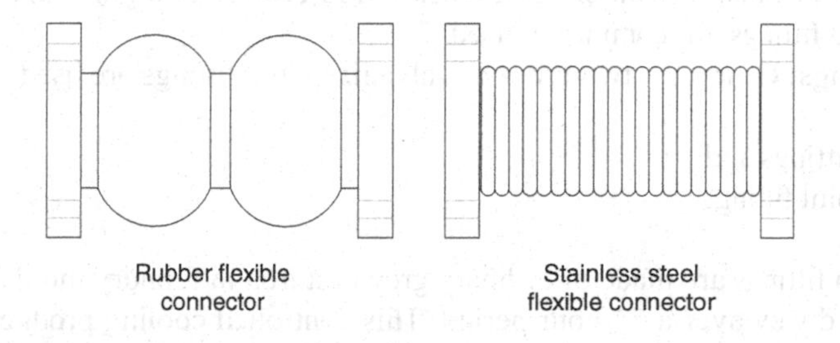

Figure 3.8 Various flexible connectors for pipes.

without impairing the efficiency of jointing. Flexible connectors also help prevent stresses due to expansion and contraction, isolate against the transfer of noise and vibration and compensate for misalignment. Use of rubber type flexible connectors is common but there are also stainless steel flexible connectors, as shown in Figure 3.8.

Cowl: A cowl, also called a vent cowl, as shown in Figure 3.9, is fitted onto the top of stacks for both supply of air into and exhaust gas off the drainage stacks. Cowl also protects stacks from unwanted matters getting inside, including rain. Openings in a cowl unit shall

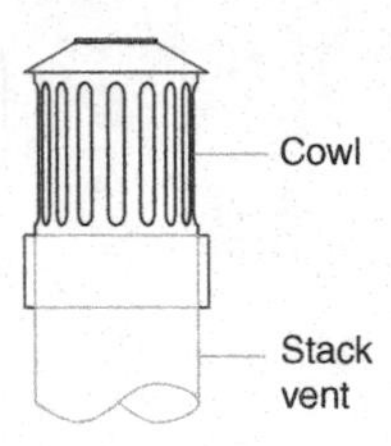

Figure 3.9 A cowl.

be capable of at least twice the cross-sectional area of the stack and shall maintain natural ventilation with a flow rate of 3028 to 4202 lpm.

3.4.3 Fittings of various materials

Cast iron pipe fittings: Bell and spigot soil pipe fittings have a bell or hub cast at the end of fittings, into which the spigot or plain end of another piece of pipe or fitting is inserted to join them together. See Figure 3.10 for various types of CI pipe fittings.

No-hub CI soil pipe fittings: In these fittings there is no hub or bell at their ends. No-hub CI soil pipe and fittings are devised to achieve a faster method of jointing the pipe and fittings mechanically. No-hub joints are made for a faster installation; they are economical and space-saving.

Standard CI threaded fittings: These fittings are made of grey cast iron in sand mould. CI threaded fittings are very economical but very brittle and may have some sand holes. So, these fittings are generally used in vent piping. These fittings are available with either a plain, uncoated finish (black) or with a galvanized iron coating. Black fittings are commonly used.

CI recessed drainage fittings: These fittings are also grey cast iron sand moulds. CI drainage fittings are cast with a recessed shoulder to provide a smooth interior surface so that when properly fitted, there is no obstruction to the flow of waste material through the pipe. These fittings are available with either a black tarred coating or a galvanized iron coating. Black tar-coated fittings are commonly used.

GI pipe fittings: Generally, two types of galvanized iron fittings are used.

1 Threaded fittings and
2 Grooved joint fittings.

Galvanized iron fittings are made of ordinary grey cast iron in foundry mould, but cooled in a very controlled way over a 72-hour period. This controlled cooling produces a change in the molecular structure of iron that makes it a tough, elastic material. After casting of these fittings, a protective galvanized coating is applied to it. These GI fittings must not be welded. The size of these fittings, which are used in water supply, varies from 12 mm to 150 mm. Various types of galvanized iron fittings are shown in Figure 3.11.

GI grooved joint fitting: These fittings are popularly used in larger diameter galvanized iron pipe, generally for water supply and distribution systems. The grooved joints have several advantages.

1 Easier and faster installation than threaded pipe joint
2 More reliable than threaded system and
3 Less expensive than other piping material.

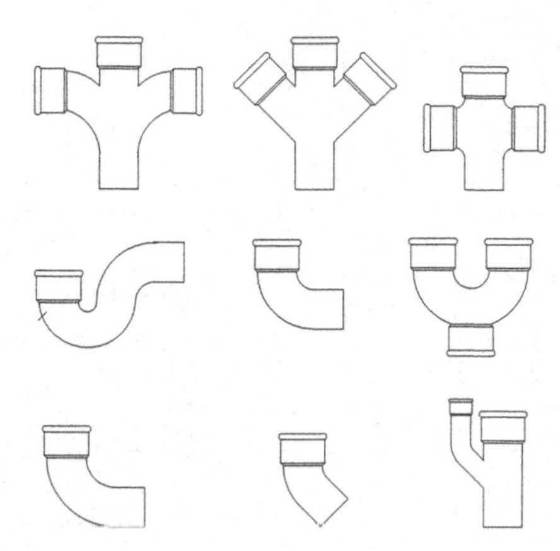

Figure 3.10 Various cast iron pipe fittings.

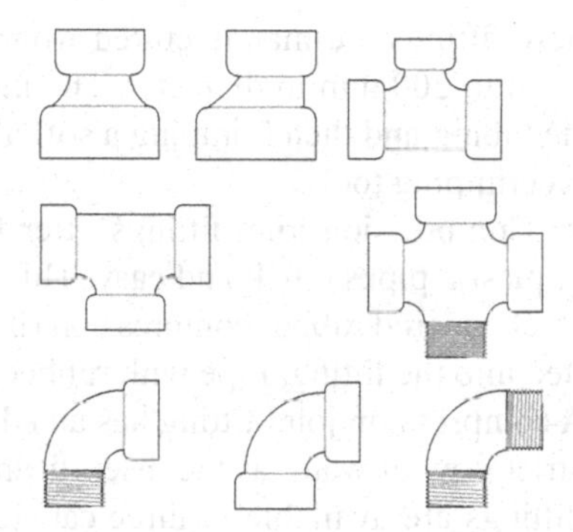

Figure 3.11 Some GI threaded fittings.

Groove joint fittings are made of malleable or ductile iron and have either an enamel paint finish or are galvanized steel-coated. The fittings used in the water supply system are mostly galvanized steel-coated. Although these fittings are available in 20 mm to 600 mm sizes, normally fittings of sizes 63 mm to 300 mm are used.

Plastic pipe fittings: Plastic pipe fittings are made of thermoplastic resins by an injection moulding process, in which heat-softened plastic is forced into a relatively cooled cavity with a shape similar to the desired fitting. The plastic is then solidified in the cavity to become the fitting. For use with various plastic pipes, fittings of a similar plastic nature are available. Plastic pipe fittings used in drainage are mostly smaller in size in comparison to the metal pipe fittings of similar diameter. Some plastic pipe fittings are shown in Figure 3.12. Characteristics of plastic fittings of different chemistry are discussed below.

uPVC fitting: These plastic fittings are light coloured and used in uPVC pipes or PVC pipes of the same nature.

CPVC fittings: CPVC plastic fittings are used in CPVC plastic pipes. The fittings are joined by solvent welding.

ABS fitting: This type of fitting is commonly used in sanitary drainage, vent pipes and storm drainage systems both above and below ground. These ABS plastic pipe fittings are used to join ABS plastic pipes with solvent welding.

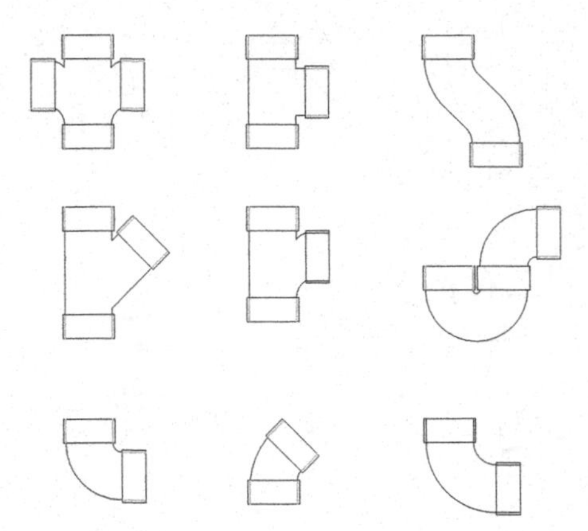

Figure 3.12 Various plastic pipe fittings.

Plastic insert fittings: These fittings are manufactured from grey acetate plastic. Most available sizes vary from 12 mm to 200 mm in diameter. The insert fitting joint is made by inserting the fitting into plastic tubing and then crimping a soft aluminium or copper clamping ring over the tubing with a crimping tool.

Compression joint fitting: Compression joint fittings offer fast and simple installation techniques to make joining of plastic pipes quick and easy. This type of fitting consists of a plastic body with hubs into which tapered rubber compression ring fits. A rubber ring is fitted onto the pipe end, to be inserted into the fitting. Pipe with rubber ring is then pushed into the fitting to make a tight joint. A compression joint fitting has an advantage over an insert joint fitting because it does not restrict flow of water as the insert fitting does.

Steel pipe fittings: Steel fittings are available in three categories: 1. cast steel, 2. semi-steel and 3. forged steel. These are mostly screwed fittings, but cast steel fittings with flange are also available. Cast steel fittings are suitable for 680 kg hydrostatic pressure, and semi-steel fittings are suitable for 907 kg hydrostatic pressure. To withstand more pressure, forged steel fittings are to be used.

Copper tube fittings: Copper tube fittings are made of either bronze (an alloy of tin and copper) or wrought copper. Bronze fittings are manufactured by casting molten bronze in sand moulds. Wrought copper fittings are formed by the hammering process using copper mill products. There is a little difference between these two types of copper tube fittings. The factors governing the choice between cast bronze and wrought copper fittings are cost, installation practice and availability. Figure 3.13 illustrates some copper fittings.

Copper fittings are joined with copper tubing by either a soldered joint or a flared joint. Soldered joint fittings are available in the following two categories.

Solder joint pressure fittings: These fittings are used in L and M type (hard) copper tubing for water supply systems. Copper pressure fittings have short radius or sweep on elbows, tees and other fittings and deeper solder sockets than drain–waste–vent (DWV) copper fittings. These fittings are available in 32 mm to 200 mm.

Solder joint DWV fitting: These fittings are used in DWV hard copper tubing, used for drainage waste and venting mainly above ground. Solder joint DWV fittings are characterized by long sweep or radius in elbows, tees and other fittings and shallow solder sockets. These fittings are available in sizes 32 mm to 200 mm. In Figure 3.13, some solder joint fittings are shown.

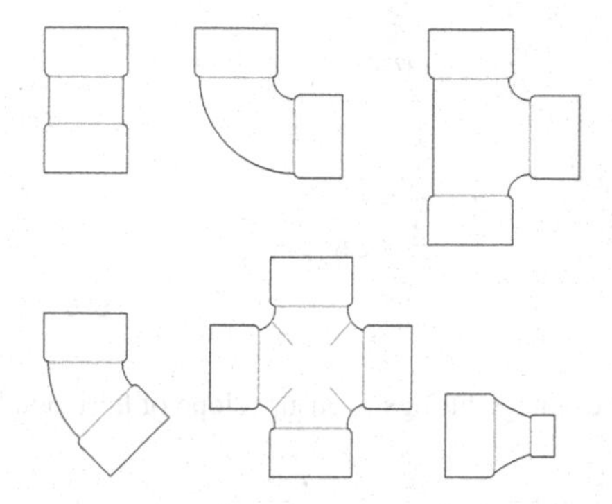

Figure 3.13 Some solder joint fittings.

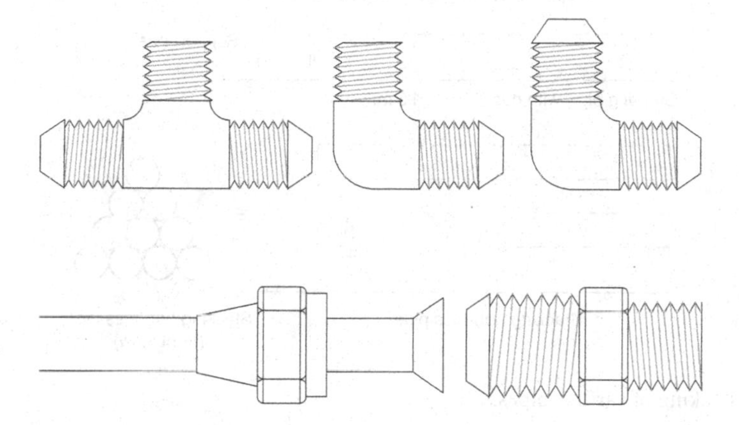

Figure 3.14 Some flared joint fittings.

Flared joint fittings: These types of fittings are mostly used with K and L soft copper tubing, normally in water service lines below ground. These are available in 10 mm to 75 mm. In Figure 3.14, some flared joint fittings are shown.

3.4.4 Drainage fitting angle

Drainage pipes are to be installed while maintaining a proper slope for smooth flow of wastewater in the pipe. If fittings like T-Y and elbow of exactly 90° bends are used, it will be difficult to install drain pipes that maintain the slope of the pipe. So, these fittings of 87.5° radius or sweep, as shown in Figure 3.15, are to be used.

3.5 Storing pipes and fittings

Storage of pipes and fittings shall be made with proper care so that the quality of these items, in terms of shape and strength, is not impaired. Pipes therefore need to be stored with sufficient care.

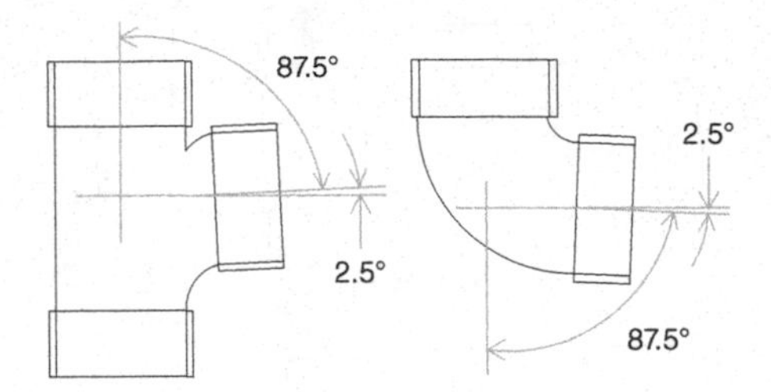

Figure 3.15 Correct angle of drainage fittings to ensure slope of horizontal drain pipes.

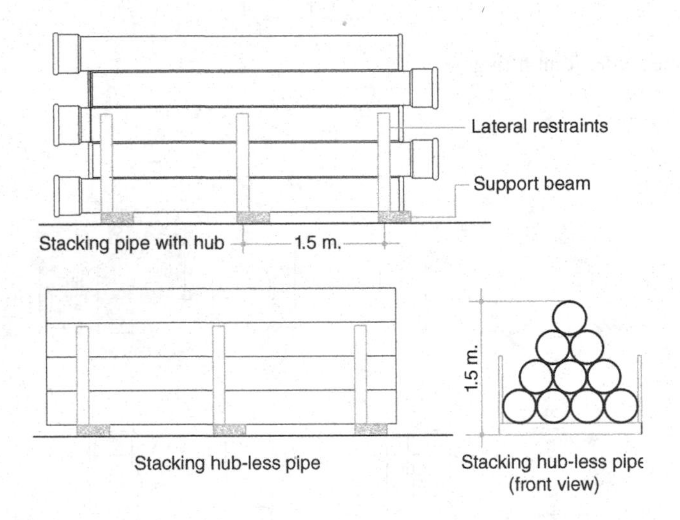

Figure 3.16 Stacking of various pipes.

3.5.1 Storing pipes

The easiest way to stack circular pipes is layer over layer. Pipes stacked one above another form different tiers or "stacks". The pipes of the upper tier sit in the grooves created in the bottom tier. Continuing to stack pipe in this way forms a trapezium-shaped stacking pattern. Pipes can be stacked in rectangular form by providing side-supports along the pipe-length, at two sides of stacks, as shown in Figure 3.16.

Stacked pipe barrels shall be continuously and uniformly supported with socket-ends alternately adjoining but extending beyond spigot-ends.

Following are some guidelines for stacking pipes of different materials.

Storing metal pipes: There are many hazards when it comes to metal pipe storage; they may potentially be hazardous by rolling, slipping, sliding or falling, so it is important that these materials be properly restrained and stored.

Racking and shelving of metal pipes can be done, but racks should be cross-braced to prevent the event of collapsing.

Storing plastic pipes: Plastic pipes, being weaker than metal pipes, shall be stored taking extra care, as mentioned below.

50 mm and smaller diameter plastic pipes can be stored in racks. Pipes on racks shall be blocked or strapped to prevent the pipes from rolling or falling off the rack. Pipes larger than

50 mm in diameter should be stacked with spacing strips between each row. To prevent pipes from rolling off the pile, each row of stacked pipe shall be arranged and blocked.

The bottom layer of pipes shall be securely chocked to prevent pipes from rolling or stack from collapsing. The width of bottom stack layers should not exceed 3 m.

Make sure that all blocks are made of material that would not deteriorate in weathering actions.

In outdoor storage, the area should be a relatively smooth, level surface without any bumps, free of stones, debris or other materials that could damage the pipe or fittings.

Pipes should be stacked in layers with socket-ends placed alternately at opposite ends of the stack and with the sockets protruding.

The bottom layer must be kept on horizontal supports, preferably of wood, at least 75 mm wide, spaced not more than 1.5 m centre-to-centre, to provide even support.

In rectangular pattern stacking, vertical supports are to be provided at intervals of 3 m along the pipe length.

For storage longer than 3 months, the maximum free height of stacks shall not exceed 1.5 m. The heavier and larger diameter pipes should be on the bottom.

Never set or keep any heavy, sharp or rough elements on top of the piping.

Plastic pipe should not be stored close to heat sources or hot objects such as heaters, boilers, steam lines, engine exhaust, etc.

Avoid covering the pipes with unventilated black tarps. In case of prolonged sun exposure, protect the pipes with an opaque material of preferably white colour, maintaining with natural ventilation to prevent overheating.

Storing concrete pipes: Concrete pipes are comparatively heavy, and larger diameter pipes are mostly used. Concrete pipes are rarely stacked except the smaller diameter pipes. For safety purposes, concrete pipes should not be stacked in greater numbers of layers than shown in Table 3.2. The height of stack should not exceed 2.0 m high in general.

3.5.2 Storing pipe fittings

Huge numbers of pipe fittings, differing in types and sizes, are used in plumbing development. Storing or keeping these fittings in a very disorganized or haphazard manner causes not only damage to fittings but also causes of time to pick up the right fittings from the lot, particularly in finding some special fittings occasionally used.

The best practice of organizing pipe fittings is to keep fittings of similar type and sizes in separately particular boxes or crates, or by making a good number of rectangular or square compartments in an area, to keep fittings of similar type and sizes in particular compartments, as shown in Figure 3.17. For storing fittings of sizes less than 50 mm, box height may be 300 mm; for sizes 50 to less than 100 mm, box height may be 600 mm; and for larger sizes,

Table 3.2 Allowable number of layers of concrete pipes that can be stacked [7].

Sl. No	*Diameter of pipe (mm)*	*Number of layers*
1	150–225	6
2	300–375	4
3	450–600	3
4	675–750	2
5	900 and above	1

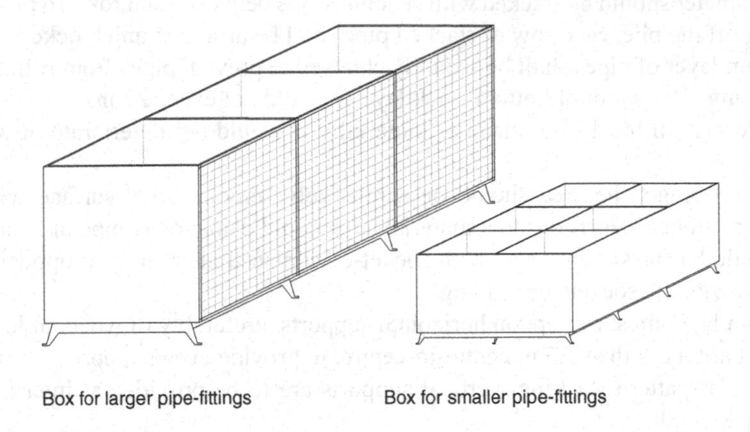

Figure 3.17 Boxes for storing pipe-fitting of various sizes.

box or compartment height should not be more than 1 m for quick and easy picking up of the desired fittings. For quick locating of the right compartment or box of particular fittings, the front may be made of suitable mesh.

References

[1] Mass Flow on-line (2020) "Where Do NPS or DN Stand for" Retrieved from www.massflow-online.com/faqs/where-do-nps-or-dn-stand-for/ on 25 December 2020.

[2] Bonds, Richard W., Barnard, Lyle M., Horton, A. Michael, and Oliver, Gene L. (2005) "Corrosion and Corrosion Control of Iron Pipe: 75 Years of Research" *Journal (American Water Works Association)* 97 (6), pp. 88–98. doi:10.1002/j.1551-8833.2005.tb10915.x. JSTOR 41312605, Retrieved from file:///C:/Users/pm/Pictures/ESTexs%20Workshop/awwa-75-years-of-research.pdf.

[3] National Capital Region Planning Board (NCRPB) "Characteristics of Various Plastic, DI and GI Pipes" Retrieved from http://ncrpb.nic.in/NCRBP%20ADB-TA%207055/Toolkit-Resources/Annexure%2014_Characteristics%20of%20plastic%20DI%20&%20GI%20pipes.pdf on 28 October 2015.

[4] Technical Learning College (TLC) (2018) "Valves and Fittings" *Continuing Education Professional Development, Course Manual*, p. 40.

[5] National Capital Region Planning Board (NCRPB) "Characteristics of Various Plastic, DI and GI Pipes" Retrieved from http://ncrpb.nic.in/NCRBP%20ADB-TA%207055/Toolkit-Resources/Annexure%2014_Characteristics%20of%20plastic%20DI%20&%20GI%20pipes.pdf on 28 October 2015.

[6] Municipal Sewer and Water (2014) "There's a Perfect Pipe for Every Water and Wastewater Project" *By Jennifer West*, Retrieved from www.mswmag.com/editorial/2014/08/theres_a_perfect_pipe_for_every_water_and_wastewater_project on 25 December 2020.

[7] Tracy Concrete "Specification, Handling & Storage – Lifting Concrete Pipes" Retrieved from www.traceyconcrete.com/site/pipes/specification-handling-storage on 25 December 2020.

4 Piping installation and testing

4.1 Introduction

The major portion of plumbing is attributed to piping. Piping installation is therefore a major job in plumbing. Complete piping installation involves various activities, starting with pipe jointing, installing various accessories and equipment on pipes and finally installing in proper position with suitable supporting techniques. After installing the pipes, all joints on piping must be checked for leakage, if any. After completion of piping installation along with all accessories, disinfection of water supply piping is done before using water.

Durability and safety in plumbing primarily depend on proper installation of piping. Success in this part of the job mainly depends upon the good workmanship of the plumbers, which is ensured by going through several tests. Bad workmanship causes manifold troubles. Good workmanship needs proper application of detailed knowledge of the work. So, in this chapter, various aspects of pipe installations and testing are discussed.

4.2 Pipe installation

Piping installation is the on-site job of plumbing, which involves various activities, starting from pipe jointing, fitting various accessories and appurtenances on and with fixtures or equipment and, finally, properly fixing in position by suitable supporting techniques. Pipe installation must meet all the requirements for durability, safety and thermal performance. For meeting all these requirements, various guidelines are to be followed, as mentioned below.

1 General requirements for pipe installation:

Pipe work must comply with the following.

a Relevant building or plumbing code
b Relevant occupational safety health and environment rules and
c Pipe manufacturer guidelines where applicable.

The most common guidelines regarding pipe installation are as follows.

a All pipes shall be laid to lines and grades and as closely as possible to the supports like wall, ceilings, columns and other structural elements, to occupy the minimum of space.
b All exposed piping shall be rigidly supported by pipe hangers and supports. Pipes shall be checked for soundness and cleanliness before installation.
c All pipes shall be carefully inspected for defects before installation.

DOI: 10.1201/9781003172239-4

d Any field cutting of the pipe shall be done only with approved guidelines.
e Pipe shall be cut using an appropriate cutting machine applicable for particular pipe. Joints shall be made up square with even pressure upon the gasket and shall be made watertight or gas tight as required.
f Any damage inflicted on the exterior coating and interior lining shall be repaired before jointing or fixing. Adequate care shall be taken to prevent scuffing of the surface during handling and transporting of pipes, plastic pipes in particular.
g Pipe shall be wrapped in flexible material or sleeved when penetrating masonry or concrete structural elements.

2 Location of laying pipe-work:

Pipes may be laid in the following locations.

a Over a floor
b Under a floor slab
c In a roof space under roof slab
d By the side of wall or column and
e Below ground.

3 Accessibility for maintenance and replacement of pipes:

Pipes installed in restricted locations that are difficult to access should meet the following requirements.

a Pipe passing through a structural element must be sleeved into a pipe of at least one size larger diameter that is open at both ends to allow inspection or maintenance.
b The gap between pipe and structural elements passing through shall be filled by fireproof intumescing acrylic mastic, cement filler and mortar etc.
c If installed in a chase or duct, it must provide ready access and sufficient space for manoeuvring, without compromising structural integrity.

4 Protection against electric hazards:

Where there is chance of having electric shock from metal pipes, pipe work must follow the following guides to avoid the potential of electric shock.

a Pipe work must be connected to an earth electrode using earth-bonding conductors.
b Keep pipe concealed where occupants are exposed to direct contact to pipes.
c Metallic fixtures must be bonded to the metal pipe by insulated fittings.
d Safety precautions must be followed when cutting metal pipes.

5 Pipe insulation:

Exposed pipes shall be properly insulated in the following conditions.

a Possibility of heat loss in hot water pipes
b Possibility of heat gain in cold water pipes exposed to sun, heating elements nearby etc. and
c Possibility of freezing of water.

4.3 Pipe cutting

The first job in pipe installation might be the cutting of pipe for jointing or installation. Pipes, based on material and size, are available at particular length. For installation purposes, if smaller pipe is needed, then it requires cutting of the pipe to proper length. Theoretically, pipe cutting, also termed pipe profiling, is a mechanized process that basically removes material from pipe or tube to create a desired profile. Typical profiles in pipe cuts include straight complete or partial cuts, mitres, saddles and midsection holes. There are pipe cutting guides for cutting pipe at different angles. These pipe guides can be used for cutting pipe at 22.5°, 45°, 90° and saddle cuts.

There are various methods of cutting pipes; the factors involved in choosing a particular method or technology for cutting tube or pipe are material, wall thickness, squareness of ends, end-conditioning requirements etc. Among various methods of cutting technology, the most common abrasive method, using abrasive saw, is generally applied for plastic and metal pipes. The most common pipe cutting tools are as follows.

A Manual hand saws:

1 Rip cut saw or hand saw
2 Hack saw for cutting small diameter pipes and
3 Pipe cutter.

B Electro-mechanically powered saws:

1 Power hack saw or band saw (portable)
2 Chop saw
3 Mitre saw and
4 Chain saw etc.

4.4 Pipe jointing

Pipe jointing covers the major part of plumbing and is one of the vital works in pipe installation. Ensuring proper jointing of pipe is the main objective of pipe works.

Imperfect joints will not only allow the fluid content of pipes to escape, causing loss of water or spreading of wastewater, but also create various other problems around the leakage point. So, success of plumbing predominantly depends on efficient installation of piping with proper joints. Good workmanship is the primary need of efficient piping installation.

Problems associated with pipe leakage: Leakage of pipe causes various problems, starting from moistening around the leaking point to property loss in huge amounts, depending upon the damage to property due to long-lasting moistening. Various problems associated with various natures of leakages are mentioned below.

1 Leakage of water supply pipes will cause the following:

i Property damage
ii Wastage of water and
iii Infiltration of polluting agents, particularly in underground pipes.

2 Leakage of underground soil or wastewater pipe will cause the following:

i Subsoil pollution
ii Underground water pollution and
iii Entrance of soil and vegetation roots into the pipe, causing stoppage of flow.

3 Leakage of gas pipe will cause the following:

   i Odorous and polluted environment and
   ii Fire or explosion of flammable gas when it comes into contact of any open flame.

Jointing various types of pipes: The jointing methodologies of pipes are primarily dependent upon the materials of pipes to be jointed. Mostly pipes of the same materials are jointed except in very exceptional cases where pipes of different materials are needed to be jointed. In this discussion only the pipes of the same material of various jointing options are described.

4.4.1 CI pipe joints

Bell and spigot CI pipe joints are either caulked or mechanical joints. Of these two, caulked joints are mostly practiced. Various CI pipe jointing methodologies are discussed below.

Caulked CI pipe joint: Caulked CI pipe jointing systems can again be performed in two ways, using two different sealing materials like lead and cement mortar.

A. Lead joint: This type of CI pipe joint can be made by molten lead and oakum with caulking yarn. Oakum is hemp, which is treated with pitch. The lead, available in 11.34 kg bars, is heated in a pot with lead furnace. Lead should be heated until it does not stick to the ladle, but not red-hot. Care must be taken that lead is not damp or frozen. In such a case, lead should be preheated before melting. The amount of lead and oakum needed in making joints of various diameters of CI pipes are furnished in Table 4.1. The steps in making a caulked lead and oakum joint are as follows:

1 Make the hub (bell) and spigot ends of the pipe dry and clean. Wet hub may cause a small explosion when molten lead is poured because the trapped moisture may turn to steam.
2 Assemble the pipes by inserting the spigot end into the bell to its full depth.
3 Yarn and pack oakum into the annular space around the pipe inserted into the bell, to a depth of 25 mm from the top of the bell, as shown in Figure 4.1. No loose oakum should protrude.
4 For vertical pipe jointing, pour the molten lead with a ladle into the remaining space around. Fill the space in one pour. In horizontal pipe jointing, a contraption called a joint-runner is placed around the pipe and clamped tightly at the top, forming a passage for pouring lead, as shown in Figure 4.2. A wad of oakum is placed under the clamp to retain the lead up to the top of the bell. The lead is then poured into the passage, filling the joint to the top. The running rope is removed when the lead solidifies.

Table 4.1 Requirement of lead and oakum in CI pipe lead joint. Converted in SI unit from source [1].

Sl. No	*Size of pipe mm*	*Lead depth mm*	*Oakum gm*	*Lead Kg*
1	50	25	50	0.567
2	75	25	71	0.794
3	100	25	85	1.02
4	150	25	99	1.36
5	200	32	198	2.721

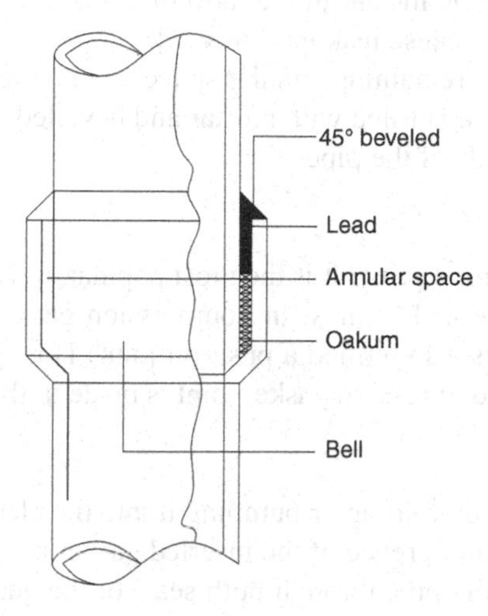

Figure 4.1 CI pipe lead-joint.

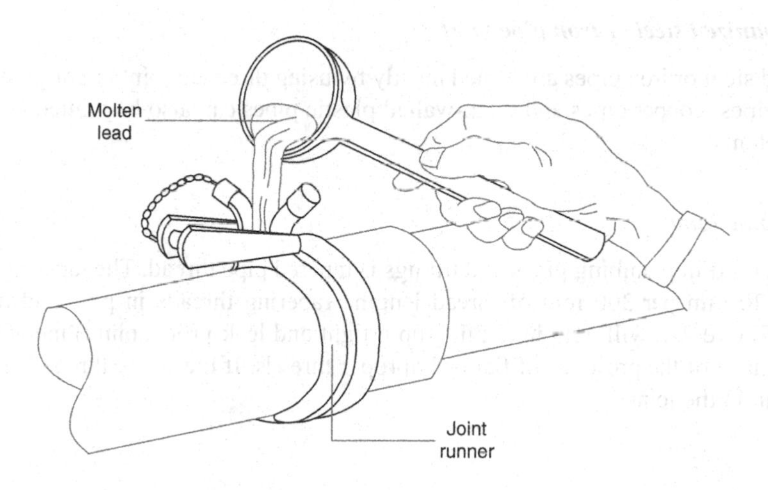

Figure 4.2 Horizontal CI pipe jointing.

5 As the molten lead hardens, it is caulked with light taps to compensate for shrinkage of lead as it cools. The neck of excess lead left in the pouring is cut off with chisel and hammer.
6 The remainder of the joint is filled with lead and bevelled off smoothly at an angle of about 45° with the outside of the pipe.

B. Cement mortar joint: The following steps are followed in making a cement mortar joint.

1 The spigot and bell to be joined are cleaned and moistened.
2 A gasket of oakum rope soaked in cement grout is inserted into the annular space inside the bell for the lower third of the circumference of the joint depth.

3 A paste of cement and sand mortar in the ratio of 1:4 is to be made. Mortar of a higher cement to sand ratio may cause leaking joints [2].
4 Insert the mortar into the remaining annular space around the joint and caulk firmly.
5 The remainder of the joint is filled with mortar and bevelled off smoothly at an angle of about 45° with the outside of the pipe.

Mechanical CI pipe joint

Compression gasket CI pipe joint: It is the most popular, quickest and easiest-to-assemble flexible joint for CI pipe and fittings. In compression gasket joints, a gasket made of neoprene-rubber is used. It is also termed a push-on joint. This type of joint is preferred in underground services. The compression gasket joint is made in the following way, as shown in Figure 4.3.

1 Insert the rubber gasket by folding or bumping it into the cleaned hub.
2 Lubricate the entire circumference of the inserted gasket by a brush.
3 Push the spigot end of the pipe through both seals of the gasket up to the full depth of the bell.

4.4.2 Galvanized steel or iron pipe joint

Galvanized steel or iron pipes are joined mostly by using threaded joint or cut grooved joint. Cast iron pipes, copper pipes and thick-walled plastic pipes can also be jointed in these two joining systems.

Threaded pipe joint

The thread used in plumbing pipes and fittings is tapered pipe thread. The tapering of thread is usually 20 mm per 300 mm of thread length. Tapering threads in pipe and fittings, as shown in Figure 4.4, will help in making up a tight and leak-proof joint. Pipe is often discarded because of the presence of flat or improper threads. If the entire thread is flat, it will cause a leak in the joint.

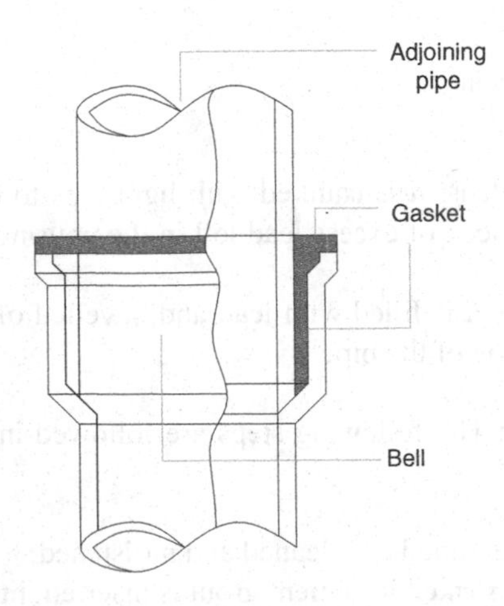

Figure 4.3 Compression gasket joint.

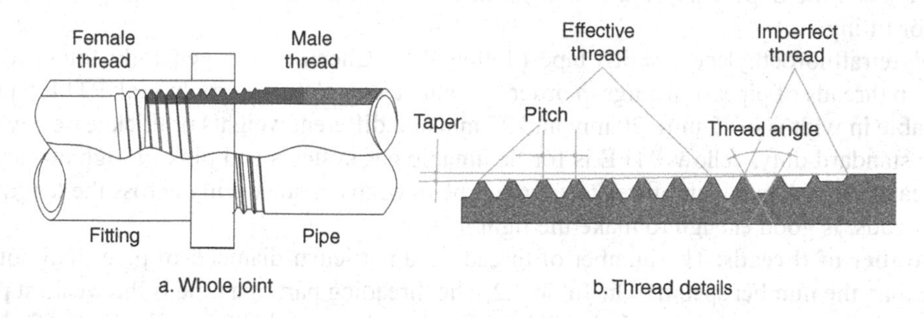

Figure 4.4 Thread joint details.

Table 4.2 Diameter and number of threads in different sizes of pipe, according to British Standard Pipe (BSP) threads [3].

BSP size mm	*Max. diameter mm*	*Min. diameter mm*	*Thread pitch (mm) TPI*
12	20.955	18.633	14
16	22.911	20.589	14
20	26.441	24.120	14
25	33.249	30.292	11
32	41.910	38.953	11
40	47.803	44.846	11
50	59.614	56.657	11
63	75.184	72.227	11
75	87.884	84.927	11
100	113.030	110.073	11
125	138.430	135.472	11
150	163.830	160.872	11

All pipe fittings are manufactured with female (internal) threads. So, for jointing male (externally threaded) pips, thread is to be made on the pipe in the following way.

1 Cut the pipe square by properly securing the pipe in the vise jaw.
2 Ream the inside of the cut to remove the burr often left by the cutting tool.
3 Thread the pipe with a die to the proper length while lubricating the die. Table 4.2 provides the proper thread length for various sizes of pipe.
4 Remove the die after threading, and wipe the thread to clean.
5 Apply the pipe joint compound to the male threads only.
6 Twist the fitting onto the pipe thread by hand and then tighten the joint with the help of pipe wrench.

Threaded joint sealants: Pipe thread joint compound or sealant, also called pipe dope, is used to make thread joints airtight. Thread joint compounds are made of filler materials held together by grease, oil and resinous or plastic binder. Linseed oil is used as a binder in the preparation of some thread compounds. Calcium carbonate, silicates, lead or barium oxide powder are suitable for many applications. Barium oxide is chemically inert within the compound. Pipe dope is applied with a brush on the threads and then screwed into another pipe or fitting.

Polytetrafluoroethylene (PTFE) tape (Teflon™ by Chemours) is generally found to be used on threads of pipes or fittings in order to obtain a complete watertight seal. PTFE tape is available in widths of 12 mm, 20 mm and 25 mm and different weights or thicknesses: white is for standard duty, yellow PTFE is for flammable gas systems and pink of high density is for heavy duty. A maximum three to four complete wraps, made evenly across the length of the threads, is good enough to make the tight joint.

Number of threads: The number of threads in a particular diameter of pipe shall not be more than the number specified in Table 4.2. The threading part of a pipe is the weakest portion of pipe, prone to corrosion in iron pipes. So, thread length shall be made as specified for the particular diameter and thickness of pipe.

Advantages and disadvantages of threaded joints.

Advantages:

1 Installation is comparatively easier and faster.
2 Specialized installation skill requirements are not extensive.
3 Leakage seldom occurs for low-pressure, low-temperature installations where vibration is insignificant.

Disadvantages:

1 Differential thermal expansion between the pipe and fittings due to rapid temperature changes may lead to leakage.
2 Vibration can result in fatigue failures due to the high stress concentration caused by the sharp notches at the base of the threads.
3 Vulnerability to fatigue damage is significant, especially where exposed threads are subject to corrosion, particularly in hazardous environments.

Cut groove pipe jointing

For this type of jointing, a groove configuration is made around the pipe end which is called a cut groove. In cut groove jointing, a coupling is used which is composed of a housing, a gasket, bolts and nuts. A coupling body has a two-piece housing-segment that fully encloses a resilient rubber gasket. The housing keys engage into the joint as the housing segments are tightened with the help of bolts and nuts. It is very essential to process the pipe grooves properly with a view to ensuring a tight, leak-proof joint for optimum performance.

To make a cut grooved joint, as shown in Figure 4.5, the following steps are to be undertaken.

1 Cut the groove to the proper depth and width by using a grooving tool and make the groove clean.
2 Disassemble the coupling components.
3 Lubricate the gasket lips and its complete exterior portion.
4 Position the gasket on one end of the pipe or fitting, being sure that the gasket lip does not hang over the pipe end.
5 Butt the fittings or pipe ends together, slide the gasket and centre it between the grooves.

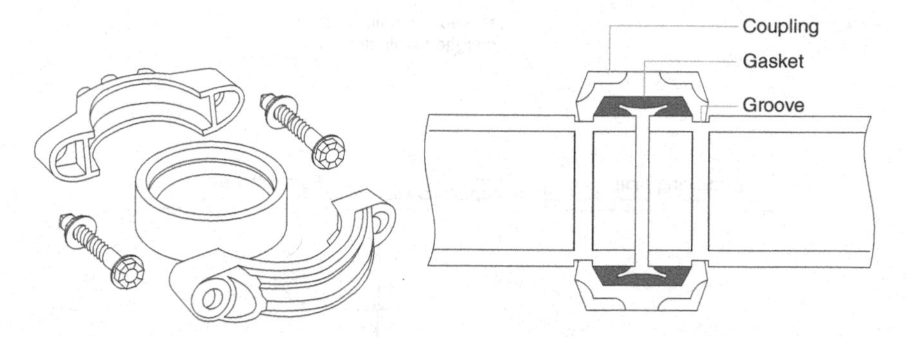

Figure 4.5 Cut groove jointing and the coupling.

6 Assemble the housing segments over the gasket, making sure the housing keys engage into the grooves on both sides.
7 Insert the bolts with the oval head fully into the housing segment and tighten the nuts uniformly.

Cut grooved joints can be made rigid or flexible. To make the joint rigid, the jointing pipe ends shall be kept closed, and to make a flexible joint, there shall be slight gap in between jointing pipe ends.

4.4.3 Copper pipe joint

Copper tubes are joined by using one of the following three most common methods.

1 Solder and braze joint
2 Flared joint and
3 Compression joint.

Solder and braze joint

Solder joints with capillary fittings, as shown in Figure 4.6, are used particularly in water supply and drainage systems. Brazed joints with capillary fittings are preferred in refrigeration piping and where greater joint strength is required or where service temperatures are as high as 177°C.

Soldered joints work on capillary action drawing free-flowing molten solder into the gap between the fitting and the tube. Flux is used as a cleaning and wetting agent to permit uniform spreading of the molten solder over the jointing surfaces.

Two types of solder are used most often. These are 95–5 and 50–50 wire type solders. The 50–50 tin–lead solder is suitable for moderate pressures, whereas for higher pressures or where greater joint strength is required, 95–5 tin-antimony solder is preferred. There are also lead-free solders. A 95–5 solder is composed of 95 percent tin and 5 percent antimony. It melts at a slightly higher temperature than 50–50 tin-lead solder. It also has a very narrow pasty range of less than 5–6°C. Pasty range is the temperature range between which solder is neither completely liquid

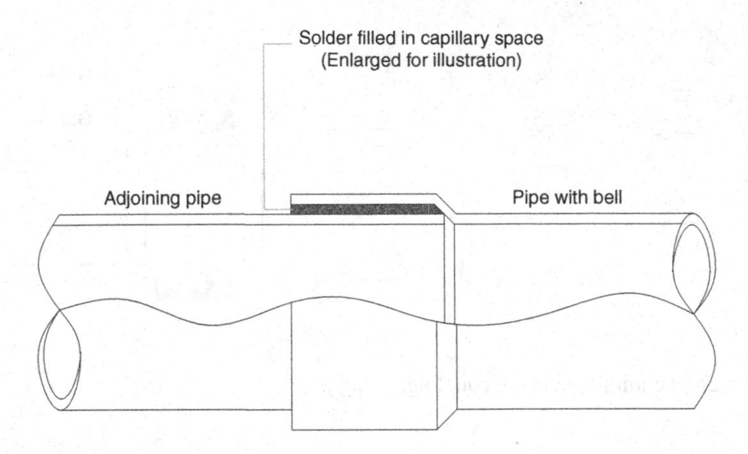

Figure 4.6 Solder joint.

nor solid. So, it is the working temperature range for the particular type of solder. However, 50–50 tin-lead solder is composed of 50 percent lead and 50 percent tin. Due to the presence of lead, it is not preferred for water supply works. But it has a 33.3°C pasty range.

Before soldering, flux is to be applied uniformly over the surface to be soldered first, which acts as a wetting agent and permits uniform spreading of molten solder. The fluxes best suited to the wire solders are mildly corrosive liquid or petroleum-based pastes containing chlorides of zinc and ammonium. Most of the fluxes are self-cleaning. The purposes of using flux are as follows.

1 Dissolving residual traces of oxides
2 Preventing oxides from forming during heating and
3 Floating out oxides ahead of the solder.

Paste type solder is also available, which is a suspension of finely granulated solder in a paste flux. If paste type solder is to be used, the following are to be observed.

1 Wire solder must be applied in addition to the paste to fill the voids and assist in displacing the flux.
2 Paste mixture must be thoroughly stirred before using. Table 4.3 can be used for estimating the amount of solder and flux required for different sizes of tube joint.

For making a solder joint, the following steps are to be carried out.

1 Cut the pipe end square to the required length with the help of a tube cutter or a hacksaw.
2 Ream the cut end to remove burrs with either the reamer or a file.
3 Clean the cut end of the tube and fitting socket using some cloth or steel wool or brush.
4 Apply flux to the tube end and fitting socket.
5 Assemble the pipe and fitting and rotate to spread the flux.
6 Apply heat by using an oxyacetylene or propane flame to the tubing first and then to the fitting until the solder melts when placed in the gap between the fittings and the tube at the joint.

Table 4.3 Approximate consumption of solders and flux for 100 joints of different tube-size; Converted in SI unit from source [4].

Sl. No	*Tube size (mm)*	*Solder requirement at 0.25 mm clearance (gm)*	*Flux requirement (gm)*
1	6	44.00	18.14
2	10	72.12	27.22
3	12	118.39	45.36
4	16	176.45	68.04
5	20	248.57	95.25
6	25	388.28	145.15
7	32	505.76	190.51
8	40	671.32	254.01
9	50	1079.55	408.23
10	63	1462.84	553.38
11	75	1966.32	743.89
12	88	2624.49	988.83
13	100	3377.45	1274.60
14	125	5167.32	1950.45
15	150	7173.56	2707.95
16	200	12226.58	4626.64
17	250	15351.83	5792.37
18	300	20682.00	7801.79

7 Remove flame and feed solder to the joint at one or two points until a ring of solder appears at the end of the fitting.
8 Remove excess solder with a cloth while the solder is still pasty, leaving a fillet around the end of the fitting, as it cools down.

Flared Joint: Impact or screw type tools are used for flaring copper tube. The flared jointing system is as follows:

1 Cut the tube square and remove all burrs.
2 Slip the coupling nut over the end of the tube.
3 Flare the tube in one of the following ways.

  A *By hammer:* Insert the flaring tool shown in Figure 4.7a into the tube and drive it by moderately light hammering strokes. Hammering is continued until desired flaring is achieved.

  B *By screw type flaring tools*: Clamp the tube in the flaring block in such a way that the end of the tube is slightly above the face of the block. Then place the yoke of the tool in such a position so that the bevelled end of the compression cone is just over the tube end, as shown in Figure 4.7b. Now turn the compressor screw down firmly to form the flare.

4 Assemble the joint by placing the fitting squarely against the flare. Engage the coupling nut into the fitting threads by hand. Tighten the nut and fitting with two wrenches.

Copper compression joint: A compression fitting is basically a coupling used to connect two pipes or a pipe to a fixture or valve. A mechanical compression joint consists of the following parts.

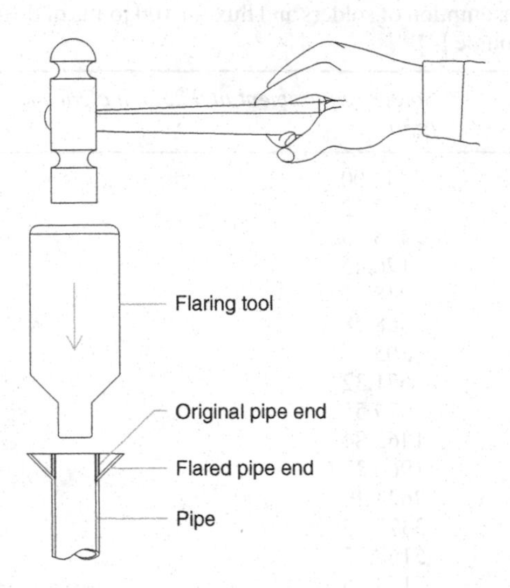

Figure 4.7a Flaring pipe using hammering flaring tool.

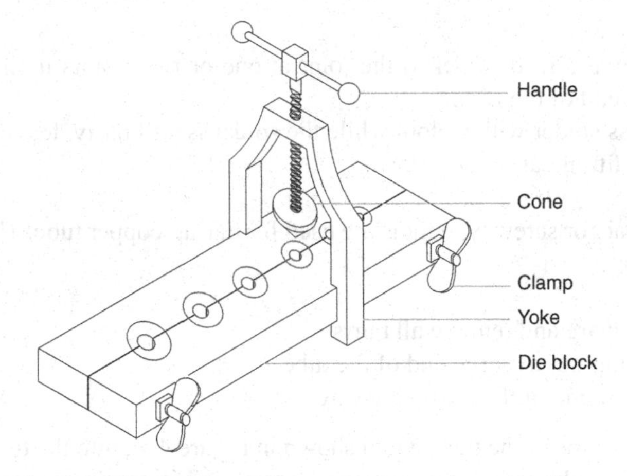

Figure 4.7b Flaring pipe using screw-type flaring tool.

1 Compression joint fitting
2 Compression ring or ferrule and
3 Compression nut.

To make a mechanical compression joint in copper tubing, the following steps are to be performed.

1 Slide the compression nut and ring onto the tubing
2 Slide the mechanical joint fitting over the tube end and

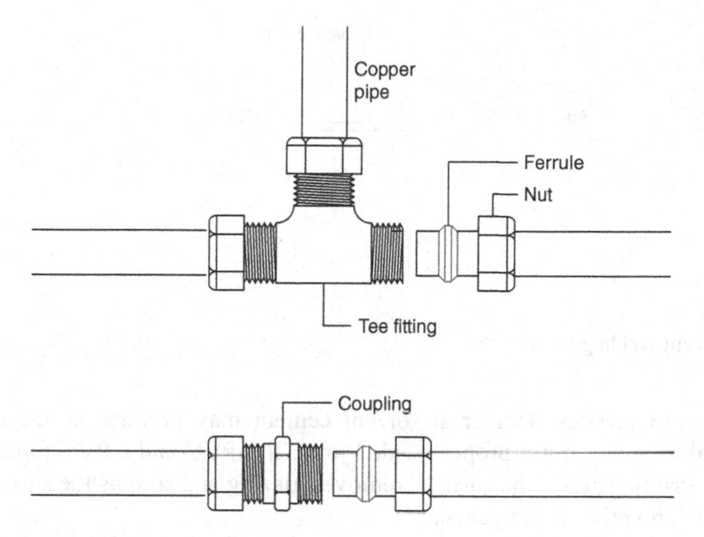

Figure 4.8 A mechanical compression joint.

3 Tighten the compression nut onto the fitting with a wrench thus jointing the seal, as shown in Figure 4.8.

The compression rings or ferrules virtually make the seal between the copper pipe and the nut seat when tightened.

4.4.4 Plastic pipe joint

Various type of plastic pipes, discussed earlier, are joined together with one of the four following methods, depending on the physical conditions in which the particular pipe will be used.

1 Solvent weld joint
2 Heat fusion joint
3 Insert fitting joint
4 Compression fitting joint and
5 Threaded joint.

Solvent weld joint

A solvent weld joint, shown in Figure 4.9, is made by applying commercially available solvent cement solution to get solvent bonding in the jointing surface. The bonding is much like a metal welded system. To achieve a good joint in plastic pipe and fittings, the following steps are to be well performed. The steps are elaborated below.

1 Choose the right primer and solvent. It is very important to use the right primer and or solvent cement for a particular type of plastic. There is no universal solvent satisfactory

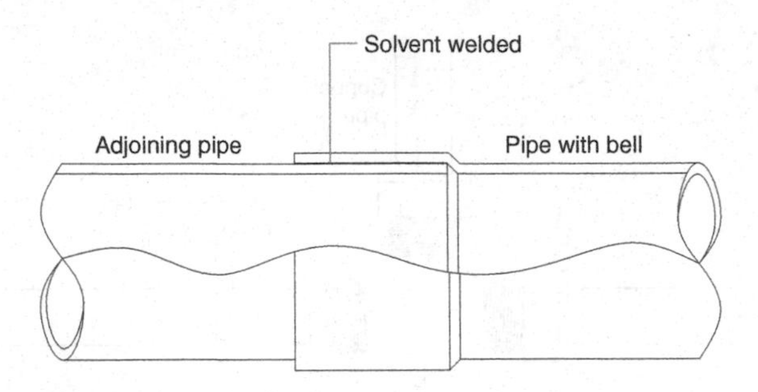

Figure 4.9 Solvent welding joint.

to all type of plastics. Universal solvent cement may produce a surface-to-surface superficial bond but not a proper welded joint. In uPVC and CPVC pipes, priming is essential. Priming clears the surface, removes glazing and softens the surface. For ABS plastic pipe, no priming is necessary.

2 Make a good interface fit. To ensure proper interface fit in plastic pipe jointing, the walls of plastic fitting sockets are tapered so that the inserted pipe contacts with the side of the fitting halfway to reach the seat of the socket. After application of primer and or solvent cement, the pipe can be forced to insert into the pipe to the full depth of the socket. It will create a tight bond and weld together.
3 Correctly prepare the pipe for jointing. To ensure a leak-proof permanent joint in a plastic piping system for drainage, waste disposal and venting, the following techniques should be followed.

   i Cut pipe squarely using a meter box or a sharp tube cutter with a special blade for plastics.
   ii Remove all burrs by using a de-burring tool or knife and make the end smooth.
   iii Check for good interface fit and select the ideal fittings.

4 Make the jointing surface of pipe and fittings clean and dry. Apply appropriate primer first to the inside of the fittings and then to the outside of the pipe end portion, which will remain inserted into the fitting. Be careful so that no puddle remains at the bottom of the socket of the fitting. After a while, apply appropriate solvent cement to the pipe at the same depth as that of the primer.
5 Insert the pipe into the socket immediately, before the solvent cement evaporates. While pushing, make a quarter turn of the pipe, which will ensure an even distribution and absorption of the solvent cement. Next, hold the joint together firmly for about 10 seconds to allow the solvent to start bonding the two surfaces.
6 Wipe off the excess solvent cement with a clean rag, leaving an even fillet all the way round. If the excess cement does not come out, then the bead looks incomplete and there is a chance of leakage.

Heat fusion method

In the heat fusion method, the joining surfaces of both pipe and fitting are heated to designated temperatures, generally at moulding temperature for the material of the pipe. For polypropylene, the required temperature is about 260°C (±10), and for HDPE, it is 204–232°C. Jointing of pipe is made by inserting melted end into other melted surface of the pipe or fitting at that temperature. The inserting force causes the melted materials to flow and mix, thereby resulting in the fusion. As the temperature cools down to ambient temperature, jointing is completed.

Heating of pipe can be done by using a electrical heating device, in which heat is created by electric current; this jointing method is termed the electro-fusion method. Heat fusion joint may be of three types: Butt, Socket and Saddle fusion.

O-ring push-fit insertion joint

O-ring push fit jointing is the simplest, yet is very common in plastic pipe jointing. This type of joint also makes a seal, preventing any risk of leakage from the water inside. It requires low insertion forces and allows the pipes and fitting to rotate.

An O-ring is a doughnut-shaped loop designed to prevent the passage of fluids. O-rings can be made from plastic or metal, but mostly O-rings made of rubber or elastomeric material are used. The O-ring is fitted along the groove made in the socket of the pipe or fitting.

When the pipe or fitting end is pushed into the socket with the O-ring, a seal is created by the mechanical deformation of the O-ring being pressed by the inserted end of pipe or fitting, which creates a barrier against any leakage between two tightly mated surfaces, as shown in Figure 4.10. In this type of jointing no adhesive or special tool is required, so this is cost and time saving. To make the joint, the steps mentioned below are to be followed.

1 Clean the socket, especially the ring groove, and spigot end of pipes to be jointed.
2 With the finger check that the rubber ring is correctly seated, not twisted, and that it is evenly distributed around the ring groove.
3 Apply jointing lubricant to the spigot end up to the insertion depth.
4 Align the spigot with the socket and apply a firm, even thrust to push the spigot into the socket. It is possible to joint 100 mm and 150 mm diameter pipes by hand.

Insert fitting joint

Insert fitting joint is made generally in PE plastic tubing used in low pressure application. The following three steps are to be followed for making such a joint.

1 Cut the tubing squarely with a plastic tube cutter and remove burrs.
2 After making the stainless steel clamping ring loose, slip it onto the tubing end. Push the insert fitting into the tubing ends up to the fitting shoulder.

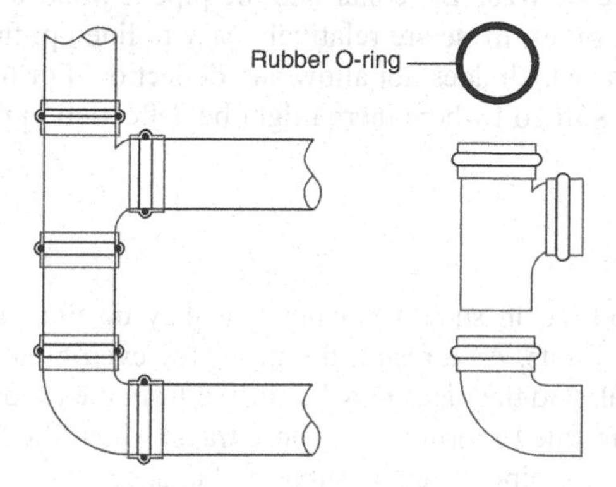

Figure 4.10 O-ring push joint and fittings.

3 Slide the clamp over the serrated portion of the fitting and tighten the clamp with a screwdriver.

Compression fitting joint

The PB plastic compression joint fitting consists of a plastic fitting body onto which a brass lock nut containing a plastic locking sleeve is threaded. The jointing procedure is as follows.

1 Cut the tubing squarely and remove all the burrs.
2 Push the tubing firmly into the fitting until it bottoms out.
3 Tighten the lock nut until it bottoms against the fitting shoulder.

Threaded joints

Though making good threads in plastic pipe is difficult, these pipes can also be joined by threading. Only schedule 80 uPVC and CPVC pipe can be safely threaded. Schedule 40 uPVC and CPVC pipe should not be threaded. Due to the reduction in wall thickness at the region of threading, the pressure rating of the pipe is reduced by about 50 percent. Therefore, threaded connections in plastic pipe are not recommended for high pressure applications. In some cases, metal threads are moulded in plastic pipe for making thread joints.

4.4.5 Concrete pipe jointing

Concrete pipes are generally jointed in two methods: 1. spigot and socket joint and 2. butt joint with collar.

Spigot and socket joint: Concrete pipes, joined in this method, have external shoulders that form a bell in one end, and the other end the spigot forms with a notch to house an O-ring or other rubber gasket, as shown in Figure 4.11. In absence of a rubber gasket, jute oakum about 19 mm thick is caulked into the bell. The remaining annular space is filled by cement mortar prepared by using one part cement, two parts sand and sufficient water to dampen thoroughly to such consistency that prevents the grout from running into the pipe. In soft soil and watertight joints under low head applications, this type of joint is recommended.

Butt joint with collar: This joint uses a precast or cast in-situ concrete collar to connect the pipes. Jointing pipes end are brought aligned face to face and put just in the middle of the collar. The annular space between the collar and the pipe is filled with cement mortar. For small diameter concrete pipes, those are relatively easy to line up; this type of joint can be done. This is a rigid joint which does not allow any deflection. For this reason, such a joint is not recommended for soft soil where there might be deflection in pipe resulting cracks in the joint.

4.5 Supporting pipes

All pipes shall be supported in such a manner that they do not sag. Sagging will cause stresses in pipes and in joints. As a result, the pipe may even break or crack between the joints, the joints may leak and the pipes may be shifted from the proper grade or pitch, causing a portion of the drain line to form traps. These traps, when filled with solid wastes will cause blockage of flow. So, pipes must be supported at specified intervals depending upon

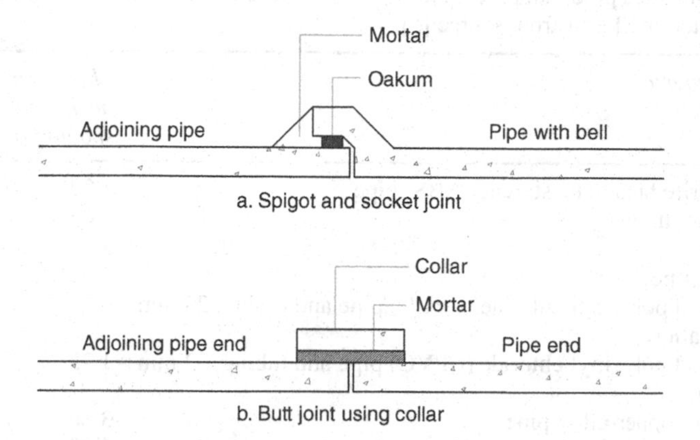

Figure 4.11 Concrete pipe joints.

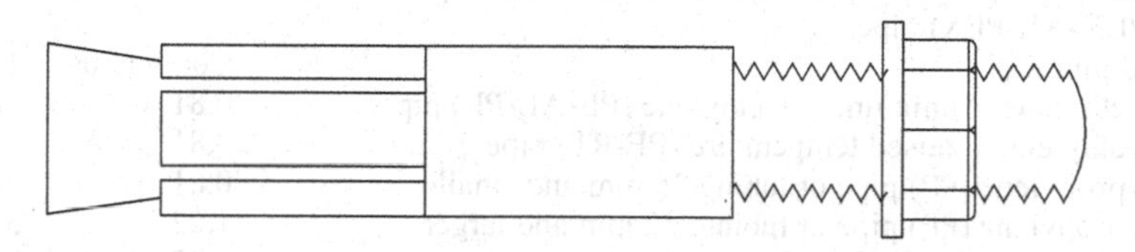

Figure 4.12 Expansion anchor.

the material, size and content of the pipe. Pipes above the ground mostly either hang from the ceiling by appropriate hangers or are attached to the vertical elements of a structure. In very few cases, pipes are run just over the floor or ground. Pipes are generally supported using anchors and hangars.

Anchors: An anchor is a device used to secure the piping to a structural element. A metal insert in the concrete of structural elements provides an anchor. Inserts are installed prior to placing the concrete. There are other methods of anchoring as well. For example, concrete is drilled to such a depth that it can receive an expansion metallic bolt. As the bolt is inserted into the drill with a hammer, the bolt expands and becomes firmly secured with the concrete of the structure, known as an expansion anchor. Another common method is to attach the anchor by a power actuated shot that inserts the anchor to the correct depth. An expansion type anchor is illustrated in Figure 4.12.

Pipe support: A variety of pipe hangers and supports are available to hold pipes in either the horizontal or vertical position. Those supports shall be tightly secured in the structural elements and sufficiently strong to carry the load of the pipe along with its content.

4.5.1 Supporting vertical pipes

Vertical pipes must be secured at sufficiently close intervals to keep the pipe in alignment and to withstand the weight of the pipe. The spacing of supports for vertical pipes of different material and sizes are given in Table 4.4. Figure 4.13 illustrates the use of riser clamps to support vertical pipe on the floor, while the use of two other types of support, called bracket and clip, support vertical pipes on walls or any vertical surfaces.

Table 4.4 Recommended pipes supports spacing for pipes of different materials and various diameters. Converted in SI unit from source [5].

Sl. No	*Piping material*	*Maximum horizontal spacing (m)*	*Maximum vertical spacing (m)*
1	Acrylonitrile butadiene styrene (ABS) pipe	1.22	3.05[b]
2	Aluminium tubing	3.05	4.57
3	Brass pipe	3.05	3.05
4	Cast iron pipe	1.52[a]	15
5	Chlorinated polyvinyl chloride (CPVC) pipe and tubing, 25 mm and smaller	0.9	3.05[b]
6	Chlorinated polyvinyl chloride (CPVC) pipe and tubing, 32 mm and larger	1.22	3.05[b]
7	Copper or copper-alloy pipe	3.66	3.05
8	Copper or copper-alloy tubing, 32 mm diameter and smaller	1.83	3.05
9	Copper or copper-alloy tubing, 40 mm diameter and larger	3.05	3.05
10	Cross-linked polyethylene (PEX) pipe	0.81	3.05[b]
11	Cross-linked polyethylene/aluminium/cross-linked polyethylene (PEX-AL-PEX) pipe	0.81	1.22
12	Lead pipe	Continuous	1.22
13	Polyethylene/aluminium/polyethylene (PE-AL-PE) pipe	0.81	1.22
14	Polyethylene of raised temperature (PE-RT) pipe	0.81	3.05[b]
15	Polypropylene (PP) pipe or tubing 25 mm and smaller	0.81	3.05[b]
16	Polypropylene (PP) pipe or tubing, 32 mm and larger	1.22	3.05[b]
17	Polyvinyl chloride (PVC) pipe	1.22	3.05[b]
18	Stainless steel drainage systems	3.05	3.05[b]
19	Steel pipe	3.66	4.57

For imperial: 1 inch = 25.4 mm, 1 foot = 304.8 mm.

a. The maximum horizontal spacing of cast iron pipe hangers shall be increased to 3 m where 3 m lengths of pipe are installed.

b. For sizes 50 mm and smaller, a guide shall be installed midway between required vertical supports. Such guides shall prevent pipe movement in a direction perpendicular to the axis of the pipe.

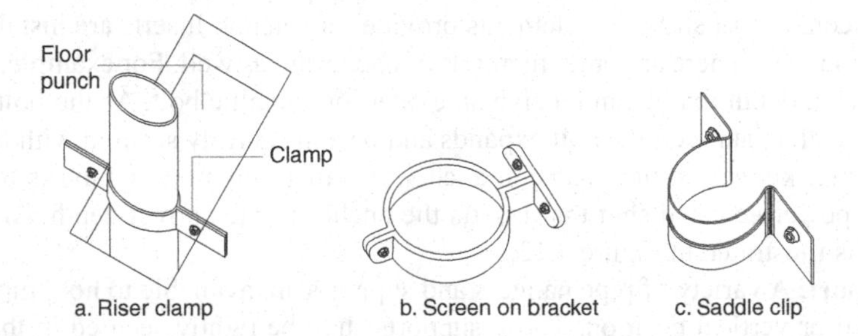

Figure 4.13 Various vertical-pipe supports.

4.5.2 Supporting horizontal pipes

The horizontal piping must be supported at sufficiently close intervals for the following purposes.

a to keep pipe in desired alignment and grade
b to prevent sagging and
c to carry the proportionate weight of piping and its content.

The items used for supporting horizontal pipes are anchors, hangers, floor supports, wall brackets etc.

Pipe hangers: Hangers are devices secured in structural elements, from which one or more pipes are suspended. The length of hanger rods is adjustable. When two or more pipes are suspended by a pair of hangers, it is called a trapeze. The hanger rods should be sufficiently strong to support the proportionate share of the load of piping. Different types of hangers are pictured in Figure 4.14, for support of different piping. Figure 4.14a shows hanging of a single pipe; Figure 4.14b illustrates how multiple pipes can be hanged. In Figure 4.12c, a pipe hanger is used in conjunction with a vibration isolator to reduce the transmission of pipe vibration into the structure.

Supporting on floor: Large pipes are sometimes supported on the floor by floor-mount saddle, as shown in Figure 4.15.

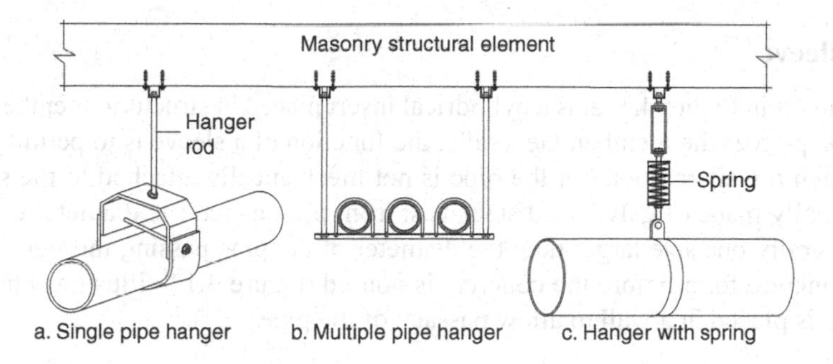

Figure 4.14 Various horizontal-pipe hangers.

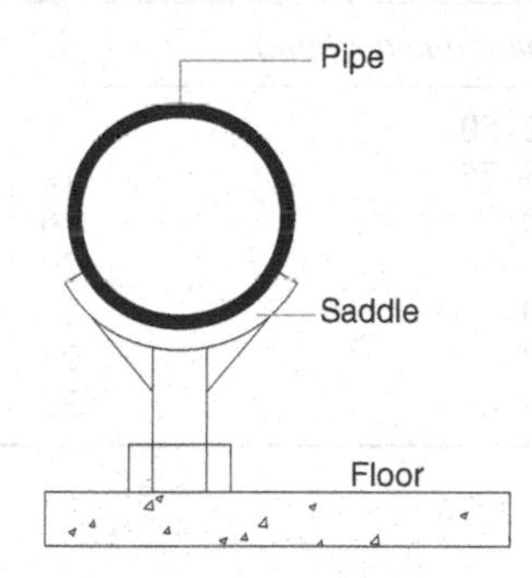

Figure 4.15 Pipe supported on floor.

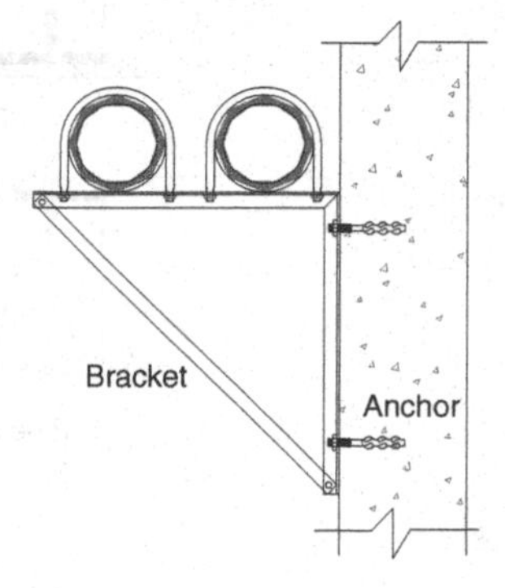

Figure 4.16 Wall bracket supporting pipe.

Wall bracket: Wall brackets, shown in Figure 4.16, are sometimes used to support horizontal piping running by the side of any wall or series of columns in a row.

4.5.3 Spacing of pipe supports

The spacing of supports depends primarily on piping materials and their sizes. The recommended spacing of supports for various pipes is given in Table 4.4. Furthermore, the supports must be located near the load concentration. i.e. near heavy valves and near all joints. Extra support is to be provided within 450 mm of pipe joint.

Size of hanger rods: Hanger rods shall be sufficiently strong to withstand the pipe, including its content. For this reason, the size of hanger rods is proposed for iron pipes of different sizes in Table 4.5.

4.6 Pipe sleeve

A sleeve, shown in Figure 4.17a, is a cylindrical insert placed in structural members, through which a pipe pierces the member. Generally, the function of a sleeve is to permit passage of a pipe through a wall or floor, but the pipe is not mechanically attached to the sleeve. The sleeve is usually made of galvanized steel, cast iron, plastic etc. The diameter of the sleeve is made generally one size larger than the diameter of the pipe passing through. A sleeve is placed in concrete form before the concrete is poured. Figure 4.17b illustrates how a fabricated sleeve is placed in a wall to allow passage of the pipe.

Table 4.5 Size of hanger rods. Converted in SI unit from source [6].

Sl. No	*Pipe diameter (mm)*	*Size of steel rod (mm)*
1	12–50	8
2	63–75	12
3	100	16
4	125–200	20
5	250–300	22
6	350–450	25
7	500–600	32

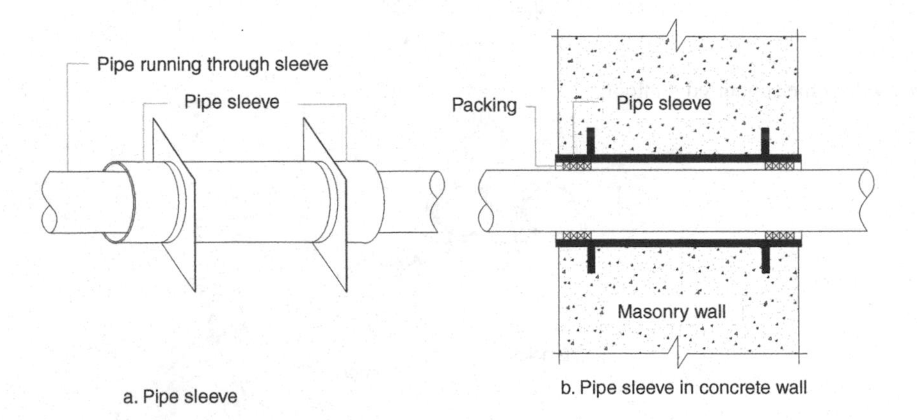

Figure 4.17 Pipe sleeve details.

4.7 Installing underground pipe

In underground pipe installation, a trench of sufficient depth and width is to be made. The trench must be made safe; otherwise it may turn to a dangerous place to work in. To make a trench safe, the following measures should be taken before digging a trench.

1. Try to locate all possible underground utilities that may be encountered.
2. Remove all objects along the trench side that may create a hazard during excavation.
3. Do not pile the excavated soil and materials within a distance from the trench side equal to trench depth.
4. Trenches in loose soil must be shored or sloped to the angle of repose of the soil.
5. Trenches more than 1.5 m deep must be sloped properly.
6. Portable trench boxes or sliding trench shields may be used where applicable.
7. While working in trenches more than 1.2 m deep, provisions for a ladder must be kept for entering and leaving the trench.
8. Nobody should be allowed to work in a trench alone.
9. Any underground water encountered in the trench shall be pumped out to a safer distance before laying the pipe.

After making sure that the trench is safe, the following measures should be followed for pipe laying.

1. Make the bottom of the trench plain, free from large lumps of dirt and bats etc.
2. With careful handling, lower the pipe into the trench.
3. Dig an appropriate-sized hole at the trench bottom so that the pipe can rest on its barrel and not on its bell.
4. Check for proper grade of pipe and make the pipe joint.
5. Backfill the trench, tamping dirt around the pipe first, and then cover the pipe in the same way. Backfill and tamp the fill materials in layers of 300 mm. Figure 4.18 illustrates an underground sewer covered by its various depth of backfilled materials.

4.7.1 Structural safety of underground piping

The forces acting on a cross-section of a subsurface rigid pipeline arise from three main sources, as follows.

1. Weight of overlying fill, including any dead or surcharge live load
2. Traffic and other transient loads on the surface transmitted through as soil pressures on the pipe and
3. Supporting reaction of soil below the pipe.

In general, pipelines are laid in trenches and then covered by back filling. On the back filling, structures may be built subject to possessing the structural capacity of pipe and the soil capacity below. So, the buried pipes are designed to make it capable of withstanding the backfill load and traffic loads and, when the diameter is 600 mm or more, part of the water load within the pipe. In determining structural safety of the reinforced cement concrete (RCC) buried pipe, the following factors must be considered.

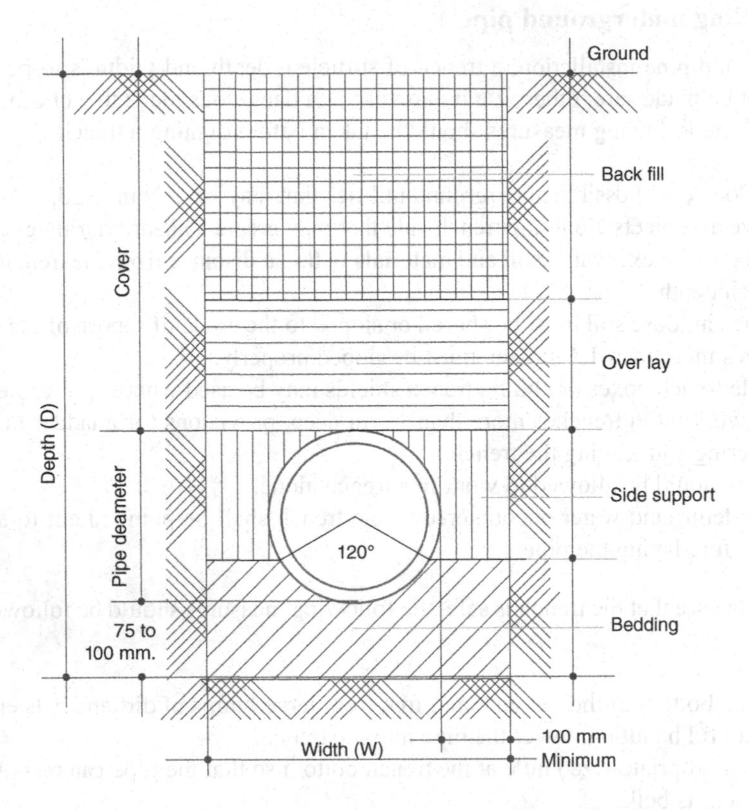

Figure 4.18 Underground sewer covered by its various depths of backfilled materials.

Bedding class and factors: The load-carrying capacity of a concrete pipeline is dependent on both the flexural strength of the manufactured pipe and the support provided by the bedding developed. The bedding factor (F_m) is the term used to denote the ratio of the strength of the laid pipe to its laboratory crushing strength [7]. The higher the bedding factor, the greater the load bearing capacity of a given concrete pipeline. The bedding factor values for different classes of bedding are furnished in Table 4.6.

Laboratory crushing strength: The crushing load (W_t) of a concrete pipe is the minimum load which the pipe will sustain without collapse. The crushing test loads in kilo-Newton per meter (kN/m) of effective length for concrete pipes of different diameter is given in Table 4.7.

Safe supporting strength: For an underground buried pipe, safe supporting strength must be checked. Safe supporting strength of a pipe must be greater than all the external pressures working on the pipe. The supporting strength of a pipe is the product of the pipe crushing strength and the bedding factor. The total external load acting on pipe is the summation of the backfill load, surcharge or traffic load and for pipes greater than 600 mm the equivalent water load.

The crushing strength of a concrete pipe can be determined by using the following formula [9].

Minimum crushing strength = $W_t > W_e \times F_s / F_m$ 4.1)

Table 4.6 Bedding factor values for different class of bedding [8].

Sl. No	*Bedding class*	*Bedding description*	*Bedding factor*
1	D	Flat sub grade	1.1
2	F	45° granular bed	1.5
3	B	180° granular bed	1.9
4	S	360° granular bed	2.2
5	A – unreinforced	120° unreinforced concrete cradle	2.6
6	A – reinforced	120° reinforced concrete cradle	3.4

Table 4.7 Minimum crush test loads of concrete pipe as stated in I.S. 6: 2004 [9].

Type of concrete pipe	*Nominal pipe Œ (DN) in mm*	*Minimum crushing load (Fn), kN/m (Fn = Wt)*
Unreinforced or plain concrete pipe	225	27
	300	36
	375	45
	450	54
	525	63
	600	72
	675	81
Reinforced concrete pipe	750	90
	900	108
	1050	126
	1200	144
	1350	162
	1500	180
	1650	198
	1800	216
	2100	252
	2400	288

Where

W_e = total external load (kN/m)

F_s = factor or safety taken as a minimum of 1.25 for concrete pipe, and 1.5 for reinforced concrete pipe [8] and

F_m = bedding factor.

4.8 Testing of plumbing system

The primary objective of testing the plumbing system is to ensure a strong and leak-proof jointing system. The other objective is to detect defects like leaks, weak joints, any flaws etc. It is necessary to test all new plumbing works and parts of the existing plumbing system which have been altered, repaired or extended, so that the work is non-defective, non-leaking and strong. Unless a plumbing system has been inspected, tested and accepted by the concerned authorities, no part of the piping system shall be covered or concealed. The testing should be conducted in the presence of the concerned approving authority.

4.8.1 Testing of water supply piping system

Water supply piping systems are tested in two stages.

i In the rough plumbing system and
ii On completely installed systems, including faucets.

Testing the rough piping installation

The following steps should be followed in testing the water supply piping system before the final connection of faucets and other equipment.

1 Install bib cocks at the two ends of the piping system to be tested. One end bibcock is for filling water into the piping, and the other is for evacuating air from the piping.
2 Cap and make watertight all other openings made for connecting faucets or fixtures with the piping.
3 Fill the piping with potable water through one bibcock, keeping the other bibcock open. When the system is full of water, all air will be forced out; the end bibcock is to be closed.
4 Depending on the size of piping, connect a hand pump, as shown in Figure 4.19, or a centrifugal pump to create pressure in the system. Connect a pressure gauge to read the pressure in the system.
5 Increase the pressure in the system so that it is not less than the following.

   i. One and a half times the pressure at which the piping system will be operated and
   ii. 1035 kPa.

6 When the desired pressure level is attained, close the bibcock. Sustain the pressure with the system for about three hours. If the pressure remains the same, then the system is watertight. If the pressure decreases, then there is leakage in the system. After locating the leakage, it is repaired and tested again until the pressure remains constant.

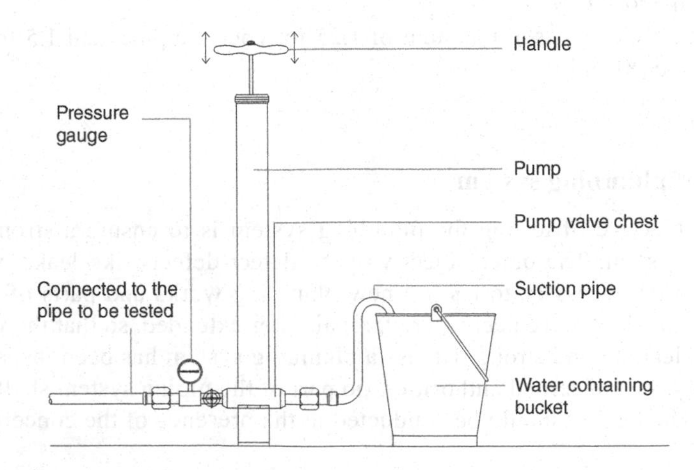

Figure 4.19 Pressure pumping set used in water supply piping.

Testing the complete plumbing system

When the rough piping installation has been found to be watertight, then the following steps are to be carried out.

1 Connect all valves, fixture faucets etc.
2 Connect the water piping system to the water supply and
3 Subject the entire system to the hydrostatic test as stated earlier.

4.8.2 Testing of drainage and vent piping system

Testing of drainage and vent piping system is carried out in two methods. These are as follows.

1 Water test and
2 Air test.

These two methods of testing are described below.

Water test

In the water test, it is necessary to seal off the flow in the drainage piping to be tested. There are two types of test plugs. One type of test plug, shown in Figure 4.20a, is made of rubber or some other elastomer. The plug is attached to a valve and air compressor. When it is fully inflated by feeding air using an air compressor, it seals the pipe watertight. The valve is then closed and the air compressor is removed.

The other type of test plug is illustrated in Figure 4.20b. The plug is put in the open end of the pipe to be tested. Turning the handle causes the gasket to expand against the interior wall of the pipe, forming a watertight plug.

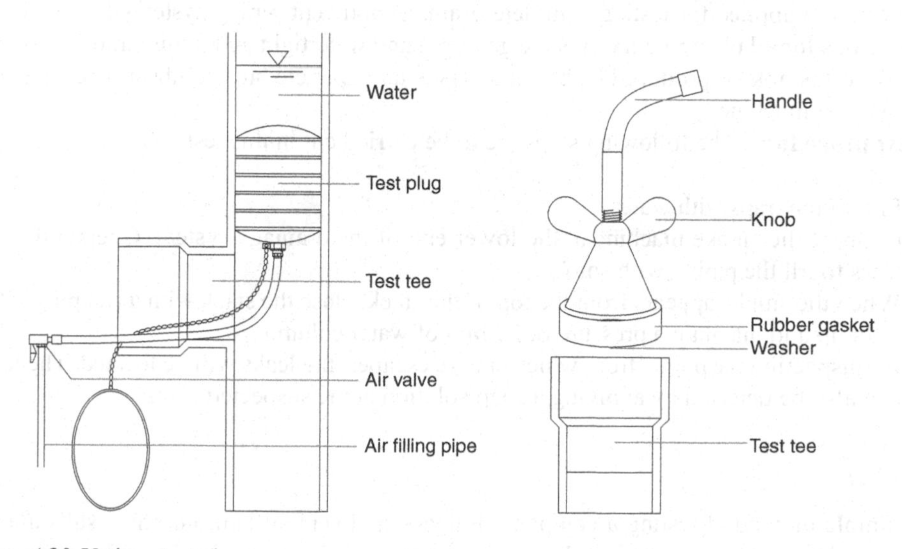

Figure 4.20 Various test plug.

Test procedure: To run the test on a drainage or vent piping system, the following steps should be performed.

1 Install a test "Tee" at the base of the stack.
2 Close all openings tightly except the highest one and fill the system with water until it overflows from the stack. The height of water column in the stack must be less than 3 m.
3 The water should remain in the system for at least 15 minutes before the inspection starts.
4 The dropping of water is an indication of the presence of a leak in the system. The entire piping system must be inspected, including the test plugs. The leaks shall be repaired.
5 When the test is complete, all test plug assembly is to be removed.

Air test

Air testing has the advantage of applying pressure uniformly throughout the system under test, whereas water testing imposes higher pressure on the lower portion of the piping system. Care must be taken so that there is no large drop in temperature, which may result in a drop in air pressure. As a result, one may get confused with the presence of a leak.

Test procedure: The following steps should be taken in performing an air test.

1 Close all outlets and inlets within the system.
2 Connect an air compressor to the system. Isolate it from the system.
3 Increase the pressure up to 34.5 kPa in the system.
4 The pressure should be maintained for at least 15 minutes.
5 If the pressure drops, detect the leaks by inspection and by applying a soap solution at probable leaking points.
6 After the leaks are located, repair and repeat the test until the system proves to be airtight.

Smoke test

Smoke test is applied for testing complete drainage and vent piping systems to ensure that connections for all plumbing fixtures are gas-tight and watertight and fixture traps are sound. For this test, smoke is produced by burning waste, tar paper etc. in a combustion chamber of a smoke test machine.

Test procedure: The following steps are to be carried out in this test.

1 Fill all the traps with water.
2 Connect the smoke machine at the lower end of the drainage system. Operate the bellows to fill the piping with smoke.
3 When the smoke appears from the top of the stack, close the stack with a test plug. Then built up and maintain a pressure of 25 mm of water column.
4 By inspecting the points from which smoke escapes, the leaks will be located. The leaks can also be detected by applying a soap solution at the suspected spots.

Odor test

It is simple method of testing a complete drainage and vent system, but the results are not always satisfactory because it is difficult to locate the leaking spot after the leakage has been detected.

Table 4.8 The disinfection time with chlorinated water of chlorine concentration 1 mg/l for various pathogens [10].

Pathogens	*Disinfection time*
E. coli 0157 H7 bacterium	<1 minute
Hepatitis A virus	about 16 minutes
Giardia parasite	about 45 minutes
Cryptosporidium	about 9600 minutes (6, 7 days)

4.9 Disinfecting piping

Before using water from newly installed or partly replaced plumbing systems, the full water supply piping system, including all machinery, accessories and appurtenances, shall be disinfected to prevent contamination of water. In addition, after the repair or maintenance of water supply systems, there is a possibility of extraneous materials getting into the inside service; the inside service shall therefore be cleaned and disinfected before water supply is resumed for consumption.

Cleaning and disinfecting the system is suggested by chlorination, made by applying 50 mg/l chlorinated water for 24 hrs, and 200 mg/l for 3 hrs. That is, the whole water supply system shall be kept filled with water having a chlorine concentration of 50 ppm for 24 hrs or 200 ppm for 3 hrs [11]. If a bleaching powder of 25 percent available chlorine is the agent chosen, 200 grams of the compound will be required per 1000 litres of solution. After chlorination, the total system shall be flushed until the removal of the chlorine from the piping.

Bleaching powder ($CaOCl_2$) is produced by directing chlorine through calcium hydroxide (CaOH). When bleaching powder dissolves, it reacts with water to produce hypochloric acid (HOCl) and hypochlorite ions (OCl^-). Disinfection time of water varies with chlorine concentration and type of pathogen present. The disinfection time with chlorinated water of chlorine concentration 1 mg/l (1 ppm) at pH = 7.5 and T = 25°C to kill various pathogenic organisms, is furnished in Table 4.8.

The maximum chlorine concentration allowed is 5 mg/l. According to WHO guidelines, the free residual chlorine concentration in drinking water should be between 0.2 and 0.5 mg/l [12].

References

[1] Cast Iron Soil Pipe Institute (2006) "Cast Iron Soil Pipe and Fittings Handbook" p. 48 Retrieved from www.cispi.org/wp-content/uploads/2017/01/CISPAF_Handbook.pdf, on 26 December 2020.

[2] Indian Water Works Association (IWWA), "Cement Mortar as a Jointing Material for Cast Iron Pipes" Retrieved from http://iwwa.info/91-decade/1971-1980/1971/apr-jun-1971/265-cement-mortar-as-a-jointing-material-for-cast-iron-pipes on 26 December 2020.

[3] The Hosemaster (UD) "How to Measure BSP Threads" Retrieved from www.thehosemaster.co.uk/bsp-pipe-threads on 17 December 2020.

[4] Copper Development Association Inc (2020) "Copper Tube Hand Book" Retrieved from www.copper.org/publications/pub_list/pdf/copper_tube_handbook.pdf.

[5] International Plumbing Code (IPC) (2015) "General Regulations Chapter 3, Table 308.5 Hanger Spacing" Retrieved from https://up.codes/viewer/utah/ipc-2015/chapter/3/general-regulations#308 on 25 December 2020.

[6] Engineering ToolBox (2003) "Hanger Support Spacing–Rod Sizes Horizontal Pipes" Retrieved from www.engineeringtoolbox.com/piping-support-d_362.html on 25 December 2020.

[7] Ontario Concrete Pipe Association (OCPA), "Concrete Pipe Design Manual" Retrieved from https://ccppa.ca/wp-content/uploads/2018/11/OCPA_DesignManual.pdf on 21 May 2021.

[8] CPM Group, "CPM Concrete Drainage Systems" *Pipeline Design*, Hydraulic, UK, Retrieved from www.cpm-group.com on 27 October 2015.

[9] Condron Concrete Works, "Concrete Pipes" *Brocheur*, p. 35, Retrieved from http://condronconcrete.ie/wp-content/uploads/2015/12/Condron-Concrete-Works-Concrete-Pipe-Brochure.pdf on 25 December 2020.

[10] Lenntech, B.V. (2019) "Disinfectants: Chlorine" Retrieved from www.lenntech.com/processes/disinfection/chemical/disinfectants-chlorine.htm on 25 December 2020.

[11] Housing and Building Research Institute (HBRI) (1993) *Bangladesh National Building Code (BNBC)*, HBRI, Dhaka.

[12] World Health Organization (WHO) (2011) "Measuring Chlorine Levels in Water Supplies" *Technical Notes on Drinking-Water, Sanitation and Hygiene in Emergencies*, Retrieved from www.who.int/water_sanitation_health/publications/2011/tn11_chlorine_levels_en.pdf on 06 October 2015.

5 Pump and pumping system

5.1 Introduction

A pump is a mechanical device used for conveying fluid from one position to another by applying some form of energy. There are various types of pump. In plumbing systems, mostly centrifugal type pumps are used in water supply and occasionally in drainage systems. There are several types of pumps with different operating principles. A pump consumes energy for its operation. It is therefore necessary to know the principles of pump operation and maintenance, so that the appropriate type of pump for a particular use can be selected for its effective use at optimum energy levels. So, this chapter is dedicated to discussing all aspects of various centrifugal pump selection, operation, installation and maintenance etc.

5.2 Pump

A pump is a mechanical device used to convert energy of a prime mover or rotor, first into velocity or kinetic energy and then into pressure energy of a fluid that is to be pumped. Generally, an electric driven motor is used as the prime mover. In plumbing, mostly water is handled, occasionally wastewater, by the pump.

5.2.1 Pump type

Pumps are primarily of three types: 1. centrifugal, 2. positive reciprocating and 3. rotary type. There are many other types of pump within each category. In plumbing system development, the most common type of pump used is the centrifugal pump.

5.3 Centrifugal pump

The typical centrifugal pump, shown in Figure 5.1, in its simplest form consists of a round disk called an impeller with vanes attached to it. The vanes are curved backward. The impeller, fixed at the end of a shaft, is housed in a volute casing which has to be filled with liquid when the pump starts operation due to rotation of the shaft attached to a motor. When the impeller starts rotating, liquid is forced away from the centre of the impeller, called the hub or "eye" of the impeller; a vacuum is created and more water gets in. The liquid entering into the centre of the impeller is dragged up by the vanes, accelerated to a high velocity by rotation of the impeller and discharged by centrifugal force into the casing and ultimately out the discharge. Consequently, more water gets into the pump through the suction eye, which is centrifuged outward and ultimately forced out through the discharge outlet; there thus occurs

DOI: 10.1201/9781003172239-5

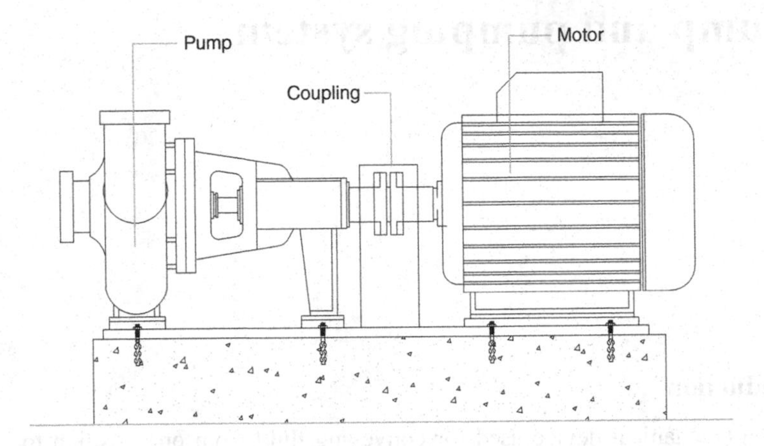

Figure 5.1 Electric motor driven centrifugal pump.

a constant flow through the pump, as it rotates continuously, driven by a motor that is mostly electrically powered.

Though centrifugal pumps are the most widely used type of pump in plumbing, for its many advantageous features, there are many disadvantages associated with these pumps, as outlined below.

Advantages of centrifugal pumps:

1 Simple operating system
2 Low investment and maintenance cost
3 Insignificant excessive pressure builds up in casing
4 Impeller and shaft are the only moving parts
5 Quiet running of the pump
6 Wide range of pressure, flow and capacities and
7 Utilizes comparatively small floor space in different positions.

Disadvantages:

1 High viscous fluids are not well handled
2 Usually not capable of handling high pressure applications in comparison to other types of pumps
3 In general, cannot develop high pressure without bringing changes in design and
4 Not suitable for high pressure delivery at low volumes, except the multi-stage pumps.

5.3.1 Major components of a centrifugal pump

Following are some major components of a centrifugal pump, some of which are shown in Figure 5.2.

Volute: Volute is the casing in which the impeller of the pump is housed. It acts as a pressure containment bowl that converts the velocity of moving liquid into pressure. It also directs the flow of liquid towards the discharge outlet of the pump.

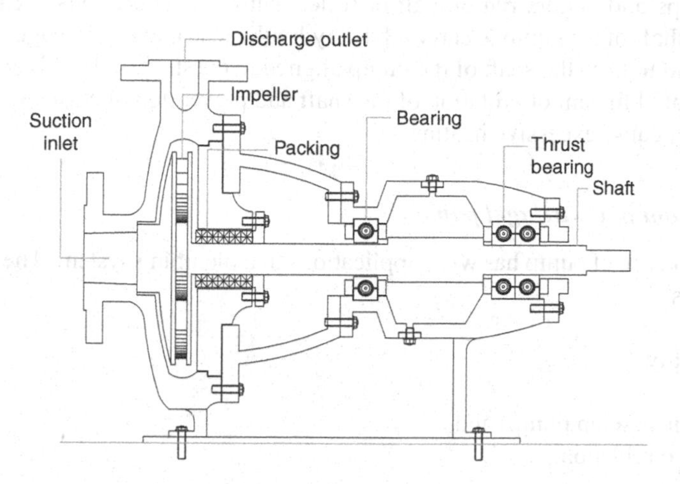

Figure 5.2 Major components of a centrifugal pump.

Impeller: The circular type part in which vanes are attached with a circular disc called a shroud. The impellers rotate at very high speed. So, the impellers are carefully designed, machined and balanced to reduce vibration and wear. To prevent corrosion, impellers can be made of non-ferrous materials such as bronze or Monel.

Shaft: The cylindrical bar used to rotate the impeller of the pump which is mounted on it. Its main function is to transmit the input power from the motor driver to the impeller of the pump. The shaft enters the pump casing through a corrosion-resistant metal sleeve that is sealed to prevent leakage. During installation of the pump, the alignment of the shaft is to be checked accurately so that it remains truly horizontal, in the case of horizontal pumps, and truly vertical for vertical pumps. Deflection of the shaft will cause a decrease in the service life of bearings, seals and wear rings.

Stuffing box: The stuffing box contains the packing materials to seal off the space (aperture) around the rotating shaft where it enters the casing through a metal sleeve called a shaft sleeve. So, liquid from the pump cannot leave and air cannot enter into the pump. Two types of seals are used: 1. soft fibre packing and 2. mechanical face seal. The shaft sleeve helps position the impeller correctly on the shaft and also protects the shaft.

Couplings: A coupling in a pumping set is a power transmission device used to connect the motor shaft to the pump shaft. Pumps can be directly coupled to the motor so that the speed of the motor and of the pump remain the same. The majority of pumps are coupled with the motor through their respective shafts using flexible couplings. Flexible couplings allow the pump to be disassembled without removing the piping and motor.

The primary function of couplings is to make a connection between the pump shaft and the motor shaft, to transmit the input power of rotary motion and torque from the motor driver to the power end shaft of the pump. The other secondary functions of couplings are as follows.

1 To accommodate misalignment between shafts
2 To transmit axial thrust loads from pump to motor
3 To permit adjustment of shafts to compensate for wear
4 To maintain precise alignment between connected shafts and
5 To allow temperature changes and movement of the shaft without mutual interference.

Bearing: Pumps and motors run on ball or roller (antifriction) bearings. Bearings provide support to the shaft of the pump to carry all the hydraulic load and assemblage weights acting on the shaft, and to keep the shaft of the pump aligned to the shaft of the driver. The bearings are kept lubricated for smooth rotation of the shaft and prevention of corrosion. Over supply of lubricant can cause excessive heating.

5.3.2 Application of centrifugal pump

The centrifugal type of pump has wide applications in a plumbing system. The following are some examples.

1 Water supply
2 Drainage
3 Cellar drains (sump pump) and
4 Hot water circulation.

5.4 Type of centrifugal pump

The types of centrifugal pumps are identified in a number of different ways based on the following factors.

1 Internal design
2 Suction configuration
3 Shape of impeller
4 Casing design
5 Motor and pump connection
6 Type of flow
7 Support of pump
8 Bearing support
9 Staging of pump and
10 Type of service.

5.4.1 Based on internal design

Volute type: The impeller of the pump is housed in spiral (volute) shaped passage called a volute casing, as shown in Figure 5.3a. The centre of the impeller is not located at the centre of the volute. The portion where the impeller periphery extends closest to the volute is termed the cut-water. Starting from the cut-water and proceeding towards the outlet, the distance between the volute and the impeller periphery increases gradually. This is essential to produce a flow of equal velocity all around the circumference and to convert velocity energy into pressure energy. Centrifuged water is then forced out of the volute, through the discharge outlet, being encountered at the cut-water.

Diffuser type: A diffuser type of pump is provided with a stationary diffuser ring around the impeller for flow guidance, as shown in Figure 5.3b. The basic function of a diffuser is similar to that of a volute. The liquid, after leaving the impeller, passes through these guiding vanes, which have a gradually enlarged passage due to which velocity of flow is reduced and pressure increases. Consequently, its efficiency becomes slightly higher. The added cost and more complicated construction of diffuser type pumps are generally not considered justified

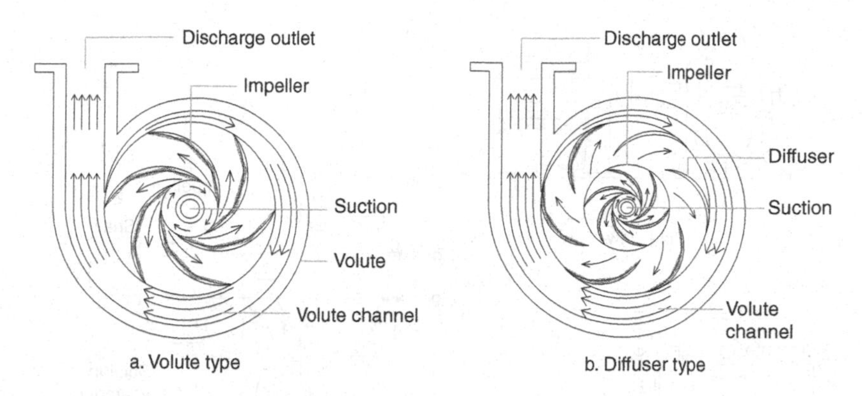

Figure 5.3 Various centrifugal pump based on internal design.

in low-pressure range operations. A diffuser pump is most commonly used in high-pressure ranges because of its wide range of capacity, pressure and efficiency. The diffuser type of pump is sometimes referred to as a "Turbine pump". But the hydraulic turbine pump differs in principle and construction from this type of pump.

5.4.2 Based on suction configuration

Single suction pump: In this pump, there exists one suction inlet only. So, water enters into the impeller from only one side. This is the most commonly used pump, and the pump in Figure 5.1 also represents a single suction pump.

Double suction pump: In this pump, water enters the impeller symmetrically from both sides. A pair of impellers is arranged back-to-back on the pump shaft. So, axial thrust is balanced, eliminating the need for thrust bearing. A double suction pump is shown in Figure 5.4. Double suction pumps are generally used when the required flow rate becomes too large for the inlet cross-sections of one impeller generally designed, or when the flow velocity in the inlet cross-section of a single impeller pump has to be reduced to prevent cavitation.

End suction: In the end suction pump, the impeller is fitted at the end; therefore, the shaft does not pass the impeller hub and overhangs on the shaft, as shown in Figure 5.1.

5.4.3 Based on shape of impeller

Open impeller: Impellers are designed curved to minimize the shock losses of flow in the liquid as it moves from the eye to the shroud. The shroud is the disk to which the vanes are attached. If an impeller has no shroud, then it is called an open impeller. This type of impeller is used for handling liquids containing suspended solids, particularly in wastewater.

Closed impeller: When shrouds on both sides enclose all vanes, then the impeller is called a closed impeller. It requires comparatively little maintenance and usually retains its operating efficiency longer than an open impeller pump. These impellers are preferred for handling relatively less turbid water, hot water etc.

Semi-open impeller: If the impeller has one shroud, then it is called a semi-open impeller. This type of impeller is used for pumping sewage water, paper pulp etc. These three types of impeller are shown in Figure 5.5.

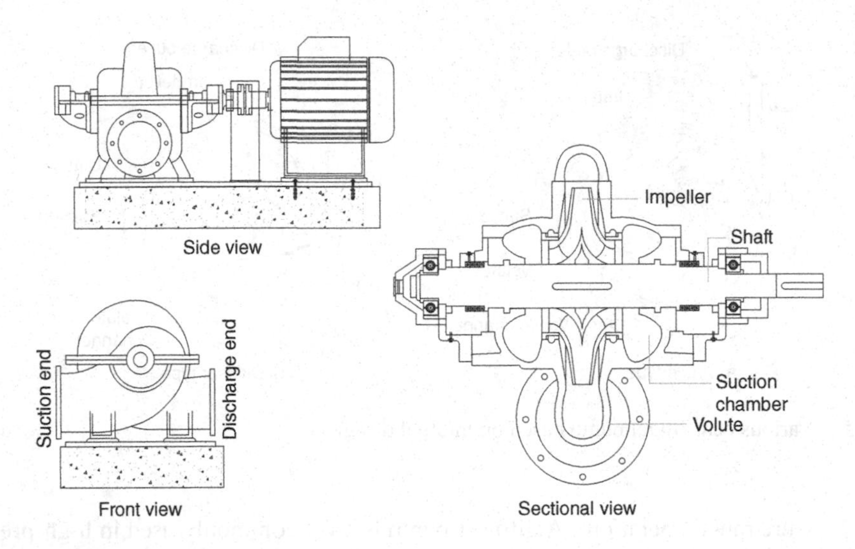

Figure 5.4 Double suction pump.

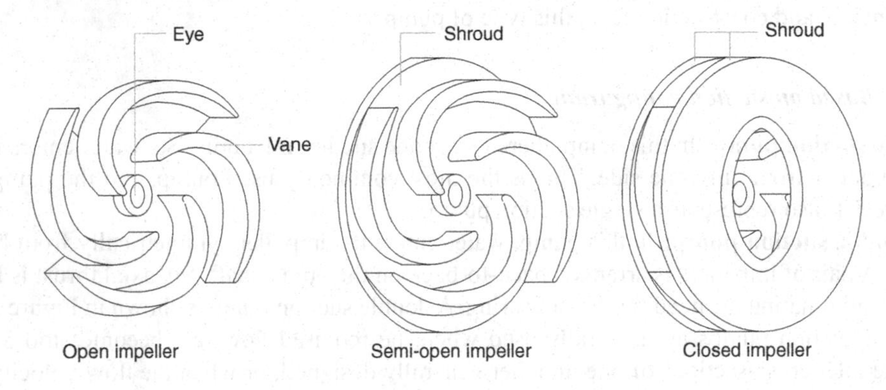

Figure 5.5 Various impellers of centrifugal pump.

5.4.4 Based on casing design

Axially split: The axially split casing is split parallel to the shaft axis, as shown in Figure 5.6a. So, the pump can be opened without disturbing the piping. This helps in servicing the pump easily. The main problem is that the pump is seldom supported at the shaft centreline and so is supported at near-centreline. This results in asymmetrical thermal expansion between the upper and lower casing parts. In intermediate pressure ranges, an axially (horizontal) split case centrifugal pump is economical.

Radially split: Pump casings are generally made with a vertical split, dividing into two halves. Splitting may also be done on the end or on the coupling side. The problem with such a system is that for having full opening of the pump, the impeller has to be removed. These pumps can be designed for centreline support, enabled for operation at higher temperatures and working pressure in comparison to axially split pumps. A radially (vertical) split case centrifugal pump should be considered only in low-pressure ranges. Figure 5.6b represents a radially split centrifugal pump.

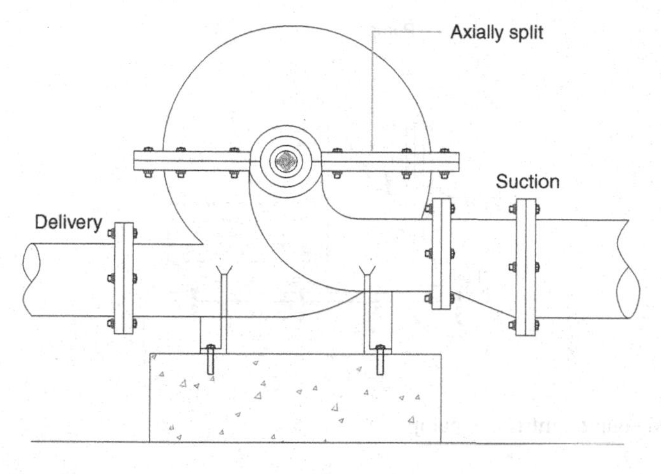

Figure 5.6a Axially split centrifugal pump.

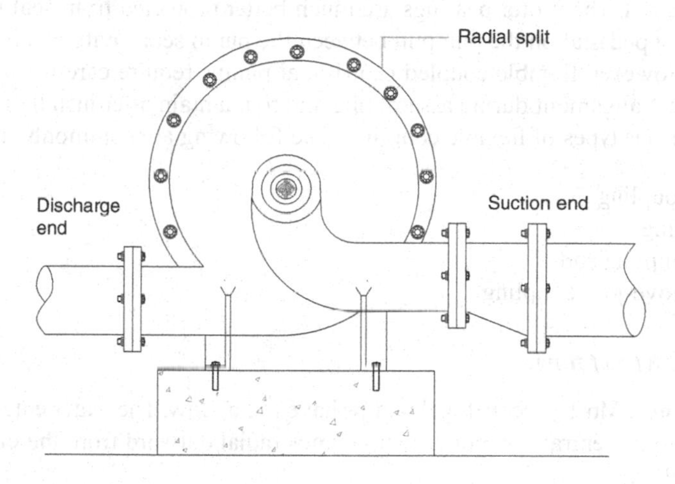

Figure 5.6b Radial split pump.

5.4.5 Based on motor and pump connection

Closed-coupled pump: In this type of pump, the same single shaft is used for both the motor and the pump, and the motor is bolted directly to the pump, as shown in Figure 5.7. As a result, manufacturing cost is relatively lower, installation is easier and problems regarding shaft alignment are eliminated. This type of pump is fairly fool-proof in operation. These pumps are not preferred for heavy duty, high power and continuous use, as there is more strain put on the common bearings and usually limited motor size. Another problem that arises is motor noise, which is transmitted to the pump and piping. These pumps can be chosen for applications where there is absolutely no possibility of misalignment.

Flexible coupled pump: In this type of pump, the electric motor drive is connected to the pump by means of a flexible coupling. The centrifugal pump in Figure 5.1 represents a flexible coupled one. Both the motor and the pump are mounted on structural base-plates to provide support and maintain shaft alignment.

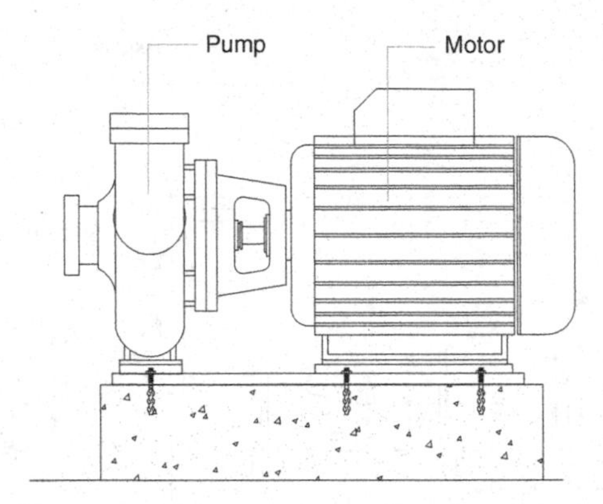

Figure 5.7 Closed couple centrifugal pump.

These pumps are very popular because the motors can be replaced by larger horsepower sizes when required. The motor bearings are much better protected from seal leaks because there is a bearing pedestal on the pump in between the pump seal cavity and the motor shaft and bearings. However, flexible coupled centrifugal pumps require careful coupling efforts to achieve perfect alignment during assembling and to maintain after installation.

There are various types of flexible coupling. The following are commonly used.

a) Pin bush coupling
b) Tyre coupling
c) Flanged coupling and
d) Jaw type (love-joy) coupling.

5.4.6 Based on type of flow

Radial flow pump: Mostly, centrifugal pumps have radial flow. The water enters through the impeller eye into its central hub portion and comes radial outward from the circumferential rim of the impeller.

Francis type: In this type, the impeller is made thicker and has lesser diameter compared to the radial flow type. The discharge diameter to inlet eyes outside diameter ratio is much less than 2, which in the case of radial is nearly 2. Running speed is much higher.

Axial flow/propeller pump: This type of roto-dynamic pump has an open type impeller with three blades. Water enters axially (i.e. parallel to shaft axis) and is propelled out axially. This type of impeller is used in the largest possible pumps to work on low heads.

Mixed flow pump: This type of roto-dynamic pump has a type of impeller in which water enters axially, but within the impeller, flow is partially radial and partially axial, resulting in a mixed type of flow. This type of impeller is designed for large pumps of medium head.

Centripetal flow: This is a self-priming pump in which water enters from the circumferential rim of the impeller and flows out through the central hub or eye of the impeller.

5.4.7 Based on type of service or duty

Lift of water: A pump can be classified according to the depth of suction lift i.e. the depth of water from which the pump can suck water without interruption. Practically, suction lift is limited to 5 m in the case of centrifugal pumps. So, if water is to be drawn from a further depth by a centrifugal pump, special means are to be adopted. According to the means adopted, pumps are named in the following way.

1 Low lift pump
2 Deep well pump

   a) Vertical turbine
   b) Eject or Jet pump
   c) Air-lift pump.

Head of water: Pumps are classified according to their total head of water as

1 Low head: up to 25 m of water
2 Medium head: more than 25–50 m of water
3 High head: more than 50–200 m of water and
4 Extra high head: over 200 m of water.

5.4.8 Based on support of pump

Horizontal dry pit: These are the most common types of pump, installed with the shaft in a horizontal position. Generally, they are installed in close vicinity to the water reservoir in a dry location, but sometimes a special pit is constructed to house the pump. The pump is supported on the floor.

In line: These types of pumps are mostly used for recirculation and boosting pressure of flowing water in water supply systems. These pumps are installed on the piping system, as shown in Figure 5.8. The piping carries the weight of the pump. Usually pump-motor assembly is mounted vertically in order to save space and centre the weight over the piping. But, some smaller pumps may be hung horizontally from the piping. To avoid penetration of water inside the motor and bearings, the motor must never be positioned below the horizontal.

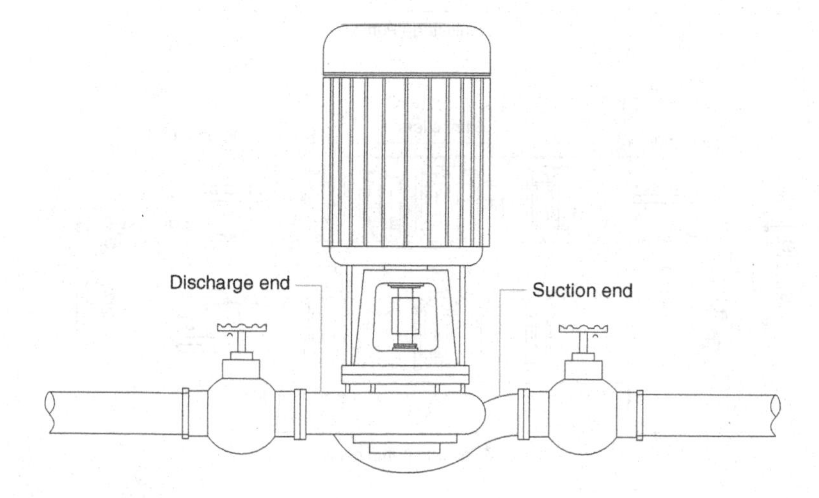

Figure 5.8 In-line pump.

Wet pits: Wet pit pumps are those that are immersed in the liquid to be pumped, contained in a pit. The pump may be supported at the base of the pit or may be suspended from the top of the pit.

5.4.9 Based on bearing support

Between-bearing pumps: This is a centrifugal pump whose impeller is supported by bearings on each side. Most double suction impellers in axially split casings are between-bearing pumps.

Overhung impeller pump: This is a centrifugal pump in which the impeller is mounted on the end of the shaft that overhangs its bearings. Inline circulating pumps are an example of this type of pump.

5.4.10 Based on staging of pump

Single stage: A single stage pump is that one which has only one impeller. Pumps for head below 90 m are generally of single stage. Single stage pumps are generally preferable for applications with high demand of flow and lower pressure head of maximum 690 kPa [1].

Multi-stage: A multi-staged pump is one which has more than one impeller connected in series. A multi-stage pump is used when the head is more than 90 m. Figure 5.9 illustrates a five-stage centrifugal pump where five impellers are placed in series.

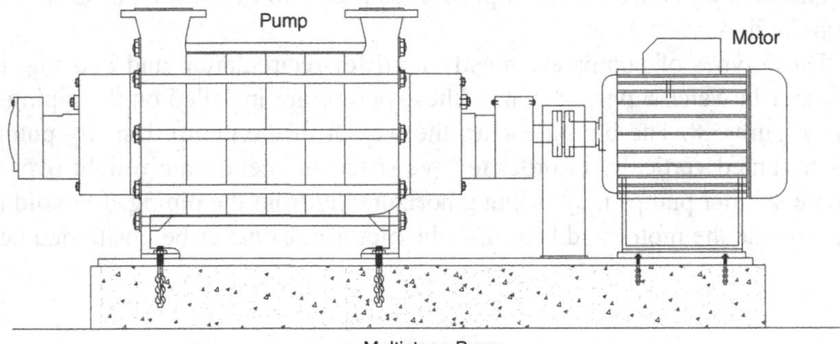

a. Multistage Pump

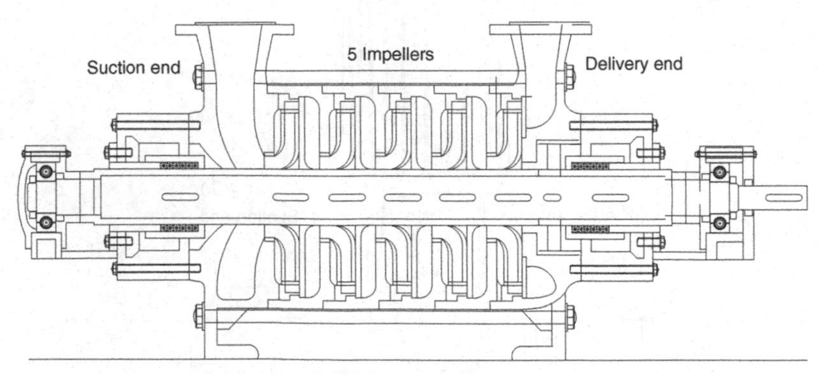

b. Multistage pump section

Figure 5.9 Multi-stage (five stages) pump.

5.5 Pump characteristics

There are various parameters responsible for determining the characteristics of pumps. The variable parameters of pump characteristics are head, power, efficiency, net positive suction head (NPSH), speed, viscosity, impeller diameter and discharge rate or capacity of pump etc. Normally the four parameters are found to vary with respect to capacity, as basic parameters which can be represented in a single graph produced by the manufacturer that represents a set of curves for each characteristic of various standard impeller diameters.

Capacity: The capacity of a pump is the rate of flow of liquid through the impeller. The capacity is generally denoted by "Q" and expressed in cubic meters per hour (cum/h) or in litres per minute (lpm) etc.

Head: The energy contained in a fluid per its unit weight is termed as head of energy, which is abbreviated as head. Head is simply energy divided by weight. For a pump, head is an indication of energy needed for the height to which a column of fluid must rise in a narrow frictionless pipe if the fluid contains the same amount of energy as is contained in one unit weight of fluid under conditions being considered. The reason behind using "head" instead of pressure to denote a centrifugal pump's energy is that the pressure to be exerted by the pump need to be changed if the specific gravity (weight) of the fluid changes, but the head will not. Head of pump is commonly expressed in meters (m).

When a pump is running, the (dynamic) total energy contained in a flowing (dynamic) fluid under pressure is the sum of its potential energy, kinetic energy and pressure, which need to be overcome by the pump. So, total head is the sum of potential head, kinetic head and pressure head.

Potential head = Potential energy per unit weight = the height of water column, also called the static head, and is expressed in terms of meters (m) of liquid.

$$\textit{Potential or Static head} = \frac{\textit{Pressure}(\textit{Bar}) \times 10.2}{\textit{Specific gravity}}(\textit{meter}) \quad 5.1$$

i.e. potential or static head is nothing but the vertical height of water column or depth in meter, above a selected datum level, as specific gravity of water is 1.0.

There are four aspects of energy to be addressed while running a pumping system, as below.

1 Pressure
2 Elevation
3 Friction and
4 Velocity.

According to Bernoulli's theorem, valid at any arbitrary point along a fluid flow is:

$$\frac{v^2}{2} + gz + \frac{P}{\rho} = \textit{Constant} \quad 5.2$$

Where
v = speed of fluid flow at a point on a streamline
g = acceleration due to gravity
z = elevation of the point above a reference plane, with the positive z-direction pointing upward, negative in the direction opposite to the gravitational acceleration

p = pressure at the chosen point and
ρ = density of the fluid.

Dividing both sides by g, it is found that

$$\frac{P}{\rho g}+\frac{v^2}{2g}+z=Constant \qquad 5.3$$

Where the first term is called the pressure head, the second term is called the dynamic (kinetic) head and the last term is called the potential (static) head.

5.6 Head calculation for centrifugal pumps

Head of a pump is the net amount of energy which is to be added to the liquid for pumping it, from one position (lower level) to another position (higher level).

The energy required to lift water from a lower level to its datum level is termed the suction head. The energy needed to deliver water from a pump's datum level to a higher level, is termed delivery head, as shown in Figure 5.10. So, total head for a pump is total suction head plus total delivery head.

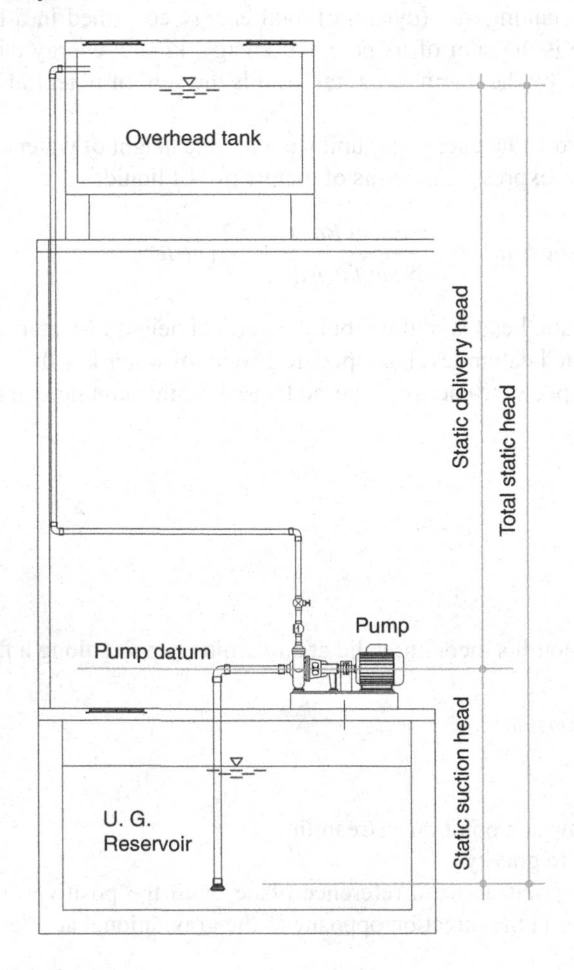

Figure 5.10 Static head of pump.

Total suction head (Hs): The suction head is the energy per unit weight on the suction side of the pump.

$$H_s = h_{ps} + h_{vs} + h_s \quad 5.4$$

Where
h_{ps} = suction pressure head
h_{vs} = suction velocity head and
h_s = suction static head.

Total delivery head (Hd): The delivery head is the energy per unit weight of fluid on the discharge side of the pump.

$$H_d = h_{pd} + h_{vd} + h_d \quad 5.5$$

Where
h_{pd} = delivery pressure head
h_{vd} = delivery velocity head and
h_d = delivery static head.

So, total head of a pump is the sum of total suction head and total delivery head. This is termed total dynamic head.

i.e. H = $H_d + (\pm H_s)$.
= $H_d - H_s$. Minus sign is to be taken when water level is higher than the pump datum line.
Therefore, $H = h_{pd} + h_{vd} + h_d - h_{ps} - h_{vs} - h_s$ (m) 5.6

When fluid flows through a pipe, there will be friction at the contact surface. To overcome this friction, some extra energy will be needed. This is one kind of head loss and is termed friction head loss (h_f).

So, total head is $H = H_d + (\pm H_s) + h_f$ 5.7

This total head is the dynamic head, which is the pressure that causes the water to flow, through the suction and the delivery pipe, up to the overhead tank.

5.6.1 Net positive suction head (NPSH)

Net positive suction head is the total suction head in meters of liquid in absolute pressure terms determined at the pump impeller, minus the vapor pressure of the liquid in meters. The net positive suction head (NPSH) required by a pump is to be determined by test. If the available NPSH for a pump is less than the required NPSH (NPSHR), there might be cavitations and the pump could fail to operate.

Cavitation: Cavitation is the phenomenon of vaporization of liquid due to pressure dropping below its vapour pressure level, which may occur at the impeller inlet of a centrifugal pump. As liquid flows from the pump inlet flange into the impeller, the head initially falls as the velocity of the fluid is increased. This drop in head may be sufficient to cause the liquid to boil. This results in cavitation. Cavitation is detectable as a rattling noise and results in low pump efficiency which may turn into high risk of damage to the pump. To prevent occurrence of cavitation in a pump, the NPSH available from the system must be greater than NPSH required (NPSHR) by the pump.

5.6.2 Factors affecting head loss

There are many reasons for which head loss occurs in pumping. If these reasons are well addressed in the planning, design and installation head loss can be reduced to a great extent. The factors to be considered, with respect to head loss minimization, are as follows.

1 Flow rate
2 Inside diameter of the pipe
3 Roughness of the pipe wall
4 Corrosion and scale deposits
5 Viscosity of the liquid
6 Length of the pipe
7 Fittings and
8 Straightness of the pipe.

Flow rate: The increase in flow rate mainly occurs due to increase in the velocity of the fluid flow in the same cross-sectional area of the pipe. The head loss is related to the square of the velocity, so the increase in head loss is very fast due to increased rate of flow.

Inside diameter of the pipe: When the inside diameter of the pipe is enlarged, the cross-sectional flow area increases and the velocity of flow is reduced, causing lower head loss due to reduced friction. Conversely, for reduction of diameter of pipe the velocity increases, so the head loss increases.

Roughness of the pipe wall: With the increase in roughness of the inside pipe wall, the thickness of the slow or non-moving boundary layer, at the interface of pipe wall and liquid, increases due to friction, resulting in reduction of flow area. As a result, the velocity of flow increases, causing increased head loss.

Corrosion and scale deposits: Scale depositions on and corrosion of the inside of pipe walls both increase the roughness and reduce the area of flow. As a result, increase in frictional resistance and the flow velocity of the liquid occurs, which ultimately causes an increase in head loss in the flow.

Viscosity of the liquid: In the flow of a liquid of higher viscosity, more frictional resistance is developed in comparison to that developed in the flow of low-viscosity liquid. So, more energy will be required for moving the liquid of higher viscosity, needing more head of pump.

Length of the pipe: Friction occurs during flow along the pipe length in contact with flowing liquid. Friction is supposed to be constant for the unit length of pipe at a given flow rate subject to having the same quality of inner surface. So, with the increase in length of pipe, i.e. flow run, the frictional resistance will be increased, resulting in increased head of pump.

Fittings: All the fittings, more or less, disrupt the smooth flow of liquid in a pipe. When the disruption occurs, this accordingly causes head loss due to friction. At a given flow rate, the losses for the fittings will be increased variably due to increase in number of fittings and type of fittings used.

Straightness of the pipe: Water tends to flow in a straight path due to momentum. This flow direction is disturbed due to any bending and offsetting of pipe, and so, water will tend to bounce off the pipe wall at the bend, and the head loss due to some sort of resistance will increase depending on the degree of bending. Sharp bending causes more friction loss than in long sweep bending.

5.7 Power

Power is required in the electric motor to spin the shaft that drives the pump. It depends on the flow that the pump needs to yield and the pressure it has to withstand.

Input power: The input power of a pump is the mechanical power in watts taken by the shaft or coupling. So, the input power of the pump is also termed break horse power (BHP). The input power (P_s) required for driving a pump varies with the capacity of the pump, as shown in Figure 5.11.

Pump BHP in imperial units is as follows.

$$BHP = \frac{Q.H.S}{367.E} \quad 5.8$$

Where
P = brake horse power (kW)
H = head (m)
Q = flow (cum/sec)
S = specific gravity and
E = pump efficiency (decimal).

Output power: The pump output power (P_w) is the useful power transmitted to the fluid handled by the pump. The value of output power depends upon three factors as mentioned below.

1 Flow
2 Head and
3 Specific gravity of the liquid.

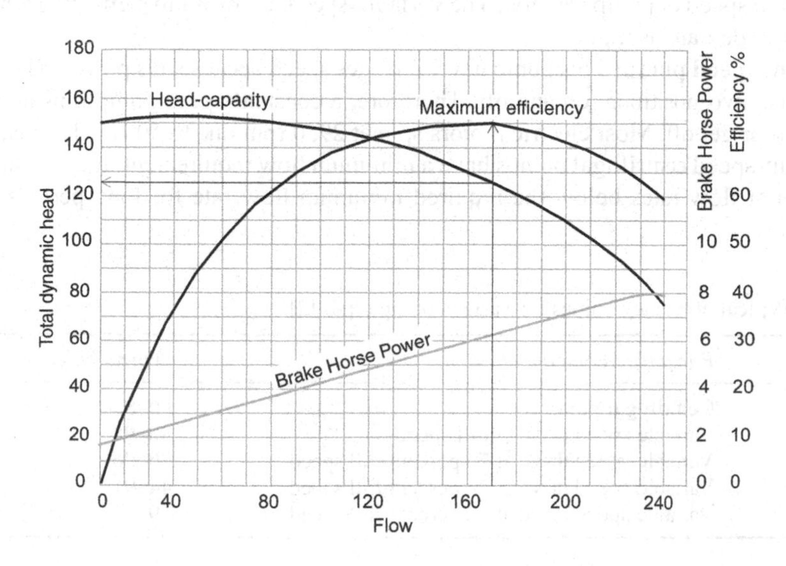

Figure 5.11 Characteristic curve of a centrifugal pump of constant speed.

The pump output power or water horsepower P_w is given below.

$$Output\ Power = \frac{Flow(cum/s) \times Pressure(Pa)}{Total\ efficiency}(Watt) \quad 5.9$$

5.8 Efficiency

Pump efficiency is obtained by comparing the power delivered to the liquid i.e. output power to the power absorbed by the pump i.e. input power. When a pump operates at 75 percent efficiency, then it only delivers 75 percent of the absorbed power; the remaining 25 percent is lost within the pump.

Theoretically, the efficiency percent with which a pump operates is the ratio of the output power to the input power multiplied by 100. It is given by the following formula.

$$Pump\ efficiency\ E_p = \frac{P_w}{P_s} \times 100$$

$$\text{i.e. } Brake\ Horse\ Power\ P_s = \frac{P_w}{E_p} \times 100 \quad 5.10$$

Efficiency varies with the capacity of pump, as shown in Figure 5.11, reaching a maximum value at one capacity where the sum of all losses is a minimum. The efficiency of a typical centrifugal pump varies within a range of 60 to 85 percent [2]. For nearly correct estimation of power requirement for various types of centrifugal pumps, particularly for small size pumps, their efficiency range needs to be known, which is furnished in Table 5.1.

5.9 Speed

Generally, a centrifugal pump is driven by a constant-speed electric motor, though a pump can be efficiently controlled by a variable-speed motor. Characteristics of a centrifugal pump vary with the speed of pump rotation. The variable-speed motor is most effective where fluctuation water demand is high.

Constant speed pump: This pump never changes speed because it is powered by a starter. A starter has two positions: on and off. Therefore, a constant-speed pump runs at its maximum speed or get off. Most electric motors spin at 2950 rpm due to 50 Hz electrical supply. All constant-speed centrifugal pumps have a minimum flow requirement. If a constant-speed pump runs at flow rates below the required minimum flow rate for long periods, various

Table 5.1 Typical efficiency values for various pump types [2].

Sl. No	*Pump type/component*	*Typical efficiency (percent)*
1	Centrifugal pump	60–85
2	Variable speed drive at full speed	80–98
3	Variable speed drive at 75 percent full speed	70–96
4	Variable speed drive at 50 percent full speed	44–91
5	Variable speed drive at 25 percent full speed	9–61

mechanical problems may occur. All drainage pumps and water lifting pumps to fill overhead water tanks shall be constant speed pumps.

Characteristics of centrifugal pumps vary with speed according to the following relationships.

$$\frac{Q_2}{Q_1} = \frac{N_2}{N_1} \quad 5.11a$$

$$\frac{H_2}{H_1} = \frac{N_2^2}{N_1^2} \quad 5.11b$$

$$\frac{P_2}{P_1} = \frac{N_2^3}{N_1^3} \quad 5.11c$$

Where
N_1 = one rotating speed (rpm)
N_2 = another rotating speed (rpm)
Q_1 = capacity at speed N_1 (lpm)
Q_2 = capacity at speed N_2 (lpm)
H_1 = total head capacity at speed N_1 and
H_2 = total head capacity at speed N_2.

The total dynamic head (TDH) of a pump is a function of its impeller's top speed, normally not higher than 213 m/min. A handy formula for peripheral velocity in m/sec is as follows.

$$V = \frac{\pi DN}{60000} \text{ m/sec} \quad 5.12$$

Where
N = impeller RPM (Revolution per minute) and
D = diameter of impeller (mm).

Variable speed pump: This type of pump can be programmed to run at many different speeds to cause variable rate of flow according to demand. Installing a variable speed pump will result in significant energy savings, likely to be between 30 percent and 50 percent, in many applications [3]. The nature of complexity of variable speed drives varies depending on the level of control required. In pressurized water supply systems for large and high-rise buildings with high fluctuating water consumption patterns, like in residential, hospital, hotel, hostel etc. buildings, a variable speed pump is suitable for energy saving.

With a variable speed pump, a pressure sensor is mounted on the outlet of the pump to measure the operating pressure of the pump. The variable speed drive (VSD) or variable frequency drive (VFD) controller is set on the motor to vary the motor speed to keep the outlet pressure of the pump constant, as shown in Figure 5.12. There are various types of VSDs. In a pumping system that requires flow or pressure control, where there might have high friction loss, the most energy-efficient option for control is an electronic VSD, commonly known as a variable frequency drive (VFD). The most common form of VFD found is the voltage-source, pulse-width modulated frequency converter. In the simplest form of VFD, the converter develops a voltage which is directly proportional to the frequency, and

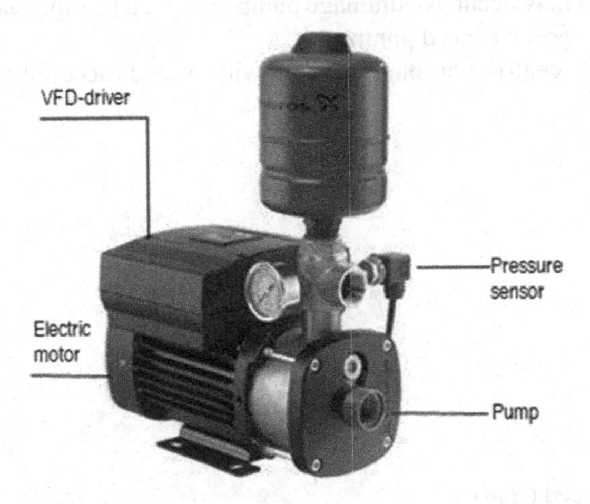

Figure 5.12 A variable speed centrifugal pump. Courtesy: Southerns Water Technology [4]

produces a constant magnetic flux in the motor. This electronic control match with the motor speed needed according to the load requirement.

A constant speed drive pump is suitable for the following conditions.

i Demand for water flow rate is more or less constant
ii Pressure boosting is required from low to medium range
iii Low cost of operation is the major factor
iv Pumps that are driven by motors of less than 40 hp and
v When fluctuation of pressure is more than 138 kPa.

Variable speed drive types are recommended for the following conditions.

i Possibility of large fluctuation of pressure in water supply systems
ii High pressure boosting is needed
iii Large variation of water demand and
iv Power requirement is more than 40 hp.

5.10 Pump application in buildings

In a building, not only are centrifugal pumps made for lifting or supplying water but specially built centrifugal pumps are also manufactured for meeting special needs in a plumbing system. Following are some special centrifugal pumps installed in a plumbing system for the purposes they are intended for.

1 Booster pump
2 Circulation pump
3 Submersible pump and
4 Sewage and sump etc.

5.10.1 Booster pump

The booster pump is used to increase capacity in terms of pressure and flow, mostly in a water supply system. It is generally installed in a water supply system where pressure is insufficient and variable. Two types of booster pump drives are available for adjusting the flow and pressure changes in the water supply system, as discussed earlier.

Capacity of pump, in terms of both flow and head, can be boosted by using multiple pumps set differently.

Flow boosting: For flow boosting, multiple booster pumps of varying or same flow capacity at the same head are added; those will be running in parallel. The delivery pipes of each pump shall be connected to the main delivery pipe of the pump set. The main delivery pipe shall be sufficient in size to accommodate added flow rate; otherwise it should be changed.

Parallel booster pump operation: In most cases, it is more efficient to operate a booster pump set with multiple smaller pumps than one or two large pumps, as water demand profiles may spike once or twice a day [1]. In parallel booster pump operation, sets of multiple pumps like duplex or triplex etc. are used for larger flow requirement with evenly or varied capacity split. In all these systems, a single standby extra pump can be added if there is a need for redundancy. Figure 5.13 illustrates duplex and triplex pumping arrangement only. In these systems, each pump can run alone while keeping others stopped; in combination with one or two; and all at a time, as sequenced.

In duplex system evenly capacity splitting needs each pump of 50 percent of total capacity or varied, as preferred, like 33 percent and 67 percent capacity splitting can also be done between the two pumps.

In triplex system evenly capacity splitting needs each pump of 33.34 percent of total capacity or varied, as preferred, like 20 percent, 40 percent and 40 percent capacity splitting can also be done between the three pumps.

Lead-lag configuration selection: In the set of multiple pumps, the pump that is chosen to run always is referred to as the "lead" pump. The other pumps are "lag" pumps. Lead pumps are prioritized with a higher priority which always runs before a pump with a lower priority. The lead and lag pump selection and prioritization can be set either by the operator or automatically.

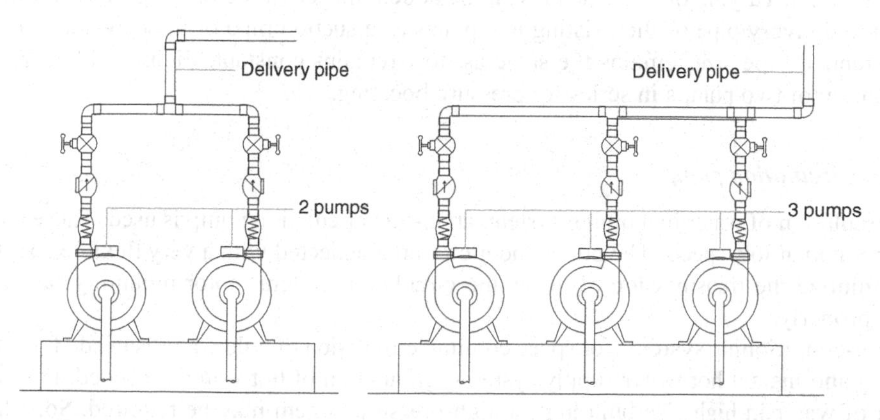

Figure 5.13 Parallel booster pump setting: a. Duplex pump set and b. Triplex pump set.

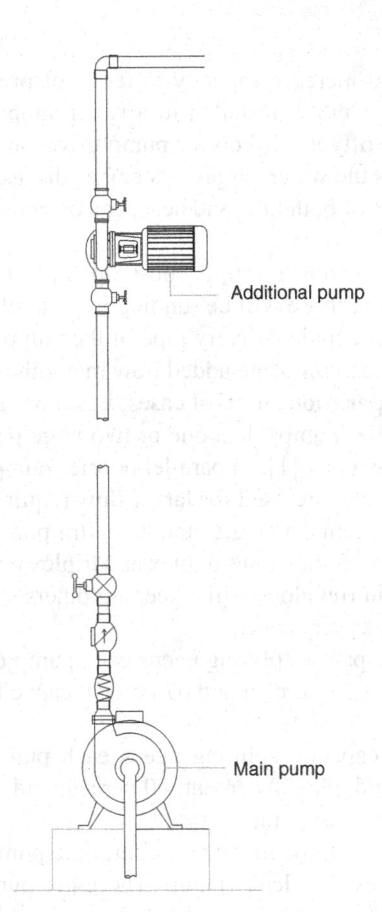

Figure 5.14 Pumps installed in series for pressure boosting

Pressure boosting: For pressure boosting, multiple booster pumps of the same flow capacity, all at varying or same head, shall be added; those will be running in series. In this case, the delivery pipe of the existing pump acts as a suction pipe of the new added pump; the extended pipe size remains the same as flow remains constant. Figure 5.14 represents installation of two pumps in series for pressure boosting.

5.10.2 Circulation pump

For circulation of water in a piping system, an in-line circulation pump is used. These pumps are operated at low speed. The pump and motor are connected with a very flexible coupling. To minimize the transmission of noise, the circulation system motor mountings are cushioned properly.

In air-conditioning systems for space cooling, circulation of cold water is needed. In space heating and instant hot water supply systems, circulation of hot water is needed. For circulation of water in high-rise buildings, a high-pressure system may be required. So, a high-pressure system steel casing circulation pump with mechanical seals capable of withstanding high pressure should be used.

Circulation pumps should have some special features; these are as stated below.

1 Shall operate quietly
2 Operates at low speed
3 Motor and pump are connected by a very flexible coupling and
4 Motor mountings are cushioned properly.

5.10.3 Submersible pump

A submersible pump is that type of pump in which the pump driving motor is submerged in the liquid to be pumped, below the pumping assembly, rather than at ground level. The pump and motor are assembled as a single unit with the motor on the lower end. Submersible pumps are usually of close-coupled construction, with the electric motor protected by a waterproof housing that permits the pump and motor to be submerged in the liquid to be pumped. The important feature of deep tube well pumps, basically a submersible pump, is that the motor is designed with a length to diameter ratio much higher than a standard motor, to permit installation in the standard well diameter for which the pump was designed. The motor thrust bearing is made strong enough to carry the full thrust developed by the pump. The bearings are usually lubricated with water or water-oil emulsion. Leakage of lubricant is prevented by providing a mechanical seal. A waterproof cable, kept clipped to the side of the rising main of the pump, is provided to feed power to the motor. Submersible pumps are mostly vertical pumps; so, these pumps are to be placed underwater in a vertical position.

Submersible pumps are less efficient because of the special design of the motor. Special advantages of this type of pump are as follows.

1 Exceptional quietness of operation because of being submerged
2 Cost of priming is not required
3 Cost of superstructure is eliminated and
4 Can be installed quickly and easily.

5.10.4 Sewage and sump pump

The sewage or sump pump is used to empty the sumps where sewage or liquid waste is temporarily stored. These pumps are specially designed centrifugal pumps, having impellers that can pump large pieces of solid matter without clogging. In addition to wide impellers, hand holes are provided in the pump casing for access to remove obstacles that may lodge in the pump. Sewage pumps perform efficiently under positive head on suction. Various sewage or sump pumps are available as follows.

1 **Solid handling pump:** Sewage containing large particles (maximum 80 percent of outlet size), abrasive and viscous wastes, fibrous and soft solids in suspension, sludge and pulpy materials etc. are pumped. Pumps are available with 8–80 m head with a flow range 5–200 lps. Hand holes are provided on the pump to facilitate removing of stringy materials lodged on impellers without dismantling the pump.
2 **Non-clog pump:** Sewage containing muddy, viscous liquid with solids in suspension can be drained easily with this type of sewage pump. An S-type semi-open impeller is used in these pumps. The size of the discharge outlet is the same as the size of the allowable-passing solid objects. For both dry and wet pit application, this pump is

available. Wet pit design pumps have a head range of 2–45 m and flow range of 4–150 lps. Column pipe is limited to 8 m.

3 **Submersible sewage pump:** Both pump and motor remain submerged in the sewage. Pump casing is a radial-split type with integral cast suction and delivery nozzle. The impeller is designed non-clog, wide channel, with two vanes. Double mechanical seals with oil lubrication are provided in between motor and pump.

Wastewater or sewage might contain oil or other viscous waste material. Centrifugal pumps are generally designed and suitable for liquids with a relatively low viscosity that might handle light oil. For a pump to be economically efficient, the maximum recommended liquid viscosity is 150 cSt (15 wt. oils) [5]. More viscous liquids than 350 to 400 cSt (30 wt. oils), at 20–21°C, will require additional horsepower for centrifugal pumps to work. For viscous liquids of more than 300 cSt (30 wt. oils), positive displacement pumps are preferred over centrifugal pumps to help lower energy costs [6].

Sump pumps are usually mounted vertically on the floor above sump, to save floor space. The supporting floor should be rigid enough to avoid vibration problem.

5.11 Pump selection factors

The following major factors affect the performance of a centrifugal pump considerably other than flow and head capacity, at its maximum efficiency of operation, and need to be considered while choosing a centrifugal pump:

Fluid viscosity: In general, a centrifugal pump is suitable for little or low viscosity fluids since the pumping action generates high liquid shear.

Specific density and gravity of working fluid: The specific density and gravity must be addressed where necessary since the weight will have a direct effect on the amount of work performed by the pump.

Operating temperature and pressure: High-temperature pumping may require special internal and body parts. Similarly, for higher pressure retaining casing needed to be required for high-pressure conditions.

Net positive suction head (NPSH) and cavitation: It is to be ensured that the system's net positive suction head available (NPSHA) is greater than the pump's net positive suction head required (NPSHR), with an appropriate safety margin.

Vapour pressure of the working fluid: The vapor pressure must be kept under consideration in order to avoid cavitation, as well as damage to the bearing caused by dry running when the fluid gets evaporated.

5.12 Pump and pipe installation

The economy, performance and operational longevity of a pump primarily depend on its accurate and correct installation in the right places. While installing a pump, the manufacturer's instructions should be followed carefully. In most cases, the following important measures should be considered.

5.12.1 Pump installation

Location: Pumps are preferably to be installed by the side of a reservoir, as shown in Figure 5.15. In such a case, a pump will not require priming. The pump shall be located nearer

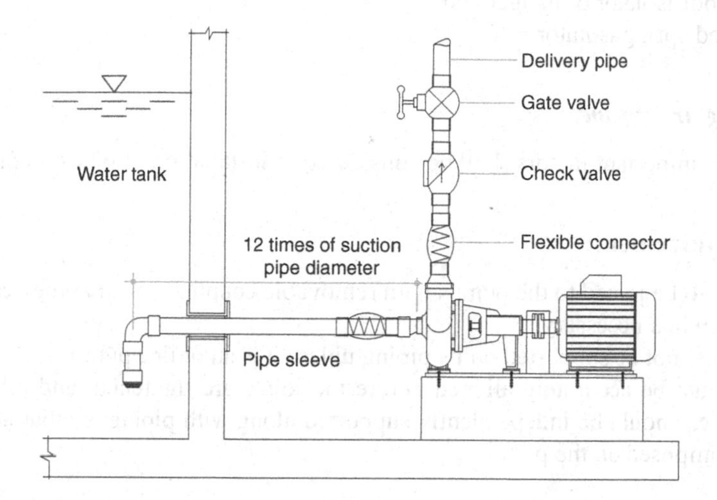

Figure 5.15 Pump and piping installation.

to the reservoir, where it is accessible and there is sufficient light to inspect the condition of the pump, specially packing and bearings.

Base: The base is a permanent support to the base area of the pump. It is made strong enough to absorb all stress and strains that may develop during operation of a pump. It is very important for vertically mounted pumps. There are three types of base, as mentioned below.

1 Type A base: no foundation; vibration isolators are attached directly to the equipment
2 Type B base: base made of structural steel rails and
3 Type C base: concrete inertia base.

A concrete inertia base is most satisfactory. Rules of thumb for centrifugal pumps suggest that the concrete base volume should be determined as three to five times the mass of the pump and driver combined [7]. The base area should be about 150 mm longer and wider than the pump base-plate dimension. The required height of the base is then calculated. Foundation bolts shall be correctly positioned according to the drawing generally supplied by the pump manufacturer.

Grouting: To provide solid bearing, 20–50 mm thick grouting prepared by cement and sand with 1:2 ratio shall be poured under bed plate. Grouting shall be allowed to set for at least 48 hours.

Levelling: Pump and motor shall be accurately aligned to minimize vibration and load on pump. Vertical pumps shall be installed truly vertical and horizontal pumps shall be truly horizontal. Alignment is checked by checking the faces of the coupling halves of pump and motor for parallelism with the help of Feeler gauge, and straightness of sides and levelling of tops by straightedge.

Vibration isolators: Pumps should not be anchored directly to any floors. In all cases, vibration isolators should be provided on pump bases. There are various types of isolators used, depending upon the pump size and supporting slab or floor condition. Following are some very common vibration isolators used.

1 Fiberglass or neoprene pad
2 Fibreglass isolation mount

3 Spring floor isolator or hanger and
4 Restrained spring isolator.

5.12.2 Piping arrangement

The following important factors shall be considered in installation of pipes on pumps.

General requirement:

1. Pipes are to be joined to the pumps with removable couplings, so that they can easily be removed when necessary.
2. Pump must not be supported on its piping unless it is an online pump.
3. Piping must be accurately aligned before the joints are tightened, and all valves and accessories should be independently supported along with piping, so that no stress and strain is imposed on the pump.

Suction pipe requirements: The following measures should be followed while installing the suction pipe of a centrifugal pump.

1. The suction pipe shall be kept as short and direct as possible and at least one size larger than the pump suction connection.
2. Where necessary, use an eccentric reducer to fit the horizontal portion of the suction pipe of larger diameter with the relatively small tapped opening of the pump, as shown in Figure 5.16.
3. For long suction pipe, it should be graded continuously downward from the pump to the source of water.
4. Where needed, use long sweep bend, and bending should be done 12 times the diameter of suction pipe apart from the suction end of the pump.
5. If strainer is to be used at the suction end, then the area of strainer openings shall be at least three times the area of the suction pipe.

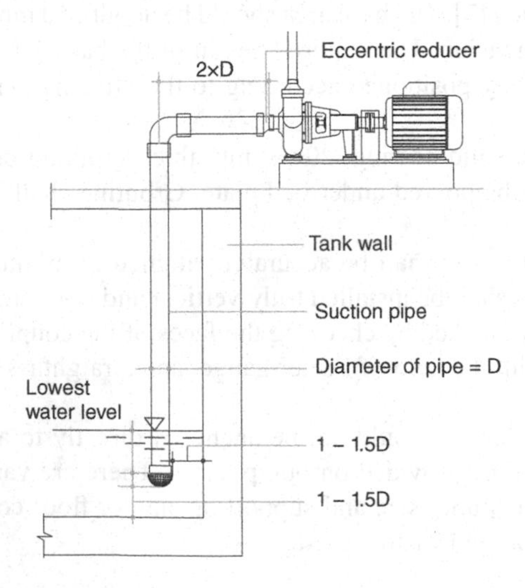

Figure 5.16 Suction pipe installation requirements.

6 All joints shall be made airtight.
7 Sufficient clearance between the suction pipe and the sides of tank and depth of water above the end of the suction pipe shall be maintained, as shown in Figure 5.16.

Delivery pipe requirements: Delivery pipe shall have the minimum number of bends. Like suction pipe, delivery pipe shall also be fitted with a pump outlet by a flexible connector. A gate valve and, when the riser pipe is more than 10 m, a check valve shall be installed on delivery pipe. Installation of a pressure gauge and water flow meter on delivery pipe is optional.

5.13 Pump accessories

Pumps need various accessories for their operation and control. All these accessories need to be chosen depending upon the requirements for pump operation and control. The major accessories for operation and control of a centrifugal pump are as follows.

1 Pump rotor and
2 Pump control panel.

5.13.1 Pump rotors

The pump rotor is the most essential accessory needed for driving the pump. Mostly, two types of machine based on source of energy are used: 1. electric motor and 2. diesel motor.

Electric motors: Generally, pumps are driven by connecting the shaft with an electric motor where electricity is available at a comparatively cheaper price. The most common type of motor used is the squirrel cage induction type, which is operated by alternating current (AC). An AC induction motor comprises two major assemblies: stator and rotor, as shown in Figure 5.17, along with other components.

The stator assemblies are composed of steel lamination shaped to form poles. Coils are wound with copper wire around these poles.

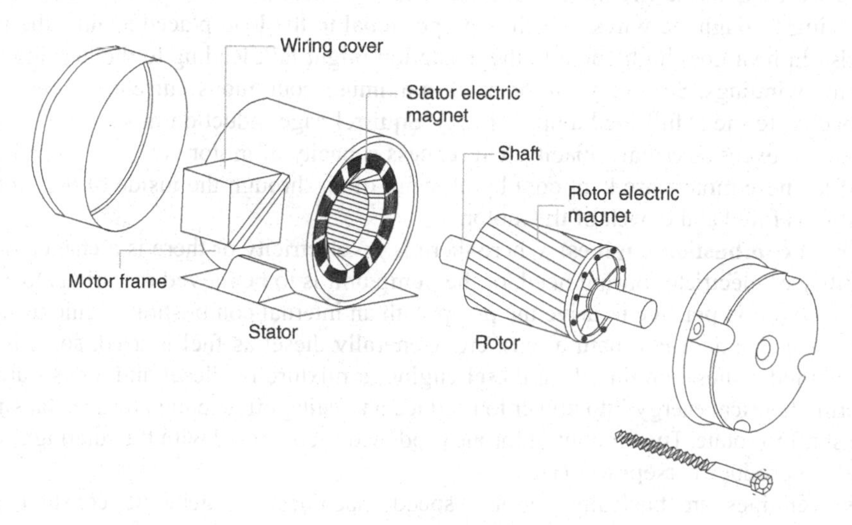

Figure 5.17 Major components of electric squirrel cage motor.

The rotor of an induction motor may be either wound rotor or a squirrel cage rotor. The squirrel cage rotor assembly comprises a steel cylinder mounted on a shaft. The surface of the cylinder is embedded with aluminium or copper conductors. The longitudinal conductive bars are short-circuited at both ends by connecting with solid plate rings, forming a cage-like shape thus named a squirrel cage.

When voltage is applied to the stator winding, electric current (amperes) flows through the wire coils and a rotating magnetic field is created. This rotating magnetic field of stator windings causes an induced voltage in the rotor bars, which are basically single-turn coils, and causes current to flow in the rotor bars. These rotor currents produce their own magnetic field, which interacts with the stator magnetic field to produce torque, causing the rotor to spin in the direction of the rotation of the magnetic field produced by the stator. The torque produced is combined with the rotational speed of the shaft to produce horsepower (HP).

The rotational speed of the shaft of the motor is closely related to the rotational speed of the magnetic field inside the motor. The rotational speed is determined by the number of poles i.e. coils of wire in the stator of the motor and the frequency of the alternating current (AC) electrical power. This is called the synchronous speed of the magnetic field, which is expressed by the following formula. Squirrel cage induction motors are obviously constant speed machines.

$$Revolutions\ per\ minute(RPM) = 120\frac{f}{p} \qquad 5.13$$

This is the synchronous speed of the magnetic field, where

f = frequency of the alternating current (AC) power expressed in cycles or hertz per second and

p = the number of poles in the motor stator.

The squirrel cage rotor rotates at a speed slightly less than the field synchronous speed. This difference in speed is called "slip", which ranges between 3 and 5 percent at full-rated power output for the kinds of motors used to drive centrifugal pumps.

The wire coils inside the motor generate heat. The heat produced is related to the current passing through the wires, which is proportional to the load placed against the motor. If the rise in heat goes high enough, the insulation might fail, leading to the burning out of the motor windings. So, every motor has a maximum continuous current (ampere) flowing capacity, termed "full load amps" (FLA). Squirrel cage induction motors are generally designed for excess thermal capacity. This excess capacity of motor is termed "service factor" (SF). These motors are kept cool by blowing of air through the inside of the motor by integral fans fitted at the back of the motor.

Internal combustion engines: Where there is no electricity or there is a chance of non-availability of electricity during need, or the pump unit is to be moved to various locations, then it is common practice to drive the pump with an internal combustion engine using fuel like diesel, gasoline, LPG, natural gas, etc. Generally diesel as fuel is used, so, it is commonly named a diesel engine. In a diesel engine, a mixture of diesel and air is burned to transform chemical energy into power to produce a twisting effect called torque, causing the engine shaft to rotate. The amount of torque produced is combined with the rotating speed of the shaft to produce horsepower (HP).

Diesel engines are basically variable speed machines. To generate constant speed, engines are supported by a "governor". When the power requirement of the pump increases, the engine begins to slow down. The governor senses the rotational speed reduction and increases the flow of fuel to increase the speed. Conversely, when the pump power requirement decreases, the rotational speed of the engine begins to increase. The governor senses this increase and reduces the flow of fuel, bringing the engine power output and speed back to match the pump power demand.

The conversion of chemical energy into power releases heat. The engine is therefore run below the maximum internal operating temperature. All engines are fitted with a switch to sense excessive temperature and stop the engine.

Diesel engines are usually supported by several accessory devices and sub-systems that make them self-sustaining. These additional devices or systems include the cooling system, cooling fan, fuel system, governor, battery-charging generator, exhaust system and the air intake system. Many of these consume power from the engine, some as much as 10 percent. Therefore, the "net continuous horse-power available at the output shaft" should be considered as the rating for engines to be used with the pumps.

Internal combustion engines are "air breathing" machines. The output power of a diesel engine is dependent on the amount of oxygen that is drawn to mix with the fuel. The amount of oxygen available to burn is determined by the density of the air (kg/cum).

The exhaust gases leaving an engine should be in a good condition which is almost invisible when operating at normal temperature with an adequate air supply for combustion. Discoloration of exhaust gas indicates a malfunction or an improper load condition. Releasing black smoke indicates that the amount of fuel passing through the engine is more with respect to the amount of air that is available to burn. The combustion process is not complete. This can be caused by a plugged air cleaner or it can be an indication of excessive engine loading, but the governor attempts to maintain the desired RPM by increasing the fuel flow. This condition will result in excessive fuel consumption and could lead to damage or wear due to heat for the engine. For the pump, this can cause breakage of internal parts.

5.13.2 Pump control panel

The pump control panel is installed virtually with the objective to regulate adequate flow rate and water pressure in the water supply system through monitoring the operation of a pump or multiple pumps. The pump control panel comprises various control equipment and necessary wires housed in an enclosure. The panel enclosure is a designed box type, made of fiberglass, steel and stainless steel sheets etc.

Common features of pump panels include disconnect, underwriters label (U/L), pump selector switches, power to pump lights and high water alarm. Optional features include hour meters, moisture sensors, temperature limiter circuits, intrinsically safe relays for pumps with explosion-proof motors, soft starters, variable frequency drives etc.

On the basis of power input, there are single-phase pump panels that operate 120VAC or 240VAC single-phase pumps up to 5 BHP, and three-phase panels for 240VAC and 480 VAC pumps of over 5 BHP. On the basis of the number of pump operations, pump-panels can be designed in the following four configurations.

1 Simplex configuration for 1 pump
2 Duplex configuration for 2 pumps
3 Triplex configuration for 3 pumps and
4 Quadplex configuration for 4 pumps.

Simplex control panels: Simplex control panels are designed to control and monitor the functions of a single pump. A simplex control panel gives many options that cannot be achieved on a direct wire installation for pump operation. The important feature is that the high water alarm and pump are put on separate circuits. So, if the pump fails and the breaker blows, the high water alarm remains functional. The panel is designed to have an "Off" float and an "On" float for the pump, which provide more flexibility in setting pumping range.

Duplex control panels: Duplex control panels are alternating pump control panel with an override function in case of either pump failing, designed to alternately control and monitor the functions of two 240/480 VAC pumps.

Triplex control panels: Triplex control panels are designed to control three pumps in water supply system.

Quadplex control panels: Quadplex control panels are designed to control four pumps, one lead pump and three lag pumps for a water supply system.

Control-panel features: Control-panel enclosure shall be made of durable and weather resistant material, sufficient in size to house all control equipment. The wiring and components shall be clean and precise within the enclosure to allow for quick and straightforward troubleshooting. All panel penetration for conduit shall be from the bottom of the panel. The pump control-panel can be an assemblage of several essential and optional components, as mentioned below.

Essential components of a pump control-panel are as follows.

1 Hand-off-auto (HOA) switch for manual activation, set in automatic float level control and disable the operation
2 Two float-switches activating a magnetic motor contactor to turn the pump "on" and "off"
3 Green pump-run-light indicator to help visualize pump is on run
4 Red pump-fault-light indicator with reset button
5 Circuit breakers to disconnect pump, alarm control and branch circuits and
6 Pump selector switches.

Optional features:

1 Single or dual power feed
2 Automatic alternation
3 Float status indicators (low and high liquid level, stop, lead, and lag)
4 Remote control and monitoring system
5 Visual and audible alarm
6 High system pressure pump shutdown with manual reset and
7 Wall-mounted, floor-mounted or pole-mounted.

Power Supply: Electricity is fundamentally defined as the movement of electrons. Electrons create charge, which is used to do work, like moving a fan or moving the shaft of a motor. Voltage is electromotive force; the measuring unit is volts and it also represents a source of electrical energy. The greater the voltage, the greater the flow of electrical current. There are two types of current: direct current (DC) and alternating current (AC). In direct current, the electrons flow (current) in one direction. In alternating current, the electrons are pushed back and forth, changing the direction of the flow several times per second. AC power is delivered in either a single-phase or a three-phase system. A single-phase AC power system transmits power through two wires: one is the phase and the other is a neutral wire. In a three-phase system, either three wires or four wires are used for transmitting power. The voltage of a single-phase supply is 230 volts, whereas in a three-phase supply it is 415 volts. The efficiency of a three-phase power supply is higher than a single-phase supply. So, recommended phases of power for operating pump based on their motors' capacity are as follows.

1 Single-phase power is allowed for motors of 5 hp or less and
2 Three-phase is required on all pumps greater than 5 hp,

Pump starter: A pump starter is an electrical component used generally to initiate the start of a pump motor.

A single-phase motor starter is used to start a single-phase motor which uses single-phase power to operate. Single-phase motors consist of two sets of windings: start winding and main or run winding. To activate and deactivate the motor running, a switch along with a capacitor is used. A single-phase motor starter may be designed with soft-starting features.

In a three-phase motor starter, a soft-start feature is unavoidable as it helps reduce the initial surge of current that flows into the windings of a motor when the power is just connected, which is also known as the inrush current. Soft starting reduces the strain on electrical circuits of the motor. Two types of three-phase powered motor starters are used.

1 Direct on-line (DOL) starter and
2 Star-delta starter.

A DOL starter, also known as an "across the line starter", is a method of starting a three-phase induction motor, being directly connected across the three-phase current supply. A DOL starter applies the full line voltage to the motor terminals. Despite this direct connection, no harm is done to the motor as the starter consists of protection and, in some cases, condition-monitoring accessories. The motor draws about 5 to 8 times higher inrush current compared to the full load current of the motor. The inrush large current gradually decreases as the motor reaches its rated speed. For motors of 7.5 hp and below, the DOL starter is mostly preferred.

A star-delta starter is the most commonly used method for the starting of a three-phase induction motor, preferably for motors larger than 7.5 hp. In star-delta starting, an induction motor is connected in through a star connection throughout the starting period. When the motor reaches about 80 percent of its full load speed, the motor is connected in through a delta connection. There is a Triple Pole Double Throw (TPDT) switch in the starter which changes stator winding from star form to delta form. The star-delta starter typically has around one-third of the inrush current compared to a DOL starter.

Main power and breaker box: Circuit breakers (single-phase) or overloads (three-phase) are used to protect controls, electrical components and pumps from over-amperage. It is installed where electrical power enters the pump house. The manual pull-down switch turns the power "on" and "off".

Pump control box: Protect pumps from low or high voltage, a drop in water level, low yield wells, clogged well screens, malfunctioning pumps and motors, and rapid cycling. Pump control boxes usually have microprocessors that monitor power-line voltage and pump motor power draw.

Level controller: Level control systems include the following.

1 Tethered float switches
2 Pressure diaphragm switches

3 Pedestal-mounted mechanical lever switches and
4 Pressure transducer level transmitters.

Float switches: A float switch is a device that monitors the liquid level in a tank or reservoir. It has a water-sealed float housing mechanism for a set of contacts such that, when the float is at the highest level, contact is released to break an electrical circuit, thereby stopping the pumps, and when the float is hanging at the lowest level closes the circuit to complete the circuit, thereby activate the pump for running. Float switches are found for three ranges of operation: tethered FS for wide angle range, vertical FS for medium range and diaphragm FS for small range operation.

Pressure switch: A pressure switch is a form of switch that cuts an electrical contact when a certain pre-set pressure has been attained on its input side. In a diaphragm type pressure switch, the water pressure moves the diaphragm up, which presses against a piston and spring and opens the contacts of the electrical circuit to stop the pump. When pressure drops it closes the contact and completes the electrical circuit, which in turn activates the pump. Again, when the set pressure is reached, this allows the contacts to open again, which turns off the pump. Pressure switches possess an adjustment screw to set the spring pressure, adjusted to change the on-and-off pressure range that the switch operates at. Proper high- and low-pressure settings allow multiple pumps to operate efficiently in lead-lag system mode.

Automatic alternation: Automatic alternation allows exercising of equal size motors on successive starts. This is generally used on equal lead or lag pumps in duplex and triplex units.

Lightning arrestor: Protects the pump motor and controls from voltage surges that might be caused by lightning, switching loads and power line interference. These do not protect against a direct strike.

5.14 Pump maintenance

Despite all sorts of care taken in operating a pump, it may go out of service much earlier than the expected designed life. Various types of problem may arise during operation of a pump. These include as follows.

1 Seal related problems like leakages, loss of flushing, cooling, quenching systems, etc.
2 Pump and motor bearings related problems like loss of lubrication, cooling, contamination of oil, abnormal noise, etc.
3 Leakages from pump casing.
4 Generation of high noise and vibration and
5 Motor related problems.

So, regular maintenance of pump is very essential. Maintenance operations for centrifugal pumps fall into two categories:

1 Routine preventive maintenance and
2 Overhaul or repair operations.

5.14.1 Routine maintenance

Routine maintenance may be expressed as work done primarily to rectify the effects of normal wear in the pump. Routine preventive maintenance operations are performed at certain intervals, usually monthly, quarterly, semi-annually and annually.

Monthly: The temperature of each bearing shall be checked in every month. In general, the temperature of any bearing should not exceed 71°C. The cause of getting heated may be for the following reasons.

i) Over lubrication of ball or roller bearings
ii) Shortage of oil for sleeve bearings or lubricant may get thick and
iii) Misalignment of the pump.

Quarterly: Once every three months, drain the oil from the sleeve-type bearings. Clean the oil rings. Replace the defective oil rings. Measure the sleeve bearings for wear. The recommended clearance is 0.002 mm per mm of shaft diameter plus 0.025 mm. Refill the bearing with recommended amount of lubricant.

Semi-annually: Check the shaft packing by observing the leakage from it. Leakage must be controlled for two important reasons: 1. to prevent excessive fluid loss from the pump, and 2. to prevent air from entering into the pump suction area. It is recommended that the box leakage should not exceed 40 to 60 drops of liquid per minute for adequate cooling. If leakage is excessive, replace all the packing in the box.

Annually: It is recommended that every year the pump should be thoroughly inspected for wear. All deposits and scales found in the pump should be removed. Check, flush and drain the sealing water, cooling water and drain piping of the pumps. Recalibrate all the instruments connected to the pump and run full tests of the pump performance. Replace all bearings and running joints whose clearance has increased more than 100 percent of the original.

References

[1] Edmondson, Chad (2020) "Domestic Water Pressure Booster Sizing Part 5: Selecting the Pressure Booster" Retrieved from https://jmpcoblog.com/hvac-blog/domestic-water-pressure-booster-sizing-part-5-selecting-the-pressure-booster on 26 December 2020.
[2] Native Dynamics, "Pump Power Calculation" Retrieved from https://neutrium.net/equipment/pump-power-calculation/ on 26 December 2020.
[3] U.S Department of Energy (2004) *Variable Speed Pumping: A Guide to Successful Applications*, Industrial Technologies Program, Washington, DC, EERE Information Center. p-12, Retrieved from https://www1.eere.energy.gov/manufacturing/tech_assistance/pdfs/variable_speed_pumping.pdf on 26 December 2020.
[4] Southerns Water Technology (2014) "Variable Speed Pumps. What Are They and How Do They Work?"by Tim Sevenson, Retrieved from https://southernswater.com.au/variable-speed-pumps/ on 18 December 2020.
[5] Nourbakhsh, A., Jaumotte, A., Hirsch, C., and Parizi, H.B. (2008) *Turbopumps and Pumping Systems*, Springer, Berlin, Germany, Retrieved from www.hindawi.com/journals/ijrm/2016/3878357/ on 26 December 2020.
[6] Power Zone Equipment, Inc, "Centrifugal Pumps – Working, Applications & Types" Retrieved from www.powerzone.com/resources/glossary/centrifugal-pump on 26 December 2020.
[7] Cahaba Media Group, Inc (2015) "Checklist for Successful Pump Installation" by Jim Elsey, Retrieved from www.pumpsandsystems.com/checklist-successful-pump-installation on 26 December 2020.

6 Plumbing accessories and appurtenances

6.1 Introduction

Plumbing deals with controlled flow of fluid in piping. A controlled flow of fluid is to be maintained for obtaining the desired flow rate in a particular pipe. For this purpose, the velocity of flow in a pipe is to be maintained within the limit. There are various other reasons which cause interruption in flow for which a desired flow rate cannot be obtained. Furthermore, various actions are needed to manoeuvre the flow in pipes. So, various accessories and appurtenances are needed to be installed in the piping systems. Valves are the major flow controlling devices, and there are many other devices to achieve different functions in the flow. In this chapter, all these accessories and appurtenances will be discussed elaborately.

6.2 Plumbing accessories and appurtenances

6.2.1 Plumbing accessories

There are various devices and equipment used in plumbing to make the system more effective and useful with a view to achieving the objective of plumbing. Common accessories used in plumbing are flow control devices. In plumbing, flow controlling devices can be put into three major groups. These are as follows.

1 Valves
2 Cocks and
3 Faucets

Various other accessories are as follows.

1 Water hammer arrestor
2 Shower head and
3 Ablution hand shower.

6.2.2 Plumbing appurtenances

Appurtenances in plumbing system can be termed those manufactured devices or assembly of prefabricated components which acts as an adjunct to the basic piping system

DOI: 10.1201/9781003172239-6

and plumbing fixtures; basically, to perform some useful functions, such as operating, maintaining, inspecting, servicing etc., of the plumbing system. These appurtenances do not add to either the water demand or the discharge load of fixtures or of the plumbing system.

The appurtenances included in water supply systems are as follows.

1 Water meter
2 Flow meter
3 Pressure gauge and
4 Water tanks.

6.3 Valves

Valves are the devices used to control the flow and pressure within a pipe conveying fluid. By definition, a valve is a lid or cover so formed as to perform any of the following functions:

1 Stopping and allowing fluid flow
2 Varying the rate of flow
3 Diverting the direction of flow
4 Regulating or reducing pressure in flow and
5 Normalizing internal condition.

There are many types, shapes and sizes of valves, but regardless of type, there are some basic parts common to most of the valves: the body, bonnet, trim (internal elements), actuator and packing. The majority of valves have inlet and outlet openings to connect with pipes by threaded, bolted or flanged joints.

The valve body is generally constructed of bronze, ductile iron with epoxy coating, cast iron or stainless steel material. Valves made of plastic are also available and are preferred for economy, light weight and corrosion resistance.

Valves can be classified into two broad categories.

1 Non-automatic valves and
2 Automatic valves.

6.3.1 Non-automatic valves

These types of valves are operated manually, mainly by hand. The valves of this group are the following.

1 On–off valve
2 Globe valve and
3 Flush valve etc.

Some of these valves can be operated automatically by incorporating electro-mechanical devices into the operating components of these valves.

6.3.2 Automatic valves

These types of valves are operated automatically for achieving some condition of services and to open or close them. The valves generally used in plumbing and falling in this group are as follows.

1 Check valves
2 Float valves
3 Pressure controlling valves
4 Safety valves and
5 Air valves etc.

6.4 On–off valves

This type of valve is used only to stop the flow in a pipe and subsequently to allow the flow by opening the gate of the valve. The closing and opening of the gate inside the valve are governed slowly by a multi-turn hand-wheel or quickly by a lever. To manoeuvre the gate for opening or closing, it is connected with the handle by a shut-off screw spindle.

On–off valves are commonly called "gate valves". A gate valve offers a straight-line flow of water, causing minimum flow resistance. These valves are best suited for services that require infrequent operation. Gate valves shall be operated keeping the disc either fully opened or fully closed. If the gate is kept partially open, then the cross-sectional area of flow through the valve is reduced, causing higher velocity of flow. Due to high velocity of flow, the wear and tear of the disk and seat will occur and eventually cause leakage. Vibration might also occur in the piping due to developing high velocity of flow for keeping the valve partially open.

Various common types of on–off valves are as follows.

1 Gate valve
2 Butterfly valve and
3 Ball valve etc.

6.4.1 Gate valves

Various types of gate valves are available. Gate valves are classified on the basis of their internal design and the stem type. With respect to the stem operation, there are two types of gate valves: 1. rising stem and 2. non-rising stem. With respect to number of discs used as a gate, there are two additional types of gate valves: 1. single disc and 2. double disc.

In non-rising stem valves, the stem is fixed to and rotates with the hand-wheel. The threads of the stem regulating the gate are exposed to the liquid being handled, so there is a chance of fouling the threads with liquid of aggressive natures. So, these types of gate valves are preferred for clean water and gases. Non-rising stem valves take up less space for manoeuvring. So, these valves are also preferred for underground piping where vertical space is limited.

In rising stem type gate valves, the stem rises with the rise of the gate when the hand-wheel is rotated. The rising stem is fixed to the gate. The hand-wheel virtually acts as a nut which is rotated around the threaded stem to raise or lower it and so the gate. The length of rising of the stem provides a visual indication of the position of the gate and height of its rise. In this valve, there is little chance of fouling the stem thread with liquid handled, so these

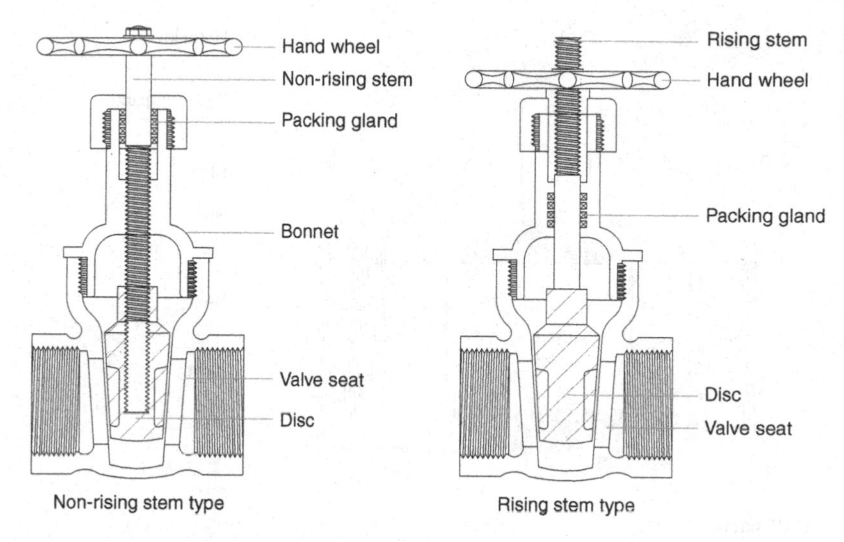

Figure 6.1 Non-rising and rising gate valve.

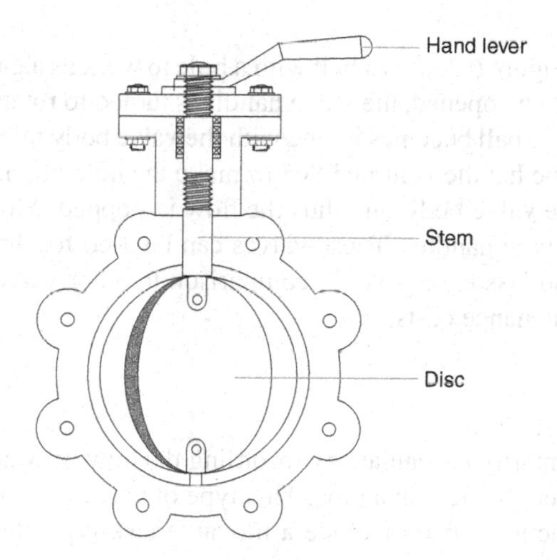

Figure 6.2 Butterfly valve.

valves are preferred for handling of foul liquids. Figure 6.1 illustrates both non-rising and rising type gate valves.

6.4.2 Butterfly valves

In these valves, a pivoted disc is operated by the handle to turn through 90° to shut off the flow of water in a pipe. These valves are comparatively easier to operate, compact in design, possess better controlling characteristics and cheaper for larger sizes. Butterfly valves, shown in Figure 6.2, are especially suitable for large flows at relatively low pressure. These valves are not preferred in water distribution mains containing solid materials in the flowing water. Butterfly valves should not be installed in pump suction piping.

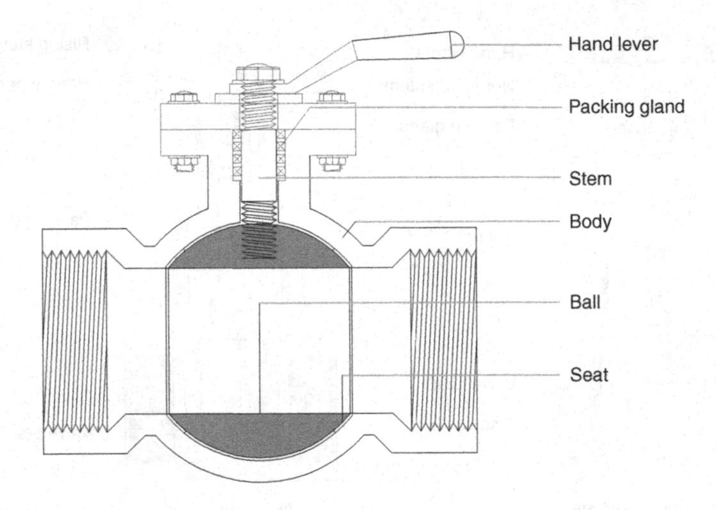

Figure 6.3 Ball valve.

6.4.3 Ball valves

A ball valve, shown in Figure 6.3, uses a ball with a hole to work as a gate to stop or start fluid flow. When the flow is to be opened, the valve handle is turned to rotate the ball to a position when the hole through the ball becomes in line with the valve body inlet and outlet. When the flow is to be shut off, the handle is turned 90° to make the hole alignment perpendicular to the flow openings of the valve body, and thus the flow is stopped. Most ball valve actuators are of the quick-acting type handles. These valves can be used for throttling purposes. Ball valves are usually found less expensive, in comparison to other valves of similar nature of use, and have low maintenance costs.

6.5 Globe valves

Globe valve is used primarily to regulate by throttling the flow of water in the pipe and secondarily to stop and open the flow in a pipe. This type of valve is also known as a throttling valve. The primary functional parts include a movable disk-type element and a stationary ring seat housed in a spherical body.

Generally, by turning the hand-wheel clockwise, attached on top of a stem, gasket or disc at the bottom is pushed down onto a seat of the valve to close the flow of water through. Conversely, the gasket is lifted up from the seat by manoeuvring the hand-wheel anticlockwise to open the flow. The flow of water inside the valve occurs in a very tortuous way. As a result, these valves cause comparatively high friction loss, so pressure in the flow is reduced to some extent at the outlet of the valve. These valves are relatively less costly.

Considering the body design for globe valves, there are three types of globe valves: Z-body, Y-body and angle globe. Considering the shape of the valve disk, there is another type of globe valve, called a needle globe valve.

Z-body globe valve: This is the simplest form of globe valve. In this type of valve, a Z-shaped diaphragm or partition across the globular body accommodates the seat in a horizontal position, to make the flow tortuous. The stem and disk move at right angles to the valve seat, as shown in Figure 6.4.

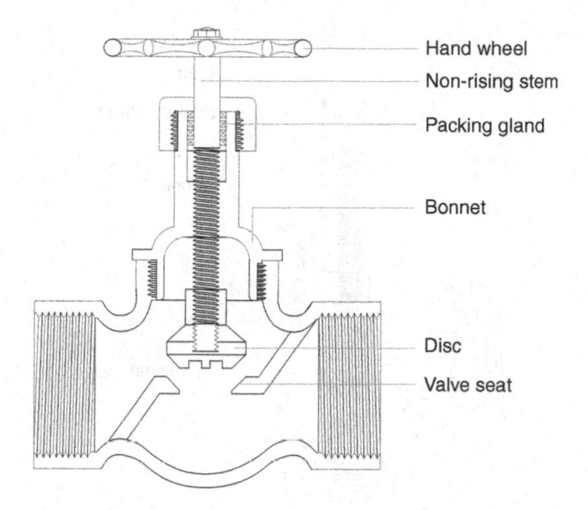

Figure 6.4 Z-body globe valve.

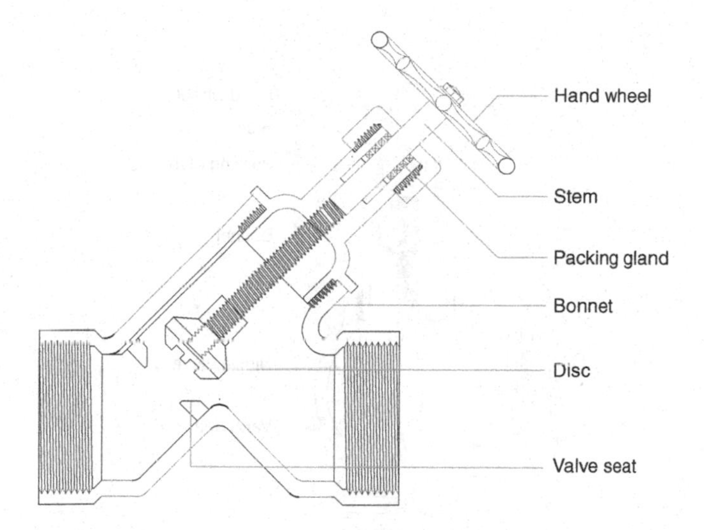

Figure 6.5 Y-body globe valve.

Y-body globe valve: In this type of globe valve, the seat and stem are angled at approximately 45°, as shown in Figure 6.5. Y-body globe valves are preferred for flow under relatively high pressure.

6.5.1 Angle globe valve

In angle globe valves, the inlet and outlet ends are made at right angles. The diaphragm is made as a simple flat plate. Water flow through the valve with only a single 90° turn and discharges accordingly. So, an angle globe valve serves as a valve to throttle the flow and change in direction of flow as elbow. This type of valve offers relatively less resistance to flow. Figure 6.6 represents an angle globe valve.

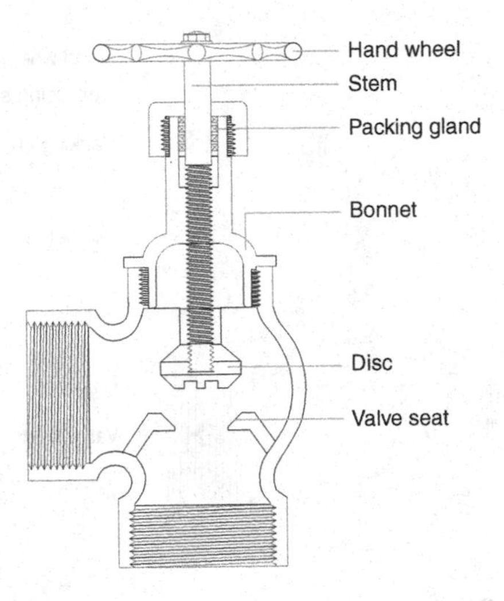

Figure 6.6 Angle globe valve.

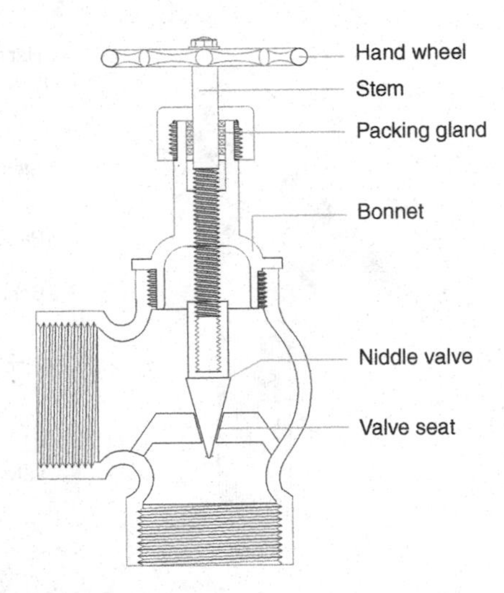

Figure 6.7 Angle needle globe valve

6.5.2 Needle valve

A needle valve is a different form of globe valve in which a threaded plunger with a needle-shaped end is used. Inside the valve, there is a relatively small port with the same tapering of the needle, forming a bevelled seat, into which the needle of the plunger exactly fits to close the flow-through. Needle valves are generally used in precise flow-controlling applications, especially when a constant, calibrated low flow rate is to be maintained in the piping for a considerable time. Needle valves are available in straight or angular form. Figure 6.7 shows an angular needle valve.

6.6 Flush valve

A flush valve, which is also known as flushometer valve, is a device to supply a predetermined quantity of water directly from the water supply pipes for flushing the water closet or urinal bowls. The details of a flush valve, shown in Figure 6.8, are discussed in a later chapter.

6.7 Check valve

A check valve is installed to permit the flow of water in a pipe in only one direction and to close automatically to prevent backflow i.e. flow in the reverse direction. So, these are also known as non-return valves or backflow preventers. Again, when this valve is used at the end of a suction pipe of a pump, then it is termed a foot valve. Foot valves prevent the draining of the water in suction pipe of a pump, thus preventing loss of prime when the pump is shut off. Check valves are available in various types as follows.

1 Swing
2 Tilting-disc
3 Butterfly
4 Lift
5 Piston
6 Stop and
7 Double check valve.

6.7.1 Swing check valve

This valve consists of a pivoted swinging flap or disc, as shown in Figure 6.9. The disc is hinged and kept suspended with a hinge pin attached to the valve body. When water flows in the desired direction, water pressure forces the disc up from its seat to keep water

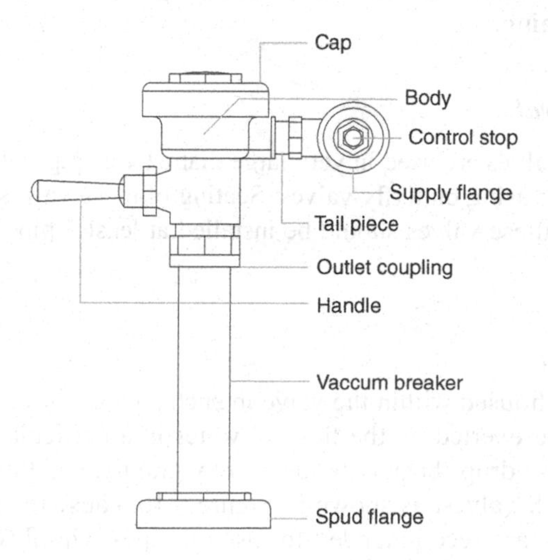

Figure 6.8 Flush valve.

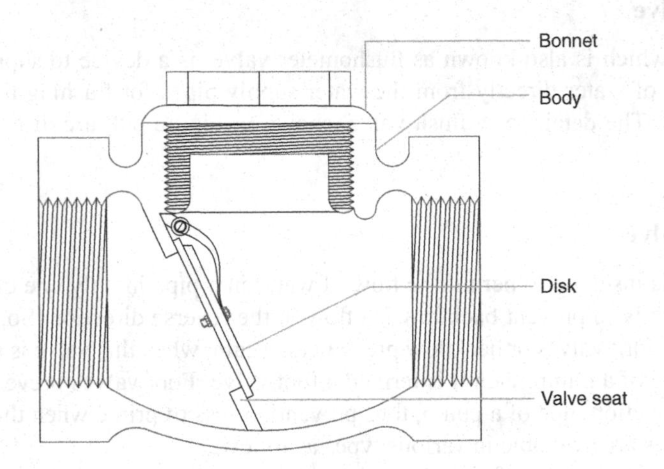

Figure 6.9 Swing check valve.

flowing through. When water flow is stopped, the disc is returned back onto its seat due to its self-weight. Thus, the backflow of water is prevented. Swing check valves offer very little resistance to the flow of water. So, turbulence and pressure drops within the valve are very low. In these valves, a metallic sound is generated when the valve is suddenly closed. To avoid such slamming, a composition disc is used instead of a metal disc. Swing check valves are available in two body patterns: 1. straight body and 2. Y-pattern.

6.7.2 Tilting-disc check valve

In this valve, an airfoil-designed disc is used which is pivoted eccentrically and so is named a tilting disc check valve. As water flows, it pushes the disc to keep the passage open, and when flow decreases, the disc starts closing and seals before reverse flow occurs. Backflow pressure moves the disc across the soft seal to fit tightly onto the seat, thereby shutting off the flow without slamming.

6.7.3 Butterfly check valve

These types of check valves are used in very large diameter of pipes. Butterfly check valves are preferred for piping using butterfly valves. Seating of these valves might be affected by turbulence in flow, so these valves should be installed at least 5 pipe diameters away from any fitting [1].

6.7.4 Lift check valve

In this valve, a disk is housed within the valve in such a way that it can be forced up from its seat by the pressure exerted by the flowing water in a particular direction. When the flow is stopped, the disc drops back onto the seat by gravity, and thus the reversal of flow is prevented. Lift check valves, as shown in Figure 6.10, cause relatively high resistance of flow of water. They are recommended for use on pipes with flows of high velocities.

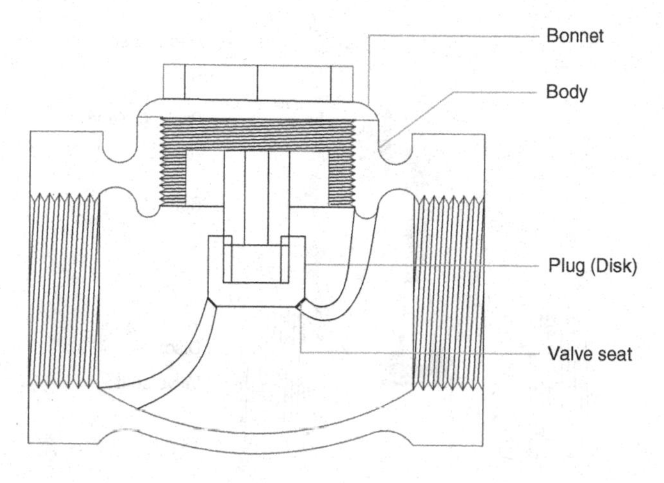

Figure 6.10 Lift check valve.

Lift check valves are again of two types: 1. horizontal lift check valve and 2. vertical lift check valve.

6.7.5 Piston check valve

A piston check valve has some similarity with a lift check valve. It possesses a dashpot consisting of a piston and cylinder that serves a cushioning effect during operation. Piston check valves are preferred where globe valves are used.

6.7.6 Stop check valves

Stop check valves serve the combined function of a lift check valve and a globe valve. These valves possess a stem, manoeuvred by a handle and a disk at the bottom. The disk is not fixed but rather slips on, as shown in Figure 6.11. When the valve is closed, the disc cannot come off the seat, thereby providing a tight seal as the globe valve does. When the stem is lifted for opening, the disc functions like a lift check valve.

6.7.7 Double check valve

Double check valves are basically two single check valves assembled in series, as shown in Figure 6.12. In this type of check valve, one check valve will continue to act even if the other is jammed wide open, and the closure of one valve reduces the pressure differential across the other, allowing a more reliable seal and avoiding even minor leakage, in order to protect any backflow. Double check valves thus create a more reliable seal against backflow in the system. It is therefore recommended for use in non-health hazard cross-connections and continuous pressure applications subject to backpressure or back-siphonage incidents, which may happen in lawn sprinklers, fire sprinkler lines, commercial pools, tanks etc., but it is not suitable for high hazard applications. Double check valves are supported by test ports on the assembly that allow a tester to determine if the check valves are watertight.

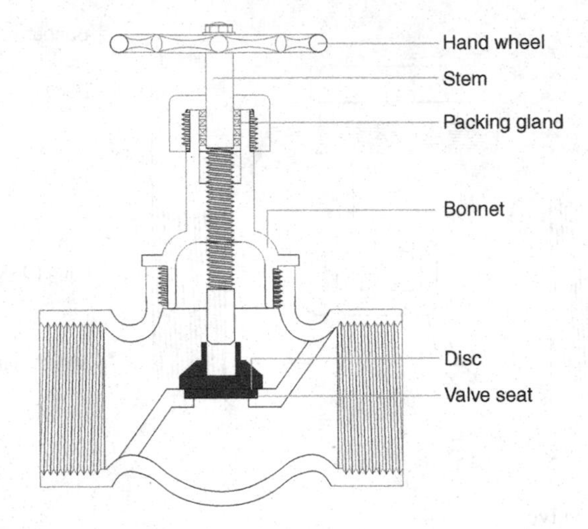

Figure 6.11 Stop check valve.

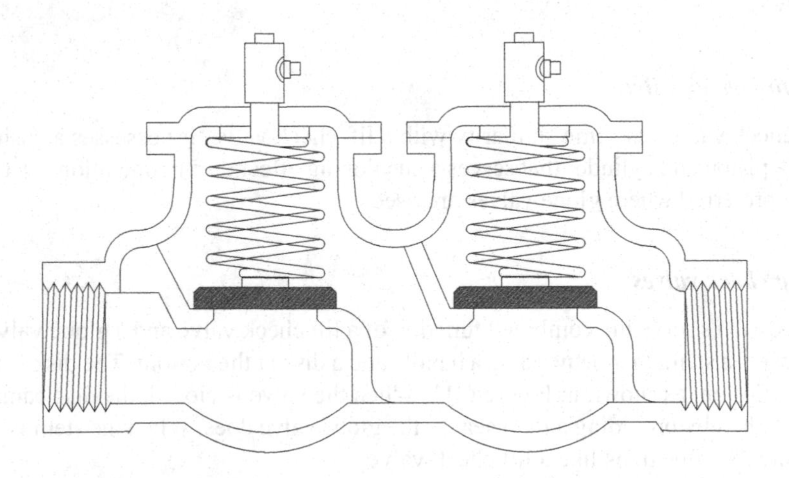

Figure 6.12 Double check valve (spring loaded).

6.8 Float valve

The function of a float valve is to shut off the water flow into a water tank or cistern when the water reaches a predetermined level. It is installed at the end of the inlet pipe to feed a water-tank. The valve is operated automatically by a ball that floats on the surface of water in the tank. As the water level rises, the position of the ball also rises, which makes the valve close to stop the inflow of water into the tank. When water level in the tank goes down, the ball also falls and thereby opens the valve to restart inflow of water to the tank, as shown in Figure 6.13.

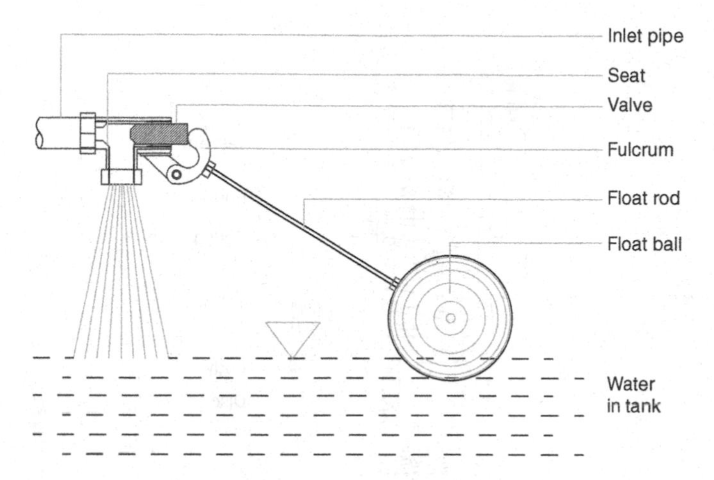

Figure 6.13 Float valve.

6.9 Pressure controlling valves

In water supply piping, there might be a need to control the pressure in the flow to maintain the desired pressure at the outlet piping. The control of pressure in a water supply system can be done in the following ways.

1 By releasing the excess pressure at a time
2 By reducing the pressure to the desired pressure and
3 By sustaining the pressure at constant level.

To perform the jobs related to controlling the pressure in the water supply piping, various valves are used. The valves used for controlling pressure are discussed below.

6.9.1 Pressure relief valve

Pressure relief valves are one kind of a safety device, as shown in Figure 6.14, designed to open when the system pressure in a vessel or pipework becomes too great and may damage the system components or endanger the lives of personnel if not relieved instantly.

6.9.2 Pressure reducing valve

A pressure reducing valve is installed on a pipe to maintain desired pre-set pressure and remain constant at the downstream of valve. When there is fluctuation of pressure and high water pressure develops in the water supply pipe, a pressure reducing valve of a certain pressure reducing capacity is installed to bring down the pressure to the desired level.

Pressure reducing valves are basically globe valves, as shown in Figure 6.15, equipped with hydraulic pilot system that is governed by pressure. The valves remain open under normal flow conditions.

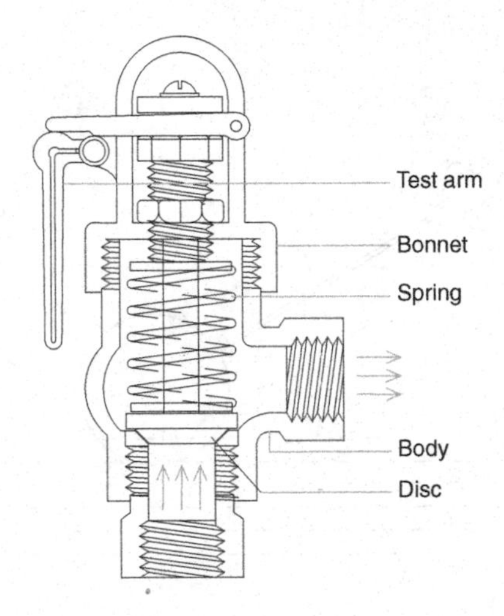

Figure 6.14 Pressure relief valve.

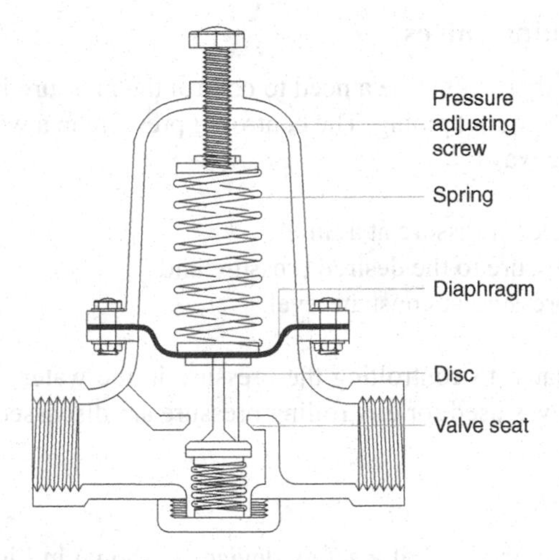

Figure 6.15 Pressure reducing valve in open condition.

The maximum pressure reduction by this type of valve can be achieved at the ratio of 10:1. But, the ratio of fall-off pressure between inlet and outlet should preferably be 4:1. Two or more pressure reducing valves can be installed in series to reduce very high pressure upstream to very low pressure at downstream. Most available pressure reducing ranges are 1380 kPa to 275 kPa and 275 kPa to 96.5 kPa. In case of high pressure dropping from installing regulators, there might be problems like noise and vibration [2]. High velocity of flow in the valve is another problem which causes damage to the valve internals.

Pressure reducing valves are susceptible to suspended particles in water. To protect the pressure reducer against damage from suspended solid particles carried through flowing water, a strainer should be installed prior to the pressure reducing valve. There are pressure reducing valves which contain strainer inside as an integral part. These strainers need to be cleaned at regular intervals. Pressure reducing valves may also have arrangements for installing pressure gauges to monitor the downstream pressure.

As a practical rule, the size of a pressure reducing valve is generally chosen to be the same as the diameter of the pipe on which the valve is to be installed. This might be true when the pipe is rightly sized for the desired flow. If the pipe is found to be oversized, then a pressure reducing valve should be chosen considering the actual flow rate in the pipe.

6.9.3 Pressure sustaining valve

Pressure sustaining valves, shown in Figure 6.16, are used to maintain a minimum pre-set pressure upstream of the valve, regardless of fluctuating flow or varying downstream pressure. These valves relieve excessive water pressure in the pipe when the pressure reaches above the maximum pre-set pressure. The valve operation may be controlled by a pilot valve, connected to the valve inlet, which can be set to the desired pressure, causing the valve to gradually open when the upstream pressure goes above the set point. Both the pressure sustaining valve and the pilot valve remain closed until the inlet pressure exceeds the setting pressure. There are also direct acting pressure sustaining valves which automatically maintain the pre-set upstream pressure required in case of sudden increase in downstream water demand or any breakage causing sudden water loss. These valves are also known as back pressure valves or regulators. A pressure gauge is installed on the valve to monitor the upstream pressure.

During operation, the valve may be in two modes: 1. operative mode, when it remains partially opened; it is gradually opened or closed accordingly to maintain the pre-set pressure on its upstream side when the downstream pressure is below this value, and 2. inoperative mode, when it remains fully opened when upstream pressure exceeds settings or the downstream pressure is above the setting, or fully closed when upstream pressure falls below the setting or downstream pressure exceeds upstream pressure.

One of the common applications of pressure sustaining valves is to accurately control pump discharge pressure with a view to protecting the pump from being overloaded and also to protect from cavitation.

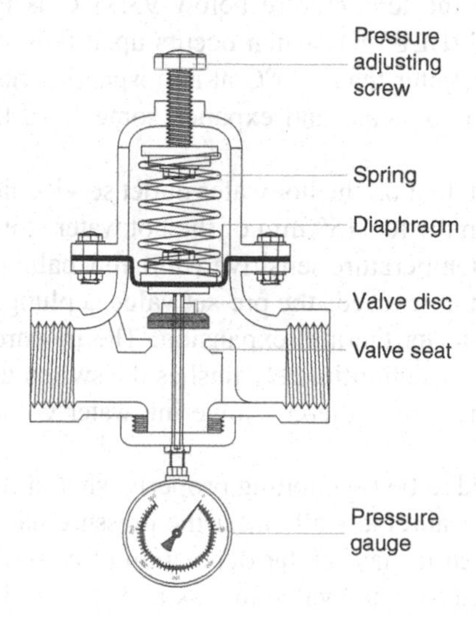

Figure 6.16 Pressure sustaining valve.

6.9.4 Constant flow valve

Constant flow valves are used to maintain pre-set flow rate in a water flowing pipe that might develop due to changes in pressure differentials between the inlet side and outlet side of the valve. The pre-set constant flow rate is maintained automatically by regulating the opening area of the orifice, inside the valve, responding to the fluctuation of pressure differentials across the valve.

To maintain a constant flow rate, the valve uses a piston type valve plug. The upstream flow exerts pressure on the upper surface of the flange on the valve plug. Similarly, the downstream flow exerts pressure on the lower surface of the flange on the valve plug. The differential pressure acting on the surfaces of the flange moves the valve plug piston either downward or upward, depending upon the direction of the force induced by the resulting pressure differential. This upward or downward movement of the valve plug piston causes the area of orifice, for fluid to flow through, to be widened or narrowed accordingly. Thus, the flow rate of the fluid through the orifice is automatically adjusted.

Care must be taken that same size of straight run of pipe is to be set with constant flow valve in back and forth of it. There are two types of constant flow valves.

1 Washer type constant flow valve and
2 Needle type constant flow valve.

Constant flow valves should be used for clean water flow services. If there is a possibility of suspended particles in water, then sediment strainer mesh is to be installed in the upstream of the valve. For more protection, installing valves with the bonnet facing downward should be avoided, regardless of vertical or horizontal installation.

6.10 Temperature and pressure relief valve

In plumbing, temperature and pressure (T&P) relief valves are generally used on water heaters to relieve pressure at 1035 kPa and temperature at 93.33°C by discharging too hot water. The reason behind keeping the temperature below 93.33°C is to prevent "boiling liquid expanding vapor explosion" (BLEVE), which occurs upon failure of a heater that contains hot water of a temperature greater than 100°C at atmospheric pressure. When a T&P valve fails, the hot water flashes into steam and expands some 1570 times its original volume, causing bursting of vessel.

T&P valves are either installed on the hot water outlet service line or mounted directly on a tank tapping located within the top 150 mm of the hot water storage tank.

In T&P valves, there is a temperature sensitive white rod, called a probe, as shown in Figure 6.17. When the temperature exceeds the pre-set value, a plunger inside the probe pushes the valve disc upward due to its thermal expansion. The upward movement of the valve pushes a spring in the valve, which ultimately pushes the switch up and causes spill-over of too hot water from inside the heater, thereby lowering water volume and reducing pressure in the heater tank.

A T&P valve is considered to be functioning properly when it relieves pressure caused by thermal expansion and resets automatically after the pressure has been relieved. If frequent relief of pressure occurs, then it may render deposition of natural minerals, predominantly calcium, on the valve seat, causing the valve to leak and become inoperative.

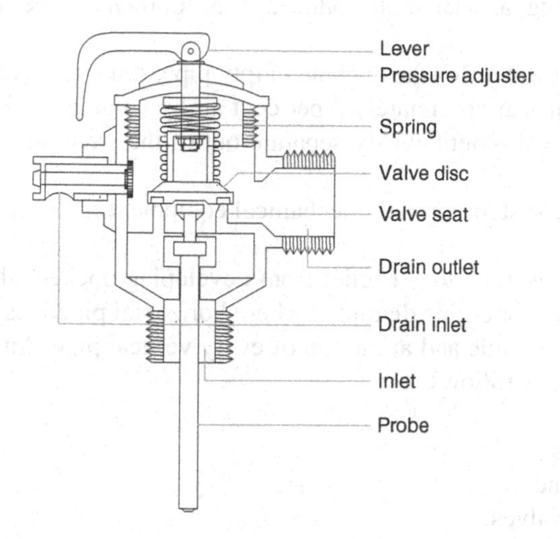

Figure 6.17 Temperature and pressure relief (T&P) valve.

According to the International Plumbing Code, "The discharge piping servicing a pressure relief valve, temperature pressure relief valve or combination thereof shall not be connected directly to the drainage system" [3]. This can create a cross-connection between the hot water in the heater and wastewater in the drain pipe. Under certain conditions, wastewater could be siphoned into the water heater.

It is recommended that the lever of the T&P valve should be lifted immediately after installation to ensure that the waterway is clear. During operation, the lever should be manually operated by lifting the lever at least once a year by allowing water to flow through the safety valve to flush out the discharge pipe of the valve.

6.11 Air valves

It is almost unavoidable to keep water supply pipelines free from having trapped air inside. The trapped air inside the pipeline ultimately accumulates at every high point in horizontal pipes where the pipeline changes grade from positive to negative and at the apex point of vertical pipes. The zones of trapped air are termed air pockets.

Accumulation of air pockets is not desirable for developing efficient plumbing systems for the following reasons.

1 Causes flow restriction
2 Increases head loss
3 Extends pumping cycles i.e. increased energy consumption and
4 Initiates corrosion.

When air continues accumulating, it restricts the flow passage, causing increased flow velocity through smaller opening in air pockets. Such increased velocity of flow can suddenly dislodge the air pockets partially or fully and push downstream instantly. Sudden dislodging of air pockets causing instantaneous increase in flow might be stopped by another nearby air pocket, which can develop a pressure surge or water hammer in pipe flow.

In water supply piping, air can get introduced from following three sources.

1 During start-up of water flow through an empty pipe, some air is trapped inside
2 Water itself contains approximately 2 per cent air by volume; during system operation, this entrained air will continuously separate out of the water and accumulate into the piping and
3 Air enters into the system through mechanical equipment like pumps, valves, etc.

An air valve must be installed to get relief from developing pockets of trapped air at every apex point greater than a one-pipe diameter, where horizontal pipelines convert from a positive grade to a negative grade and at the top of every vertical pipe. Air valves are available in three configurations as follows.

1 Air release valves
2 Air inlet valves and
3 Combination air valves.

6.11.1 Air release valve

Air release valves, also known as "small orifice" valves, as shown in Figure 6.18, continuously release air trapped in piping during system operation. As air from the pipeline enters into the valve, it displaces the water down, causing the float to drop. The air is then released into the atmosphere through a small orifice of the valve. As the air is vented out from the valve, the space is replaced by water again, raising the float to its normal position and closing the valve orifice. Air release valves are essential for efficiency in water supply system and protecting from water hammering.

6.11.2 Air inlet valve

Air inlet valves, also termed "large orifice" valves, as shown in Figure 6.19, are used to exhaust large quantities of air upon system start-up, as well as allowing air to re-enter the

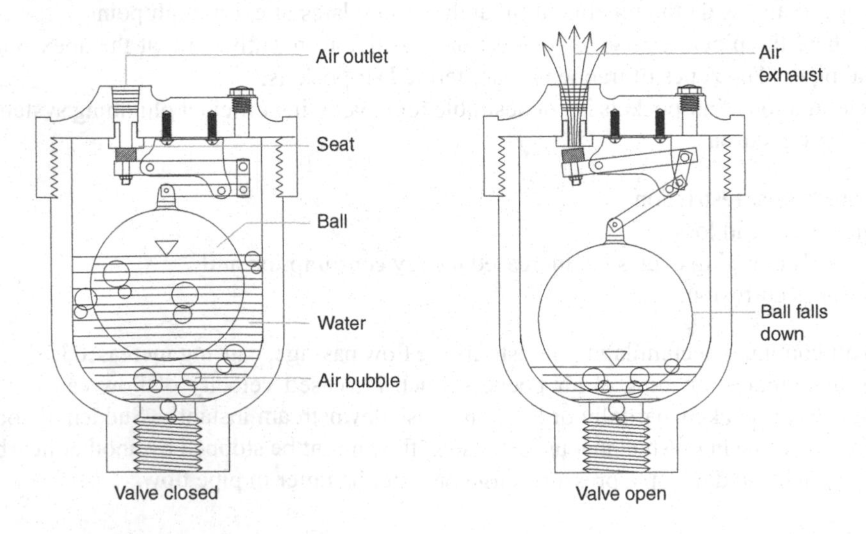

Figure 6.18 Air release valve.

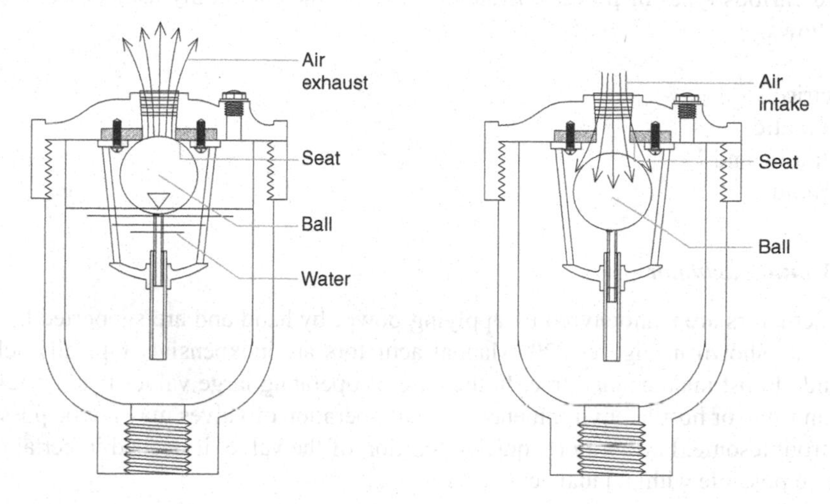

Figure 6.19 Air inlet valve.

pipe upon system shut down or system failure. As water enters the valve, the float will rise, closing the discharge port. The valve will remain closed until system pressure drops to near zero. Air inlet valves also protect the pipeline from creation of vacuum inside. If negative pressure develops, there is a possibility of piping collapse. These valves will not open to release any accumulated air while the system is under pressure.

6.11.3 Combination air valve

Combination air valves perform the function of both the air releasing valve and the air inlet valve. Combination air valves are available in single body and dual body configurations. The single body configuration is more compact and economical.

6.12 Actuators

An actuator is the mechanism, acting upon valves, that is used to regulate the flow of fluid by closing, opening or intermediate positioning of the valve. Actuators help produce torque or thrust to move or rotate the stem, and thereby the valve, for desired flow control. Automatic valves may be self-actuated but the majority of non-automatic valves are adaptable to automatic actuation. Some valve actuators possess switches or other ways to remotely control the position of the valve. Power-operated actuators are suitable for adjusting valve operation remotely and rapid operation of large valves.

Valve actuators are selected based upon a number of factors, including torque necessary to operate the valve and the need for automatic actuation. Valve actuators may be manually operated or power operated. The valve actuators require periodical maintenance or repair work. So, it should be placed in a vault and in an easily accessible location.

Manually operated valve actuators include the following.

1 Hand-wheel
2 Lever and
3 Gears.

There are various types of power-operated actuators. The commonly used power actuators are as follows.

1 Electric
2 Pneumatic
3 Hydraulic and
4 Solenoid.

6.12.1 Manual actuator

Manual actuators are manoeuvred by applying power by hand and are supported by a gear assembly, as shown in Figure 6.20. Manual actuators are inexpensive, typically self-contained and almost maintenance free. In the case of operating large valves that are remotely located, in toxic or hostile environments, manual operation of valves may not be possible or may be troublesome. Furthermore, quick operation of the valve, if needed in certain cases, may not be possible with manual actuators.

6.12.2 Powered actuator

Two types of powered actuators are available, as mentioned below.

1 Rotary valve actuators: produce the rotational motion and are used to operate rotary valves like ball, plug and butterfly valves etc. and
2 Linear valves actuators: produce linear motion and are used to operate valves like gate globe, valves etc.

Most types of powered actuators can be supported with fail-safe features to close or open a valve under emergency circumstances.

Electric actuators: The electric actuator uses an electric motor to produce torque to operate the stem of a valve supported by gears, as shown in Figure 6.21. The operation of electric

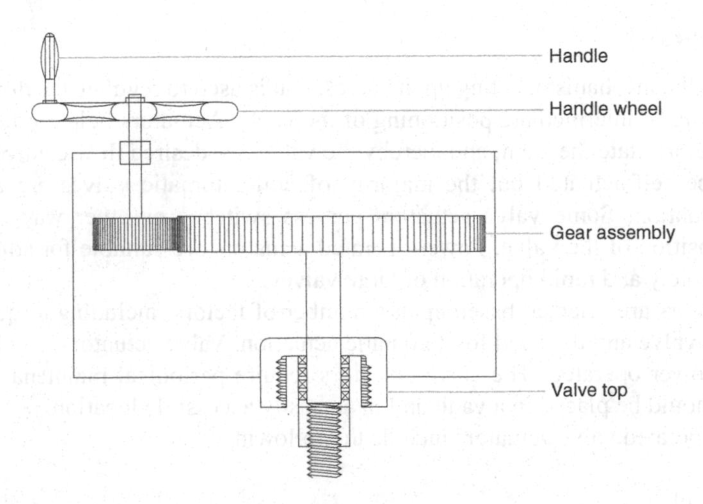

Figure 6.20 Manual actuator incorporating gear assembly.

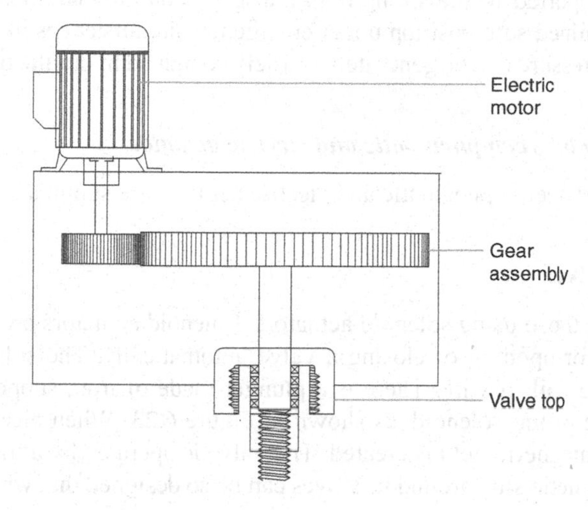

Figure 6.21 Motor actuator.

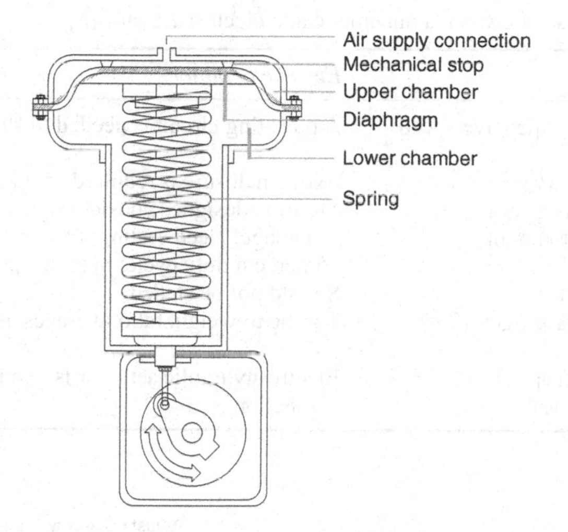

Figure 6.22 A pneumatic actuator.

actuators is relatively quiet, non-toxic and energy efficient. However, availability of electricity must be ensured, which might fail during fire or due to any electrical faults.

Pneumatic actuator: Pneumatic actuators basically use air or gas pressure as the source of power to manoeuvre the stem or valve, as shown in Figure 6.22. A central compressed air or gas provides the necessary power to make either linear motion or quarter-turn of valve stem required for operation. In linear type pneumatic actuators air or gas pressure acts on a piston or bellows type diaphragm causing linear force on the valve stem. Alternatively, in quarter-turn type pneumatic actuator the pressure acts on the quarter-turn vane, thereby producing torque to provide rotary motion to operate the rotary valve stem.

Hydraulic actuators: In hydraulic actuators, fluid pressure is used to convert pressure into motion. Like pneumatic actuators, these are also used on linear or quarter-turn valves. Fluid pressure acting on a piston causes linear movement for gate or globe valves. Hydraulic actuators can be supported with a spring return, also called a fail-safe, features to drive the valve to a predetermined safe position under emergency circumstances like power or signal failure. Hydraulic pressure can be generated by a self-contained hydraulic pressure pump [4].

6.12.3 Comparison between pneumatic and electric actuator

The comparisons between a pneumatic and electric actuator are summarized in Table 6.1.

6.13 Solenoid valves

Solenoid valves are those using solenoid actuators. Solenoid actuators are electrically operated, mostly used for opening or closing a valve automatically. The solenoid is basically a cylindrical-shaped coil of wire. There is a plunger made of iron, supported by a spring, placed at the centre of the solenoid, as shown in Figure 6.23. When electric current flows through the coil, a magnetic field is created. The valve is operated by attracting the plunger by generating a magnetic slug around it. Valves can be so designed that when power is given

Table 6.1 The comparison between a pneumatic and electric actuator [5].

Pneumatic actuator	*Electric actuator*
Simple, accurate and inexpensive speed control	A pulsating circuit is needed to slow the operating speed
Explosion proof and spark proof	Extra enclosure is required for hazardous area
Not subject to overheating; not sensitive to wet environment	Motor is designed considering current and temperature damage. Need sealing for moisture, heat etc.
100 percent duty cycle	25 percent duty cycle; may be upgraded
May be stalled indefinitely	Should not be stalled
Torque to weight ratio averages 123:1 in 170 n.m.	Torque to weight ratio averages 44:1 in 170 n.m.
Spring return (fail-safe) option is practical and economical	Electro-hydraulic actuator is good choice for fail-safe function

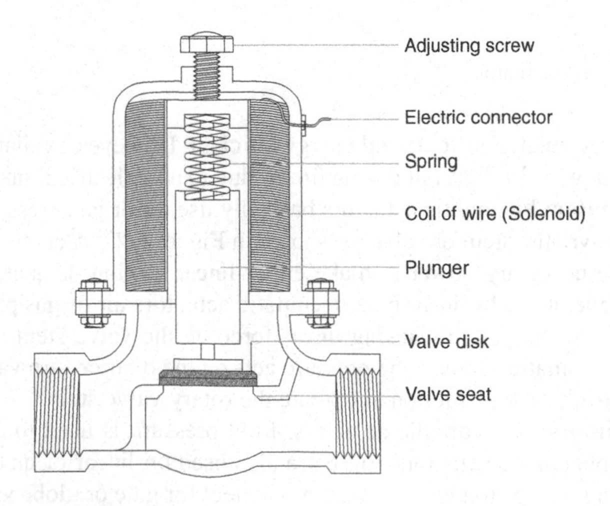

Figure 6.23 Solenoid valve: closed position.

to the solenoid, it will either open or close the valve. Conversely, when power of solenoid is put off, the spring returns the valve to the reverse position. So, by using two solenoids, both the opening and the closing of a valve can be performed by putting power to the respective solenoid.

When power is put on a single solenoid, spring pressure is used to act against the motion of the slug. Single solenoid valves are termed fail opened or fail closed depending on the position of the valve when the solenoid is de-energized. Fail-open solenoid valves are opened by spring pressure and closed by energizing the solenoid. Alternatively, fail-closed solenoid valves are closed by spring pressure and opened by energizing the solenoid.

Double solenoid valves may also fail when both of the valve positions do not change after both solenoids are de-energized [6].

6.14 Installation of valves

Efficiency in performance of a valve depends upon the perfect installation of the valve on the system. Following are the steps to be followed and ensured while installing a valve.

1. Ensure perfect alignment of piping by proper supporting and avoiding bending at the valve connection.
2. Install the valve in alignment with the pipe to prohibit excessive strain on the connection to the pipe. Tighten bolts and nuts sequentially using the crossover method.
3. All valves should be independently supported against movement and stress from the connected piping system.
4. Install valve in an accessible location with the arrow on the side of the valve body in the direction of flow.
5. Upright positioning of valve is preferable for ease of maintenance.
6. Sufficient work space around the valve is to be kept for easy removal or reinstallation of parts during maintenance.
7. Blow or flush out pipeline thoroughly before installing a valve.
8. Visually inspect the valve seating and ports for cleanliness immediately prior to installation.
9. Ensure that the valve pressure rating is compatible with the service conditions.
10. Verify that packing nuts are tight before pressurizing the system.
11. Valve must be in the closed position before installation.

6.15 Cocks

A cock is a type of on–off valve intended to form a convenient means of shutting off the flow of water from a pipe. Cocks are also termed stop cocks. There are several kinds of cocks used to meet various requirements of services, named below and shown in Figure 6.24.

1. Straight-way cock
2. Angle cock and
3. Waste or drain cock.

6.15.1 Straight-way cock

In this cock, water flows in a straight direction and so thus provides its name. The cock used to control the flow through shower head is mostly the straight-way cock.

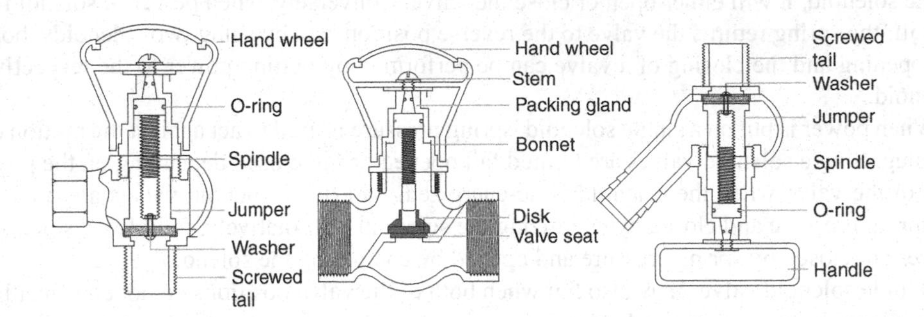

Figure 6.24 Various cocks.

6.15.2 Angle cock

In an angle cock, the flow changes flow direction about 90°. These cocks are more or less similar to an angle globe valve. So, friction loss is comparatively high. The water connection between the supply pipe and the faucet of fixtures are generally done using angle stop cocks.

6.15.3 Drain cock

Drain cocks are used at the bottom of any riser pipe or at the bottommost points of water supply pipes, water vessels etc. to drain out the accumulated water inside. It can also be called a drain valve.

6.16 Faucets

The term faucet or bib is used generally to signify an on–off valve controlling the outlet of a pipe conveying water. With respect to mode of internal operation, faucets are classified as below.

1 Compression faucets
   - a Plain faucet and
   - b Self-closing faucet
2 Fuller faucet and
3 Mixture faucet.

6.16.1 Compression faucets

In this type of faucet, a disc type washer is compressed against a seat by a stem operated by a handle to close the flow of water. Therefore, these faucets are grouped under compression faucets. Compression faucets are of two categories, as follows.

Plain faucet: In this type of faucet, a disc washer is used, which is attached at the end of a stem manoeuvred by a handle fitted on its top. There is a solid or removable seat inside. When the handle is turned off, the washer at the end of the stem rubs and compresses against the seat to close off the water flow. By turning the handle counter-clockwise, the washer is lifted up to allow the flow. Plain faucets are also called a “tap” or “cock”. There are two types of plain faucets, Pillar cock and Bibb cock, as shown in Figure 6.25 and Figure 6.26, respectively. Bib cocks are designed to be installed on vertical surfaces and pillar cocks are on horizontal surfaces.

Self-closing faucet: Self-closing faucets are spring-loaded faucets that close automatically after some time of operation. Figure 6.27 illustrates a typical self-closing faucet. The faucet head should be depressed or operated to obtain the water flow. When the head is depressed by hand, water starts flowing; after some time, the flow will be automatically stopped. Self-closing faucets are also termed non-concussive faucets. Self-closing faucets of varying running periods, e.g. approximately 7, 15 or 30 seconds after every press of handle, are available. These types of faucets are used to minimize the consumption of water by preventing wastage. So, these faucets are recommended for public toilets and in washrooms of

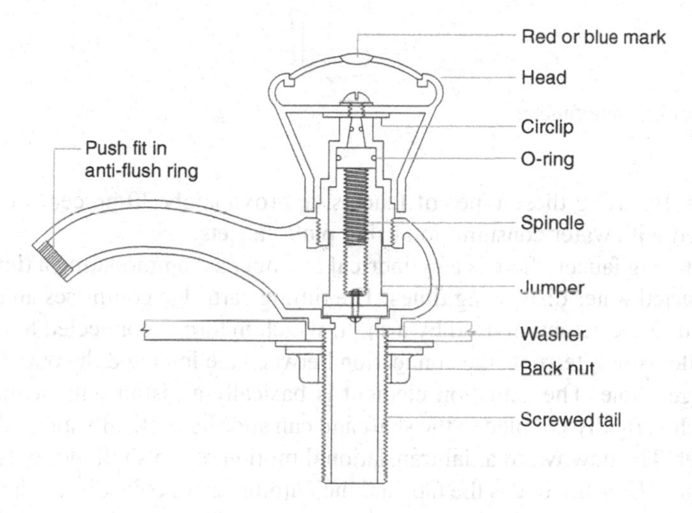

Figure 6.25 A pillar faucet.

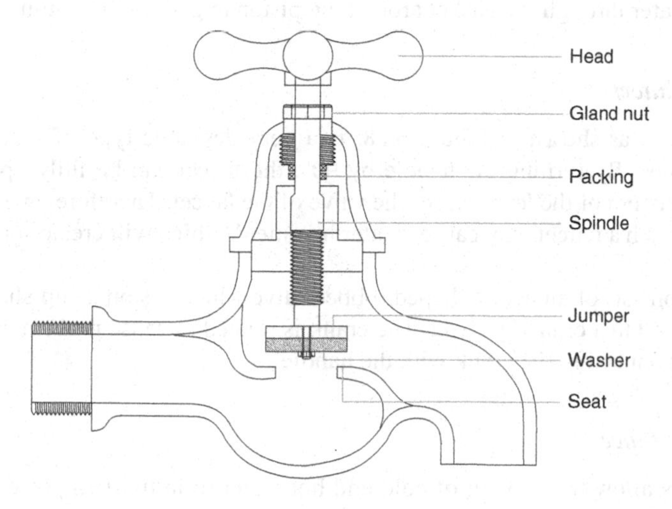

Figure 6.26 A bib tap.

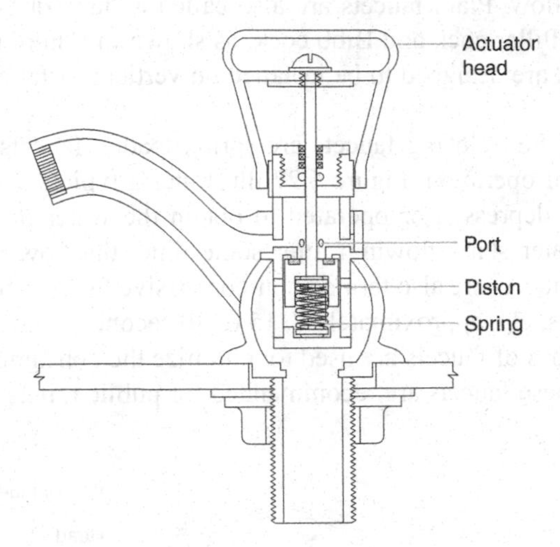

Figure 6.27 A self-closing faucet.

assembly halls. By using these types of faucets, approximately 30 percent of water can be saved compared with water consumption using plain faucets.

In the self-closing faucet, there is a cylindrical chamber accommodating a timing cartridge designed for varied water dispensing times. The timing cartridge comprises an actuation element connected to the actuation head by a shaft, which in turn is connected to a flow control element that allows or interrupts the connection between the inlet and the outlet of the faucet through a gauged hole. The actuation element is basically a piston with sealing lip gasket around it, which is rigidly coupled to the shaft and can slide hermetically along the wall of the timing chamber. The downward axial translational motion of the shaft along its axis, caused due to depressing the head, opens the tap, and the automatic opposite direction motion closes the tap by way of elastic means biased by a return spring. The opening time of the faucet is dependent on the inflow of water into the timing chamber through a gauged hole or by the infiltration of water through the gasket around the piston to get into the timing chamber.

6.16.2 Fuller faucet

The Fuller faucet, as shown in Figure 6.28, is a very desirable type of faucet for use on low-pressure lines. By turning the handle by 90°, the faucet can be fully opened quickly. A reverse quarter-turn of the lever closes the valve of the faucet. Therefore, in a high-pressure line, the use of such a faucet may cause "water hammer", which will create sound and vibration in the pipe.

The faucet consists of an acorn-shaped rubber valve which fits on a cup-shaped seat. The valve is connected to a crank by a rod. The crank is rotated to cause movement of the valve by a crank rod connecting the crank with the handle.

6.16.3 Mixture faucet

Mixture faucets allow the mixing of cold and hot water to individual preference. Instead of two separate faucets, one for cold and one for hot water flow, the mixture faucets

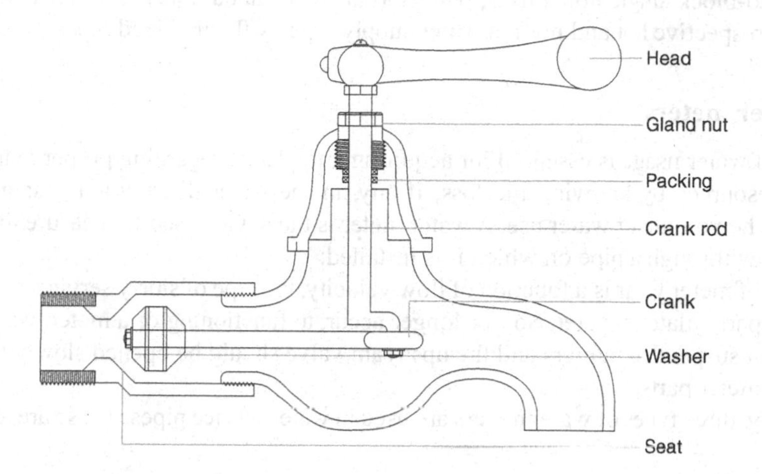

Figure 6.28 A Fuller faucet.

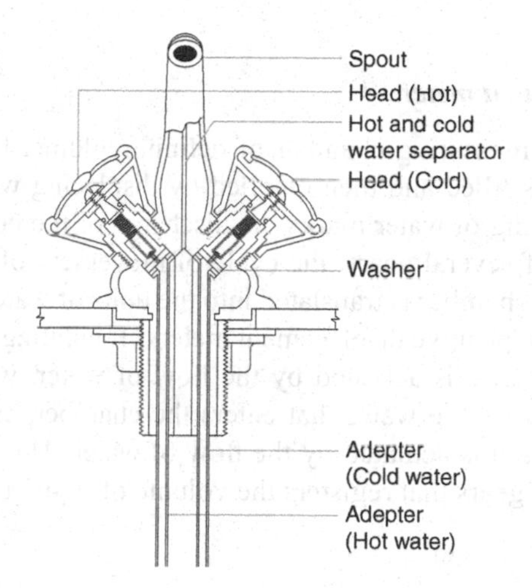

Figure 6.29 Mixer faucet.

combine cold-and-hot water valves with a single spigot to supply water of the desired temperature through a single spout. In a manually controlled mixture faucet, the temperature is controlled by manipulating each valve with a separate handle or a single lever by turning it aside. Such faucets are usually satisfactory on a basin, a bathtub, a sink and shower head. Depending on the type of fixtures, different mixture faucets are named; e.g. basin mixture for basin; bathtub mixture for bathtub and so on. Depending on the alignment of the tap, the number of holes for taps and the design of body, mixtures are termed differently; e.g. two tap-hole pillar mixture, one tap-hole mono-block mixture etc. Instead of two actuation knobs, one single lever is provided to operate the mixture faucet, which is termed a single lever mixture faucet. Figure 6.29 illustrates a mixture faucet. In mixture faucets, the pressure at the inlet of the cold and hot water should be maintained more or less the same.

In a mono-block single hole mixer, two short adapters, about 20 cm, are provided to connect to the respective hot and normal water supply pipe by flexible feed pipes.

6.17 Water meters

Metering of water usage is essential for acquiring knowledge regarding proper management of water resources by knowing the loss, if any, in the water distribution system through measuring the volume of water use. A water meter is the device used to measure the volume of water flow through a pipe on which it is installed.

The rate of meter wear is a function of flow velocity, the type of slurry service, and the percentage of particulate in water. So, for longer accurate functioning of a meter, water should be free from suspended matters and the upstream valve should be opened slowly to prevent damage of meter parts.

Generally, three types of water meters are used in water service pipes. These are as follows.

1 Positive displacement meter
2 Velocity meter and
3 Compound meter.

6.17.1 Positive displacement meter

This type of meter has a measuring chamber of definite volume. During passage of water through this chamber, it is filled and then emptied by displacing water mechanically. Each cycle of filling and emptying of water marks the discharge of the contents of the measuring chamber. With the help of several gears, the cycle of movement of the parts for displacing water from the measuring chamber is translated into the units of water volume on the register dial. There are two types of positive displacement meters: 1. nutating disc and 2. piston meter.

In a disc type meter, a disc is actuated by the flow of water, which rotates in nutating motion and thereby displaces the water that enters the chamber, as shown Figure 6.30. In a piston type meter, a piston is actuated by the flow of water. The movement of the piston drives the arrangement of gears that registers the volume of liquid to record on the dial.

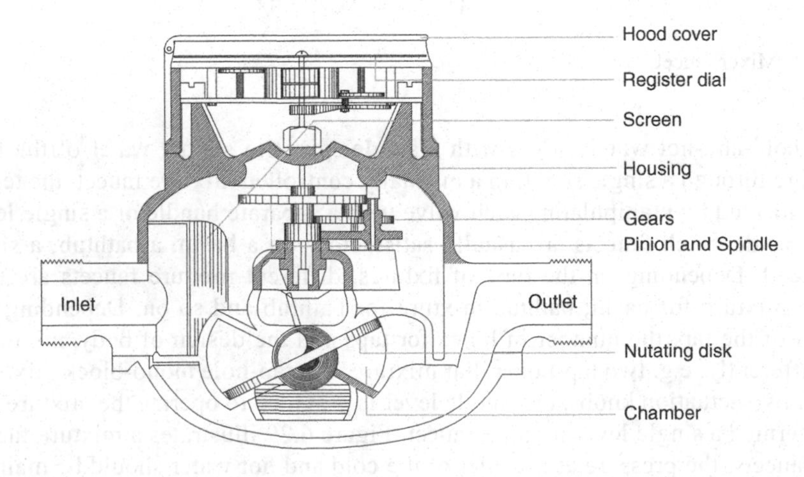

Figure 6.30 Disc water meter section.

This water meter gives accurate measurement for low flow of water, so it is usually used in pipes of 50 mm in diameter or less.

6.17.2 Velocity meter

In velocity meters, the velocity of inflowing water is considered in measuring the volume of water flowing. These meters measure the velocity of water passing through a known cross-sectional area to equate to a volume of water flowing. These turbine meters are available in sizes starting at 20 mm.

A velocity meter is used in water service pipes in which large and constant volume of flow occurs. So, turbine meters measure accurately a large volume of water flow, and in small flow, its measurement will not be accurate. These meters offer comparatively less friction loss, of about 48 kPa, at its maximum capacity. It requires comparatively less maintenance. There are various types of velocity meters, like turbine, multi-jet, propeller, ultrasonic, venturi and orifice meters.

Turbine meters: This type of meter is a volumetric measuring turbine type. This meter engages a vaned rotor which rotates at an angular velocity proportional to the flow rate. The angular velocity of the rotor results in the generation of an electrical signal (AC sine wave type) in the pickup. The summation of the pulsing electrical signal is related directly to total flow. The frequency of the signal relates directly to the flow rate. This signal is transferred to the gears of dial register by means of a vertical spindle which is electromagnetically coupled with the rotor. The volume of water is measured by the number of revolutions by the rotor.

Propeller meter: In a propeller meter, a fan-shaped rotor is introduced that spins by the flow of water. A recorder is connected to the rotor to register the readings.

Multi-jet meters: In these meters, there are tangential openings in a chamber to direct the water flow across a rotor with many vanes, to rotate it. The speed of rotor is proportional to the flow rate.

Venturi meters: This type of meter uses a venture with a section smaller than the pipe on the upstream side. The velocity, so the pressure, of flow through the venture increases for its smaller diameter "throat". The change in pressure at upstream and in the venture is proportional to the square of velocity. Flow rate is thus determined by measuring the difference in pressure.

Orifice meters: In orifice meters, there is a circular disc with a concentric hole. Flow rate is increased, and so is the pressure, while flowing through the hole of a comparatively smaller diameter. By measuring the pressure in upstream flow and through the hole, the volume of flow is measured, as in venturi meters.

Ultrasonic meters: In ultrasonic meters, sound waves are transmitted diagonally across the flow of water in the pipe. The frequency of the wave relates directly to the flow rate. Changes in the velocity of water are thus converted electronically to the change in flow rate.

Magnetic meters: In magnetic meters, an insulated section is used, through which water is allowed to flow. The flow of water induces an electrical current that is proportional to the velocity of flow. Measuring the differences in flow velocities the corresponding flow rate is measured.

6.17.3 Compound meter

A compound water meter is simply a combination of a disc and a turbine meter in one body, as shown in Figure 6.31. A compound water meter is therefore used in water service pipes

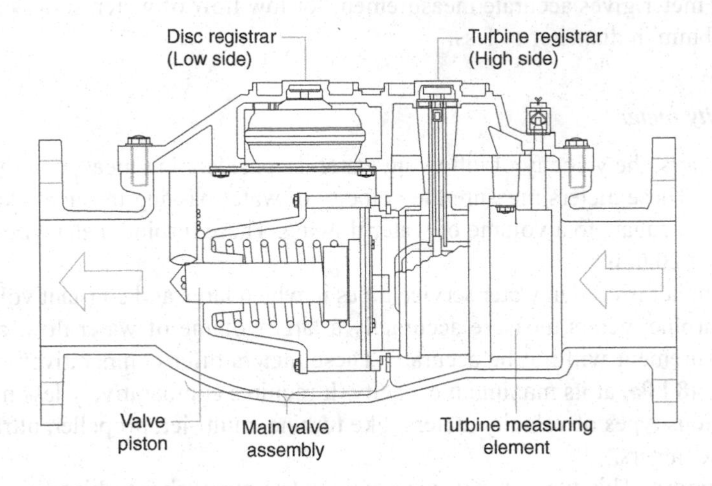

Figure 6.31 Compound water meter section.

Table 6.2 Comparison in characteristics of different water meters. Converted in SI units from sources [7, 8].

Type Size mm	*Disc type (1)*			*Compound type (2)*			*Turbine type (3)*		
	lpm		*kPa (*) at Max lpm*	*lpm*		*kPa (*) at Max lpm*	*lpm*		*kPa (*) at Max lpm*
	Max	*Cont.*		*Max*	*Cont.*		*Max*	*Cont.*	
16 × 20	76	38	104						
20	114	57	104				114 (4)		
25	190	95	104				189 (4)		
40	380	190	104				380 (4)		
50	606	303	104				606 (4)		
75				1212	606	138	1325	908	48
100				1900	950	138	2385	1590	48
150				3800	1900	138	5300	3483	48
200				6056	3028	138	9085	1600	48
300				8706	4353	138	14384	9464	48

(*) Maximum pressure loss at safe maximum operating capacity.
(1) AWWA C700 Cold Water Meters – Displacement Type,
(2) AWWA C702 Cold Water Meters – Compound Type
(3) AWWA C701 Cold Water Meters – Turbine Type, Class II, In-line (High velocity types- Pressure losses do not include strainer)
(4) Max flow for residential fire service – continuous flow and kPa loss characteristics are similar to displacement meters

where there is a large volume of water flow but high fluctuation of flow occurs. When the flow increases beyond the capacity of the disc meter, a special automatic valve opens and permits the large volume of water to flow through the turbine meter. This type of meter is available in sizes from 76 mm to 250 mm.

Compound meters give high accuracy in water volume measurement but offer maximum friction loss of about 138 kPa at its maximum capacity [7]. Compound meters need extensive maintenance for having huge number of mechanical parts inside.

Comparison in characteristics of different water meters are shown in Table 6.2.

6.17.4 Outside register of meters

Water meters are read by the meter readers of the water supply authority. In some cases, meters might be located in a pit deep below the ground. When taking a reading of the meter becomes difficult or causes disturbance to others for any valid reasons, provision can be made for remotely placing the meter register. The outside remote register provides easier access for water meter readers to go from one location to another without disturbing anybody. The advantage of using these meters is the capability of reading them from above the surface without having to open a meter pit. The meters requiring a meter pit to be opened could cause the meters to freeze due to warm air escaping from the pit.

Outside or remote register meters have a built-in generator, which sends pulses to an outside remote register. This register can be used with all types of water meters.

6.17.5 Meter installation

Meters shall be installed properly for correct and smooth operation. The following procedures shall be followed for the purpose.

1 The meter should be located in locations which are easily accessible for service, inspection and taking readings.
2 Though the meter can be installed in any position, like horizontal, vertical or inclined, it should be installed in horizontal position, as this position helps to provide accurate results in comparison to the results given in other installation positions.
3 The meter must be always full of water while operating.
4 Prior to the installation of a new meter, the pipeline must be cleaned and flushed out.
5 Prior to installation, the meter should be checked internally for foreign material.
6 The meter must be installed with the flow arrow, etched on the meter body, pointing in the direction of flow.
7 From any bends, the straight pipe section of the same diameter D as the meter, with lengths of 10D and 5D, shall be installed upstream and downstream of the meter respectively.
8 If there is likelihood of suspended particles in water, a mesh strainer should be installed upstream of the meter.
9 The meter shall be installed 100 to 150 mm above the bottom of the meter box. A clear space of at least 300 mm shall be maintained on each side and 1.2 m above the meter. The face of the meter should be between 450 and 600 mm from the top of the meter pit lid [5].

6.18 Flow meters

Fluid flow rate in a pipe is measured by the flow meters. There are a variety of flow meters for measuring flow rate of liquids, such as ultrasonic flow meters, magnetic flow meters, variable area flow meters, turbine flow meters etc. Among these, the turbine flow meter is discussed herein.

In a volumetric measuring turbine type flow meter, as shown in Figure 6.32, fluid entering the meter flows through a strainer first, where turbulence of flow is reduced and the velocity profile is improved. Then fluid passes through the turbine vanes, causing it to rotate at an angular velocity proportional to the flow rate. As the vane rotates through the magnetic field,

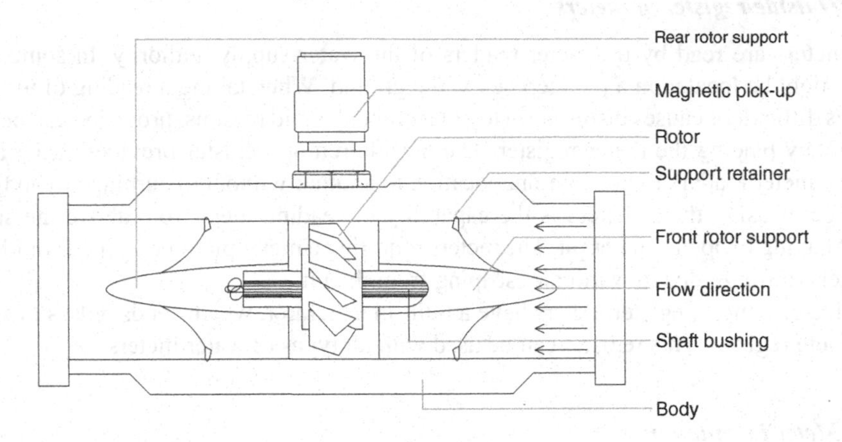

Figure 6.32 Turbine flow-rate meter section.

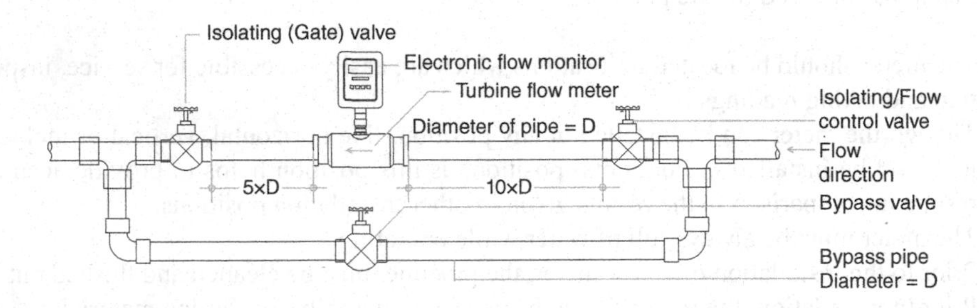

Figure 6.33 Water flow meter installations.

created at the base of the pick-up transducer, an AC sign wave type pulse is generated in the pick-up coil. These impulses produce an output frequency. The summation of the pulsing frequency signal is directly proportional to the volumetric flow through the meter. A flow transmitter in the meter processes the pulse signal to determine the flow rate of the fluid. Thus, the frequency of the signal, directly related to flow rate, is used in measuring the flow rate which is registered and shown in the electronic monitor attached to the meter.

6.18.1 Installation of flow meter

While installing a flow meter, the following instructions should be carried out for proper functioning and maintenance of the meter.

1. Install meter horizontally, keeping the magnetic pick-up facing upward.
2. If suspended matters are present in water, a mesh strainer should be installed at upstream.
3. Install bypass line to allow meter inspection and repair without interrupting flow, as shown in Figure 6.33.
4. Install straight pipe of minimum length equal to 10 pipe diameters, on the upstream side, and 5 diameters on the downstream side, of the flow meter.
5. Install the flow meter and connection cable far away from electric motors, transformers, sparking devices, high voltage lines, etc. Put cables in conduit close to those electrical items.

6.19 Pressure gauge

A pressure gauge is a mechanical device used to measure pressure and display the value in an integral unit. There is another pressure gauge which is used to measure pressure lower than the ambient atmospheric pressure, called a vacuum gauge and also known as an absolute pressure gauge.

There are various types of pressure gauges based on the technology used in measuring the pressure. These are as follows.

1 Bourdon tube pressure gauge
2 Diaphragm pressure gauge and
3 Piston pressure gauge.

6.19.1 Bourdon tube pressure gauge

Bourdon tube pressure gauges are mostly used. These pressure gauges use a flexible C-shaped or even a helix, called Bourdon tube, which acts as a pressure-sensing element. When the tube is subjected to pressure, the tube flexes and tends to straighten out or uncoil elastically or to regain its circular form in cross-section. The resulting motion of the tube is transmitted as a measurement through a mechanical movement to the pointer on the dial face, as shown in Figure 6.34.

6.19.2 Diaphragm pressure gauge

In a diaphragm pressure gauge, a flexible membrane as a diaphragm or a cascade of multiple diaphragm capsules is used. When the diaphragm is subjected to pressure, it deflects. The magnitude of deflection is dependent on the difference in pressure between its two faces. The reference face is made open to the atmosphere to measure gauge pressure. The deformation of diaphragm is measured using mechanical, optical or capacitive techniques. The resulting

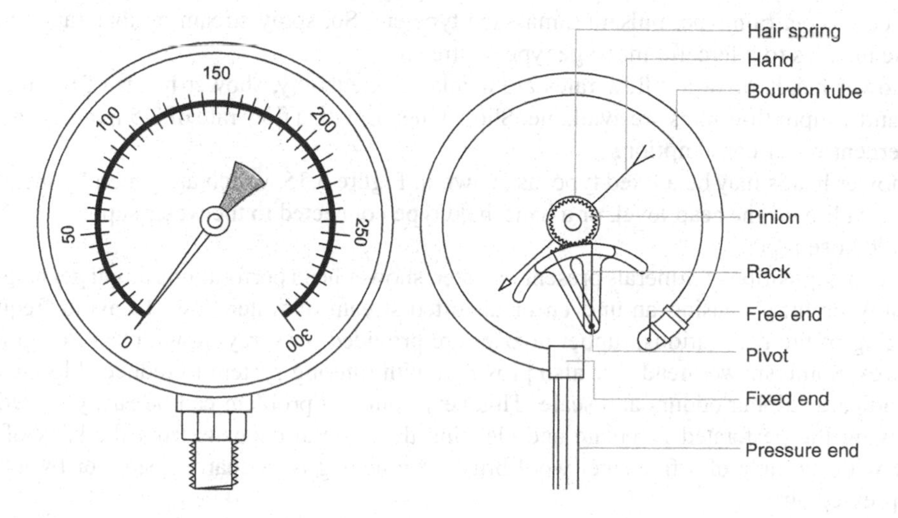

Figure 6.34 A pressure gauge and its major parts.

deflection of the diaphragm capsule is transmitted as a measurement through a mechanical movement to the pointer on the dial face. The diaphragm materials are the same as those used for the Bourdon tube.

Most of the pressure gauges of low range, less than 104 kPa, utilize a diaphragm sensing element to get accurate measured pressure. Typical measuring units used are mm of water, kilo Pascal (kPa) and bar.

6.19.3 Pressure gauge selection

To avoid damage to the internal parts of the gauge, the range of measuring pressure by the gauge should be selected considering the maximum operating pressure of the application, which must not exceed 75 percent of the measurement range selected. It is preferred that the range of pressure of the gauge should be about twice the normal operating pressure.

The wetted parts of pressure gauges shall be protected, as far as possible, from vibration, excessive temperatures, corrosive liquids and rapid changes in pressure. Normally, the tube and socket assembly of a pressure gauge, which come into contact with the fluid acting as a pressure medium, is considered as the wetted part. For air, gas, steam, water and other non-corrosive media, usually a bronze or brass Bourdon tube and brass socket assembly are preferred, and stainless steel or Monel wetted parts are suggested when the medium is corrosive in nature, or at high operating pressures or temperatures. The diaphragm seal is recommended for highly corrosive media and when there is possibility of solidifying or depositing of solids within the tube and socket assembly of the gauge.

6.20 Shower heads

Shower heads are appliances used for spraying water, mostly for taking a shower. For this purpose, it contains a face plate with innumerable, very small perforations through which water flows in a spraying manner. The spraying flow pressure of the shower head is mainly governed by the system pressure, but in some shower heads an option for powerful water spraying is provided. Shower heads of various spraying stream patterns are available, such as needle type, rain type, pulsating massage type etc. So, spray stream options range from gentle needles to a forceful massage type of stream.

Shower heads of varied flow rates are available. Generally, shower heads of 6 lpm, 7.5 lpm and 10 lpm flow rates are available. Shower heads with a flow rate of 7.5 lpm save about 40 percent water consumption.

Shower heads may be a fixed type, as shown in Figure 6.35, which are generally installed on the wall over the head level, or a hand-held type connected to the water supply pipe by a flexible hose pipe.

Due to deposition of minerals present in water, shower head perforations might get clogged partially or fully, causing an uneven or distorted stream of water flow. To avoid frequent clogging of the perforations, rubber nozzles are provided for spraying with “anti-clogging” features. Some shower heads are also provided with filtering system to reduce chlorine and to eradicate sulphur odours and scale. This very common problem can be easily solved by removing the perforated faceplate and cleaning the mineral deposits from the back of the plate with the help of a fine steel wool brush or pinching by a coarse needle or by jetting compressed air.

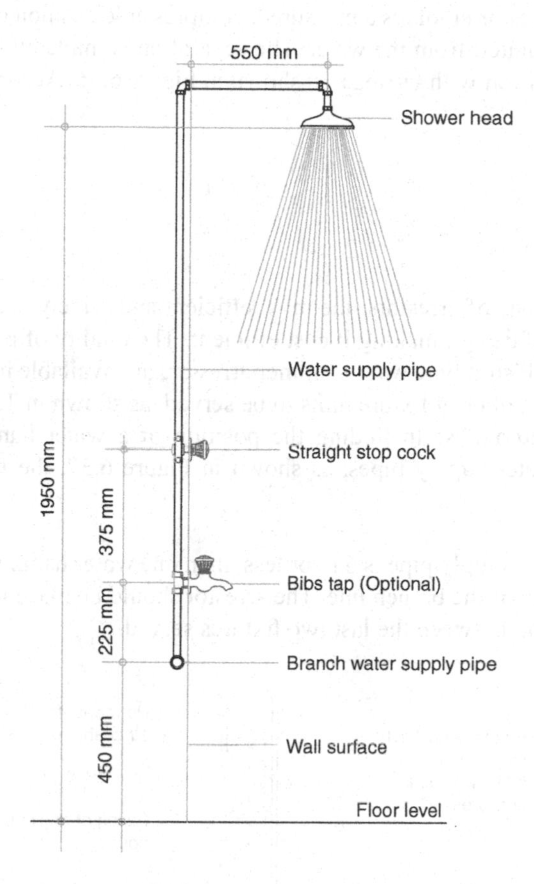

Figure 6.35 Shower head (fixed) installation.

6.21 Water hammer arrestor

Water hammer is a phenomenon, usually recognized in a water flowing pipe by a banging or generation of a thumping noise in the flow. This situation arises in a pipe when water flowing at high velocity is suddenly stopped, generally by shutting off a quick-closing valve. This sudden stop of flow results in the development of a pressure surge behind the valve, which reacts like a tiny explosion inside the pipe. These shock waves generate pressure spikes exerted all around and reverberate throughout the pipe, resulting in rattling and shaking of the pipe. The longer the pipe, the bigger the diameter, and the faster the flow, the more intense the water hammering will be. The other causes of generating water hammer are due to loose pipe installation or the presence of a worn-out washer in a faucet.

After generation of water hammer, it continues until it is absorbed somehow. Normally, trapped air in pockets of a piping system absorbs these shock waves. If water hammering is allowed to continue, it may damage the fixtures and appliances in the plumbing system. To stop water hammer, the energy of the shock wave is absorbed by installing a device called a water hammer arrestor.

A water hammer arrestor employs a measured, compressible cushion of air or gas which is kept permanently separated from the water. The separation is made by a rubber diaphragm, a metal bellows or a piston with O-rings, as shown in Figure 6.36. Accordingly, the arrestors are typed as below.

1 Diaphragm type
2 Bellows type and
3 Piston type.

Among all these types of arrestors, the most efficient and widely used is the piston type due to its simplicity of design making it cost efficient. The quality of piston parts primarily governs its longevity. Piston type water hammer arrestors are available in various sizes, to be chosen based on the number of fixture units to be served, as shown in Table 6.3.

Arrestor placement rules: In finding the position of a water hammer arrestor to be installed in branch water supply pipes, as shown in Figure 6.37, the following guidelines should be satisfied.

1 If the branch water supply pipe is 6 m or less, then one water hammer arrestor should be installed at the end of the branch line. The arrestor should be placed 2 m or less from the end of the pipe and between the last two fixtures served.

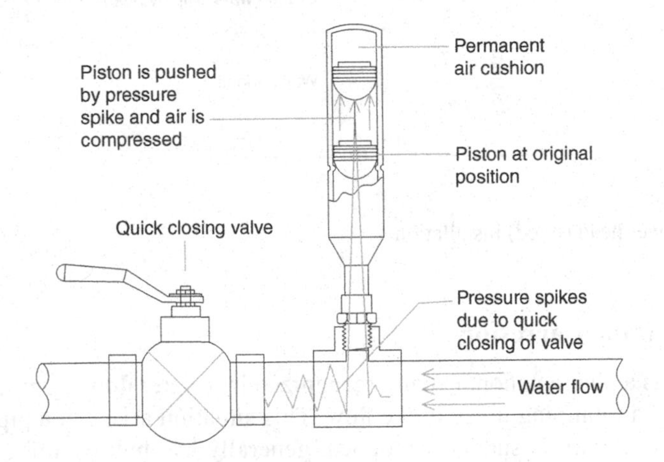

Figure 6.36 Water hammer arrestor: piston type.

Table 6.3 Type and size of water hammer arrestor with respect to fixture unit to be addressed [9].

Type PDI Units	*Connection size in mm*	*Fixture unit capacity*
AA	12	1–3
A	12	1–11
B	20	12–32
C	25	33–60
D	25	61–113
E	25	114–154
F	25	155–330

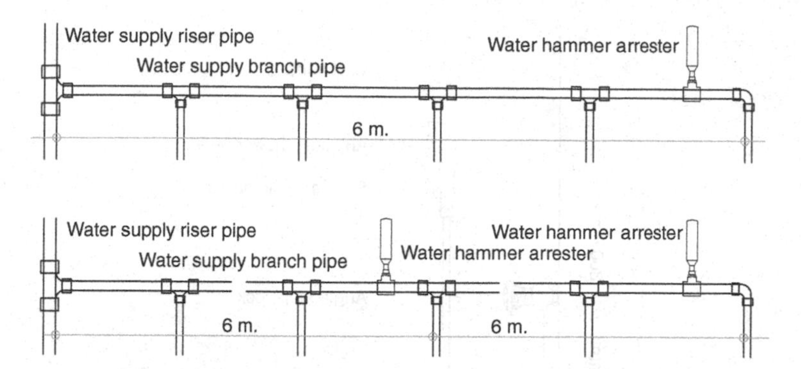

Figure 6.37 Water hammer arrestors positioning.

2 When the branch water supply pipe is longer than 6 m, an additional water hammer arrestor should be placed at the end of each 6 m section, within 2 m of the end of those sections. The sum of the fixture unit ratings of all the arrestors combined should be equal to or greater than the total fixture unit served.

6.22 Ablution hand shower

A hand shower for ablution enables maintenance of personal hygiene throughout every emptying of body waste, without the use of masses of toilet paper. The hand shower allows for the cleaning of genital parts with a gentle spray of fresh water while sitting on the toilet.

Hand showers are basically a mini-stream of water flow controlled by the trigger attached to the shower head, as shown in Figure 6.38. The mini-shower head is generally manufactured in high impact ABS with a chromium plate finish. The mini-shower head is connected to a stop cock by flexible hose pipe about 1.4 meters long and 12 mm in diameter for easy manoeuvring. The mini-shower head is held by a wall-mounting bracket fixed on the side wall.

Flexible pipe connectors: 300–500 mm long flexible braided plastic or braided metal hose pipes are used to connect basin faucets or water closet cisterns to the stop cocks installed on the water supply pipe ends near the fixtures.

6.23 Water tanks

Water sometimes needs to be stored for some period depending on the source of water, purpose of water use and type of water supply system. Water tanks, particularly their material of construction, play an important role in maintaining quality of stored water.

Storage tanks are either constructed or assembled on-site or directly installed. Generally, commercially available smaller-sized tanks (particularly in volume) made of various materials are directly installed. Masonry or reinforced cement concrete tanks that are comparatively larger in size are constructed on the site where they are to be located. Larger tanks made of iron or steel sheets are generally assembled at the location site. The selection criteria for storage tank materials should be based on economy, durability and possibility of water contamination. Tanks of various materials are discussed below.

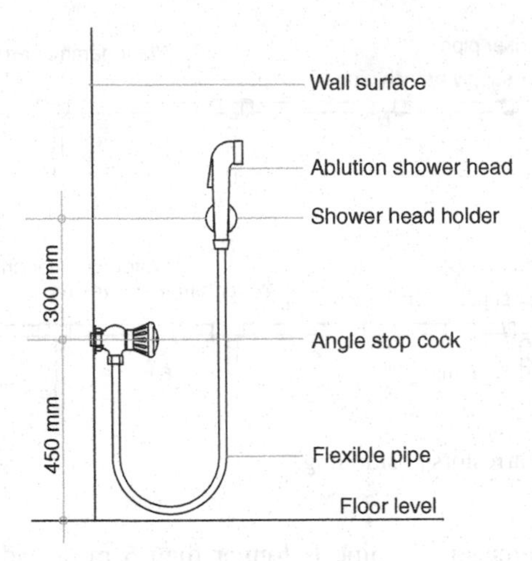

Figure 6.38 Ablution shower head installation.

6.23.1 RCC or masonry tank

Constructed tanks on the ground or roof, generally with capacities exceeding 50,000 l, are mostly made of reinforced cement concrete (RCC). The RCC tank is less expensive for tanks sized smaller than 1,000,000 litre capacity. For tanks greater than 1,000,000 litre capacity, a pre-stressed concrete tank is supposed to be less expensive by approximately 20 percent compared with the cost of an RCC tank of that size [10]. Walls of underground water reservoir of comparatively smaller depth can be made of bricks. Generally, tanks of capacities ranging from 15,000 to 50,000 litres can be made of bricks [11]. Tanks can be built in any shape and size with these materials. Masonry tanks are economical; however, making them watertight is difficult.

6.23.2 Ferro-cement tank

Ferro-cement is a cement-based composite construction material; it is modified from normal reinforced cement concrete by using mesh reinforcement instead of reinforcing bars. This type of construction is an effective and durable construction material for water tanks. Ferro-cement water tanks can be used for water storage in buildings. A commercially available ferro-cement tank of 1.2 cum area, is built by assembling 20 numbers of 525 sqmm ferro-cement plates of 12 mm thickness, made of cement and sand and reinforced with two layers of 18 BWG wire mesh. Ferro-cement tanks can be found pre-assembled and also in parts for erection at the site. Techniques for constructing ferro-cement tanks of ≤25,000-litre capacity have been developed [12]. If stored water is acidic and contains very few mineral salts, the water stored in an RCC, masonry or ferro-cement tank becomes neutral or lightly alkaline (pH between 7.5 and 8.5) and is also weakly mineralized.

6.23.3 GI tank

These tanks are fabricated of galvanized iron sheets. Smaller tanks can be made of 16–18 BWG-thick iron sheets by riveting with GI rivets at corner edges. The maximum capacity of these tanks is ≤1800 litres. Corrugated galvanized iron (CGI) tanks are also available, which are manufactured of corrugated iron or steel sheets of 20-gauge thickness. With this type of sheet, tanks of ≤300,000 l capacity can be built. This type of metal tank is popular for aesthetic reasons, fire resistance, unchanged quality of stored water and durability.

6.23.4 Stainless steel tank

These tanks are made of stainless steel sheets with thickness varying from 0.6 to 3 mm depending on the size and ranging from 200 litres to as big as 1 million [13] litres. These tanks are durable, highly resistant to corrosion and almost maintenance-free; however, they are costlier.

6.23.5 Plastic tank

These tanks are made of fibreglass-reinforced plastics, high-density polyethylene or other plastic materials. The tanks are relatively light, easy to carry and easy to install; however, they are not durable due to the deterioration of plastic quality from the sun's rays. A plastic tank of maximum size thus far manufactured has a 45,000-litre capacity. All plastic tanks used for storing water to be used for culinary or drinking purpose must be made of food-grade plastic. A vertical-type plastic tank, as shown in Figure 6.39, should be placed on a masonry base, which must be constructed in such a way that it supports the entire bottom surface of the tank.

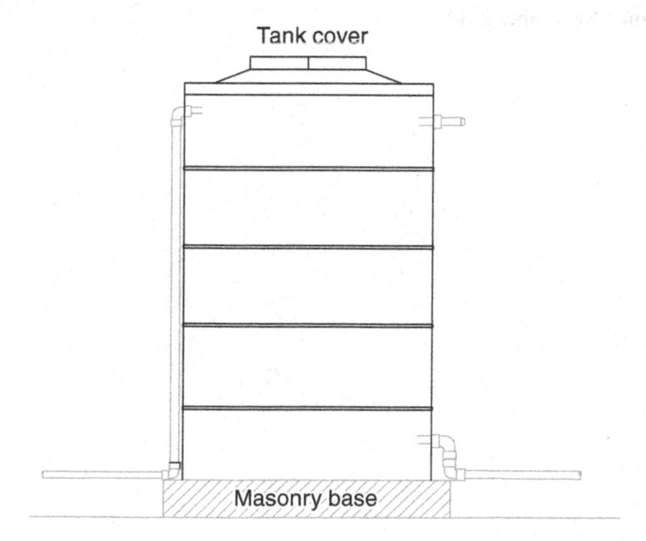

Figure 6.39 Plastic tank installations.

References

[1] Conely Company (2018) "Butterfly Check Valve" *Spears*, p. 106, Retrieved from www.conelyco.com/documents/Spears/Spears%20Butterfly%20Check%20Valves.pdf on 26 December 2020.

[2] Baboo, Prem (UD) "Problems Faced in Control Valves & Trouble Shooting" Retrieved from www.researchgate.net/profile/Prem_Baboo/publication/286879244_PROBLEMS_FACED_IN_CONTROL_VALVES_TROUBLE_SHOOTING/links/566ecfcd08ae430ab50039a4/PROBLEMS-FACED-IN-CONTROL-VALVES-TROUBLE-SHOOTING.pdf, www.physicsforums.com/threads/how-does-a-pressure-reducing-valve-work.208747/.

[3] International Plumbing Code (2015) "Chapter 5 Water Heaters, Clause 504.6" Retrieved from https://codes.iccsafe.org/public/document/code/550/9793923 on 26 December 2020.

[4] Wikipedia, "Valve Actuator" Retrieved from https://en.wikipedia.org/wiki/Valve_actuator on 26 December 2020.

[5] Ulnaski, Wane (1991) *Valve and Actuator Technology*, McGraw Hill Inc, New York.

[6] Department of Energy (1993) "Fundamentals Handbook" *Mechanical Science, Module 4, Valves*, p. 48.

[7] Lange, Garry (UD) "Domestic Water Meters" *Characteristics and Applications, LVVWD Engineering Services Division.*

[8] American Water Works Association (2018) *AWWA Manual M6, Water Meters – Selection, Installation, Testing, and Maintenance*, 5th Ed., AWWA, Denver.

[9] The Plumbing and Drainage Institute (2017) "Water Hammer Arresters Standard" *PDI-WH 201*, p. 20.

[10] Metkar, S.R. (2015) "Economics of R.C.C. Water Tank Resting Over Firm Ground vis-a-vis Prestressed Concrete Water Tank Resting Over Firm Ground" *Civil Engineering Forum*, Retrieved from www.engineeringcivil.com/economics-of-r-c-c-water-tank-resting-over-firm-ground-vis-a-vis-prestessed-concrete-water-tank-resting-over-firm-ground.html on 30 October 2015.

[11] Ammas, "Water Storage Tanks" *Response of Preeti Saxena*, Retrieved from www.ammas.com/q&a/Water-storage-tanks/q/185752on 1 November 2015.

[12] Sharma, P.C., "Ferrocement Water Storage Tanks for Rain Water Harvesting in Hills and Islands" Retrieved from www.eng.warwick.ac.uk/ircsa/pdf/12th/6/PCSharma.pdf on 06 November 2015.

[13] Bestank, "Stainless Steel Water Storage Tanks" Retrieved from http://bestank.com/products/water-tanks/on 1 November 2015.

7 Water supply systems

7.1 Introduction

Water is a necessity in almost all types of buildings, predominantly for its occupants as well as various other needs, depending on the type of use of the building. Water is first collected from a source and then conveyed to points at different locations in a building with demand for water, mostly by means of pipes. Various fixtures and appliances are installed through which water is consumed. To enable plumbing fixtures and appliances to function satisfactorily, these are provided with water in sufficient volume and at adequate pressure. To feed these fixtures, appliances or equipment water distribution pipes are installed in horizontal and vertical arrangements. For developing pressure in the supply system, various pumps are installed, and to maintain desired flow rate in the pipe, some controlling devices are installed on the piping, which were illustrated in the previous chapter. In this chapter, various methods of water distribution, pipe sizing approaches etc. are discussed.

7.2 Water

Water is one of the basic needs for human life and living. Water, in its purest form, is a combination of two parts hydrogen and one part oxygen (H_2O), but, as such, it is never found in nature. Pure water is basically a transparent and tasteless liquid. It appears in nature in three physical states: liquid, solid (ice, snow) and gas (vapor or steam).

Water in liquid form weights approximately 1 kg/cum. This is 830 times heavier than air. However, in the form of vapour, water is 133 times lighter than air. Water reaches its highest density at 4°C, freezes at 0°C and in an open container boils at 100°C at sea level, transforming into steam with the volume expanding about 1600 times. Upon freezing to ice, water expands in volume by about one tenth and exerts a pressure of 227,527 kPa.

Water is the "universal solvent" due to having the ability to dissolve solids and absorb gases and other liquids. So, all natural water contains minerals and other substances in a solution received from the air and soil through and over which it passes or flows. So, when water is full of minerals conducive to health, it is potable mineral water, which is directly consumable. Conversely, when it contains substances injurious to health, it is non-potable and needs treatment to make it potable. The properties of potable water are: it is harmless to health, colourless, odourless and clear.

For domestic use, potable water may not be directly available and so non-potable water is made potable water i.e. the quality of water must be made fit for drinking and culinary purposes. The limiting concentration of some substances in potable water is shown in Table 7.1.

DOI: 10.1201/9781003172239-7

Table 7.1 Maximum limits of some substances permitted in potable water [1, 2].

Sl. No	*Water quality parameters*	*Unit*	*WHO guideline values, 2011 [1]*	*European Union standard for drinking water [2]*
1	pH		6.5–8.5	–
2	Color (filtered)	Pt. Co. Unit	15*	–
3	Turbidity	NTU	5*	–
4	Iron (Fe)	mg/L	0.3*	0.2
	Manganese (Mn)	mg/L	0.4	0.05 mg/l
6	Arsenic (As)	mg/L	10	10 μg/l
7	Chloride (Cl)	mg/L	250*	–
8	Fluoride (F)	mg/L	1.5	1.5 mg/l
9	Nitrate/Nitrate-Nitrogen (NO3/(NO3-N))	mg/L	50 (as NO3)	Nitrate – 50 mg/l Nitrite – 0.5 mg/l
10	Total dissolved solids (TDS)	mg/L	1000 *	-
11	Total coliform (TC)	#/100 ml	00 TC/100 ml	-
12	Faecal coliform (FC)	#/100 ml	00 FC/100 ml	-

Note: * Based on acceptability.

7.3 Water supply requirements

The supply of water in any building implies the incorporation of a system by which a desired quality and quantity of water can be supplied to various points of use where there is demand for it. In buildings, water is not produced but is rather collected from various potential sources. Sources may be one or more depending upon the potentiality of the sources and according to demand. After collecting water from the source, it is supplied to the various points of use in the building through the development of a water supply system in the building. So, for water supply system development in a building, the following major jobs are to be performed.

1 Estimating the demand of water
2 Selecting sources and
3 Developing the supply system.

7.4 Demand of water

At the onset of planning for water supply system development, the requirement for water is to be determined. In buildings of different occupancies, water is demanded in varied quantities for various purposes of use. The requirement of water for those purposes may differ in nature also. When the requirement of water, for all the purposes of use in a building, is determined then its availability is to be ensured. So, total water requirement in a building is to be determined first. While planning and designing the water supply system in any building, it is necessary to identify first all the purposes of water use in the building.

There are many factors involved in estimating per capita (person) requirement of water in a building. So, it is nearly impossible to determine accurately the actual demand of water in a building, but judiciously considering those varying factors of per capita water demand, water requirement in the building could be rightly estimated.

In buildings, water is not only used by the users for their own purposes. There are various other uses of water in buildings, such as for safety, sanitation, recreational purposes etc. Following are various sectors in the building and its premises where water would be needed.

1. Occupants of buildings of various occupancies
2. Building construction, maintenance and demolition
3. Visitors
4. Swimming pools, spas, water bodies
5. Fire suppression
6. Road, pavement washing
7. Irrigation and gardening
8. Poultry or animal rearing
9. Fountains, waterfalls, water cascades etc.
10. Air conditioning and heating systems
11. Cleaning building sewers and
12. Special purposes.

In Table 7.2, the water requirement for various purposes of water use in various characters of occupancies of buildings is furnished, to help estimate the prime and basic water demand per day in buildings.

In Table 7.3, the water requirement for various purposes of water use in buildings is furnished to help estimate the optional water demand in buildings.

Table 7.2 Water demand in various building occupancy [3, 4, 5].

Sl. No	*Type of institutional building*	*Litres per day (lpd)*
1	Residential	70–400
2	Hospital (Including laundry)	
	a) Number of beds exceeding 100	455
	b) Number of beds not exceeding 100	340
	c) Health centres and hospitals.	5 litres/out-patient 40-60 litres/in-patient/day 60 litres/patient/day
	d) Cholera centres	15 litres/carer/day 30 litres/in-patient/day
	e) Therapeutic feeding centres	15 litres/carer/day
3	Hostel	135
4	Nurses home and medical quarters	135
5	Restaurant (per seat)	70
6	a) Factory where bathroom is provided	45
	b) Factory without bathroom	30
7	Offices	45
8	Hotel (per bed)	180
9	Cinema, concert hall and theatre (per seat)	15
10	School i. Day school	45
	ii. Boarding school	135
	iii. Nursery school	45 3 litres/pupil/day for drinking and hand washing.
11	Sport ground	3.5 litres/sqm
12	Vehicle	45

(*Continued*)

Table 7.2 (Continued)

Sl. No	*Type of institutional building*	*Litres per day (lpd)*
13	Shopping center	180
14	Airport and seaport	70
15	Railway and bus station	45–70
	a. Terminal	45
	b. Junction station for mail or express:	
	i. With bathing facility	70
	ii. Without bathing facility	45
	c. Intermediate stations where mail and express does not stop:	
	i. With bath	48
	ii. Without bath	23
16	Community center	180
17	Religious place	180
18	Shopping center	180
19	Group housing	180
20	Airport and seaport (international and domestic	70
21	Sports center	5–8
22	Mosques	2–5 litres/person/day for washing and drinking
24	Public toilets	1–2 litres/user/day for hand washing 2–8 litres/cubicle/day for toilet cleaning
25	a) Flushing toilets	20–40 litres/user/day for conventional flushing toilets connected to a sewer
	b) Pour-flush toilets	3–5 litres/user/day for pour-flush toilets
	c) Anal washing (bidets)	1–2 litres/person/day

Table 7.3 Water requirement in various optional purposes of uses of water [4, 5, 6].

Sl. No	*Purpose of use*	*Water requirements*
1	Horticulture	4.5 million litres/acre/day
	Small-scale irrigation	3–6 mm/sqm/day, but can vary considerably
2	Sanitary sewer cleaning	3.0–5.0 litres/head/day
3	Road washing	5.0 litres/head/day
4	Fountain	3–5 lpm/nozzle
5	Swimming pool	4 percent more than pool capacity
6	Kitchen garden	1.4 litres/sqm/day
7	Sports ground	3–5 litres/sqm
8	Park and garden	2–3 litres/sqm/day
9	Animal/cattle rearing	
	Cow or buffalo	40 to 100 litres/capita/day (lpcd)
	Horse	40 to 50 litres/capita/day [4]
	Dog	8 to 12 litres/capita/day [4]
	Sheep or goat	5 to 10 litres/capita/day [4]
	Poultry	0.37 litres/capita/day [5]
10	Vehicle washing	
	Two-wheeled carriage	30 to 40 litres/day [4]
	Four-wheeled carriage	50 to 70 litres/day [4]
	Car (self-service operation)	55 to 82 litres/vehicle [6]

7.4.1 Population projection

The population that will be using water during their stay in different types of buildings shall be calculated rationally. It is necessary to determine the gross volume of water required for various purposes on a peak day. Population size depends upon the type of usage and facilities provided in the building. Information regarding the population using a building can be obtained from the owners, users and building codes. Table 7.4 gives a basis for working out the projected population to be served in a building.

7.4.2 Variations in water demand

Water demand varies depending upon prevailing conditions and time of using water. The variation in water demand based on conditions and time of use are as follows.

1 Seasonal variations: 15 percent of the average demand of the year
2 Daily variations: the maximum daily flow = 1.80 to 2.0 times the average daily flow
3 Hourly variations: the maximum hourly flow = 2 to 3 times the average hourly flow and
4 The minimum hourly flow = 1/3 to 2/3 of average hourly flow.

Table 7.4 Population projection for various occupancy buildings [7].

Sl. No	*Occupancy type*		*Specification of occupancy*	*Population*
1	Residential	a	Normal dwelling unit (DU)	4.5–5/DU
		b	Large bungalow with guest rooms, etc.	5–7 persons/DU
		c	Add for staff quarters (single)	1/DU
		d	Add for family type staff quarters	4.5–5/DU
2	Office	a	Office workers	1 person/7.5–10 sqm of gross plinth area
		b	Maintenance employees	As per building type and size
		c	Visitors	5–15 percent of office employees or actual data for buildings of high public contact
		d	Floating population (drivers, shoppers, casual visitors etc.)	As per building type and size
3	Day schools	a	Students	Total student strength designed
		b	Teaching staff	1 per 20 student
		c	Administrative staff	1 per 100 student
		d	Visitors	2–5 percent of total student
4	Hostels of school and college	a	Students	1 per bed
		b	Warden's residence	4.5 per unit
		c	Staff residences	4.5–5 per unit
		d	Kitchen staff	-
5	Hotels	e	Guests	1/bed
		f	Employees	2.4–3/bed
		g	Restaurant meals	-
		h	Visitors	2–5 percent of guest numbers
6	Hospitals	i	Patient	1 per bed
		j	Doctors and nurses	10–15 per patient
		k	Administrative staff	25–30 per patient
		l	Visitors	0.5–1 per patient
		m	Floating population (drivers, shoppers, casual visitors etc.)	As per building type and size

7.5 Sources of water

Water to be used in the plumbing system can be obtained from various sources: natural or man-made. The natural sources are the primary sources of water. The natural sources are as follows.

1 Surface water like rivers, canals, lakes or even seas
2 Underground water and
3 Rainwater.

On the other hand, man-made water sources are as follows.

1 Surface waters like ponds and dammed water reservoirs and
2 Water supply mains developed by any water supply agency or authority.

Potentiality of various water sources: Potentiality of all types of sources is not the same in terms of quality and quantity. The potentiality of a particular source of water may even vary with time. To select the right source of water, its potentiality is to be highly judged. To understand the potentiality of water sources, all the types of water sources are discussed below.

7.5.1 Surface water

The term surface water refers to the water, on the earth's surface, that flows in streams and rivers as well as water in artificial or man-made lake, canals, ponds etc. Seawater, which is salty, also falls into the group of surface waters.

Among the surface waters, river water is found to be the best choice as the source of sweet water. The sweet nature and potentiality of available water from nature are believed to be the main reasons behind growing ancient civilizations by the sides of many rivers in the world. Dhaka, the capital city of Bangladesh, started growing by the side of the river Buriganga due to its sweet nature. With the continuous sprawling of urbanization and growth of industries, the surface waters become polluted due to discharge of treated, partially treated or even untreated wastes into it. When the pollution potential of surface waters go beyond the manageable limit, dependency on groundwater sources is high as the second alternative option.

In underdeveloped countries, most urban centers or cities situated near or by the side of a river discharge their treated or untreated wastewater and sewage into the rivers. Moreover, industries also discharge their treated or untreated wastes into nearby surface water bodies, causing heavy pollution. There are areas where the pollution level in surface waters is found to be comparatively very high, particularly in dry periods, due to having less water flow. In these circumstances, the treatment cost of water for supply gets very high, which may exceed the cost-effective limit.

For areas far away from any potential sources of surface waters, the cost for conveying water may be too high to be cost-effective. In those areas, if underground water is found satisfactory in terms of quality, quantity and cost-effectiveness, then it is considered to be the next alternative source of water.

7.5.2 Groundwater

Groundwater is an important and potential source of water for supply and commercial uses. The earth's surface is built up of different layers of soil mass like sand, gravel, clay, rock, etc. The layers of rock or compact clay, being solid or non-porous, cannot store water. On the contrary, the layers of coarse sand and gravel have many pores and cracks, which allow rainfall to enter into the soil, starting percolation from the natural surface. The porous soil layers, remaining filled with water, are called aquifers. Groundwater is extracted from these aquifers.

Groundwater may be found close to the earth's surface or at profound depths. Groundwater is supposed to be wholesome water and rich in minerals as it flows deeper. Groundwater usually gets recharged naturally by the percolation of rainwater or surface water but can also be artificially recharged, infiltrating mostly rainwater.

For groundwater as a source of water, the major concerns are depletion and fast rate of lowering the water table. In the process of urbanization, the impacts of building development on groundwater hydrology may be listed as below.

1 Increase in water demand
2 More dependence on groundwater use
3 Over-exploitation of groundwater
4 Increase in runoff, fall in groundwater levels and decline in well yields
5 Reduction in open soil surface area and
6 Reduction of green surfaces for direct and natural infiltration and deterioration in rainwater quality

Due to ever-expanding urbanization, particularly in land-hungry and densely populated areas, the major portion of the urban area becomes covered by the metal surfaces of buildings, roads, parking etc., leaving few green spaces. As a result, the scope of natural rainwater-infiltration is reduced considerably or even may not occur at all, in comparison to the rate of withdrawal of groundwater, to meet the growing demand of water, mostly in buildings. Natural systems established a condition of equilibrium in the process of precipitation, infiltration, surface runoff, evaporation and evapotranspiration. Continued and over-withdrawal of groundwater by pumping leads to a continual drop in the water table. The water table can drop at rates of up to 1 meter per year or even more. In Dhaka city, the capital of Bangladesh, at present the average rate of lowering the groundwater table is about 3 meter per year [8].

Groundwater is supposed to be wholesome and potable and is rarely found to be polluted or contaminated, though there are instances of pollution or contamination of groundwater due mostly to human interventions. Pollution of shallow groundwater is mostly caused by biological activity, much of it due to human interventions. Industries also play a considerable role in polluting groundwater. Among other causes of groundwater pollution, waste dumping on the ground or under it, land filling with contaminated soil and leaching of agricultural pollutants etc. may be considerable. There might be underground connections between surface streams and groundwater, and so, if stream water is polluted or tainted, the effects can be transmitted to the groundwater. Pollution or contamination of groundwater might also happen due to unethical and unregulated human interventions through wastewater dumping deep underground or through injecting contaminated water. In coastal areas, the groundwater is often found to be brackish or saline due to the intrusion of saline water from the sea.

In areas where both the surface water and groundwater reach such conditions that they have lost their dependability as sources of water, rainwater is considered to be the next option of a source of water that has been found reliable [9].

7.5.3 Rainwater

When dependability on water from usually available sources is considerably lost, then, as an alternative source, rainwater is considered to be a potential supplementary option for buildings. Rainwater has proven to be an alternative source of water available at building doorsteps, having the lowest risk in use. It can provide a readily accessible and reasonably reliable source of water to meet the shortfall for demand of water in a building.

Non-potable uses of rainwater have wider range and divergence. These may include flushing of water closets and urinals inside building toilets; indoor plant watering and landscape irrigation; outdoor washing such as for cars, building facades, sidewalk washing, road sweeping etc.; water for fire suppression such as fire trucks, fire hydrants and sprinkler systems etc.; supply for chilled water cooling towers, replenishing and operation of water features and water fountains and laundry if approved by the local authority. Replenishing of swimming pools may also be acceptable if special measures are taken, as approved by the concerned regulatory authority. After proper treatment and disinfection, rainwater can even be consumed confidently for drinking and culinary purposes.

The use of considerable amounts of rainwater for multiple purposes needs to be collected, stored and distributed in the buildings concerned, which falls under the subject of plumbing. This system can be set up independently or in conjunction with normal water supply systems in a building and in a building complex.

Rainwater, as a source of water, is found reliable in many parts of the world, subject to its availability and quality. Regarding rain or rainfall, the general concept for finding a rainwater utilization system feasible is that the rainfall should be over 50 mm/month for at least half a year or 300 mm/year, unless other sources like surface and ground waters are extremely scarce [10].

7.5.4 Water supply mains

Water supply mains are the piping networks developed in certain areas for supplying water to meet various needs of water after collecting it from potential natural sources of water and treating it to the desired quality if needed. In an area, the water supply network is developed by either a public or a private service provider, who is given the responsibility for ensuring availability of the desired quality and quantity of water round the clock within the area, maintaining certain pressure in the mains. The pressure maintained in the urban water distribution systems of many developed country is found to range from 10 m (1 bar) minimum to 70 m (7 bars) maximum [11].

Water service providers are held responsible for the water supplied to the boundary of the property of the building's owner. The property or building owner is supposed to make the necessary arrangement of pipework to bring water from the main into the building after getting connected to the main subject and to obtain necessary permissions in this regard, and then to develop a water distribution system to supply water to the various points where water is needed in the building. There are two major problems faced in the performance of water supply mains:

1 Non-availability or intermittent flow of water and
2 Low pressure.

Non-availability of water in the main might be due to accidental faults in the system or breakage of pipelines. Intermittent flow may occur due to water supply management during shortages of water production at the source. Low pressure in the water supply main may occur due to various reasons, as mentioned below.

1 Excessive withdrawal of water, more than the capacity of supply
2 Excessive usage of water during peak hours of water usage
3 Occurrence of extra head loss in the pipe due to expansion of piping network and
4 Leaking in the main pipes.

7.6 Water service pipe

The water service connection pipe is a pipe connecting the distribution water main in the street to the water distribution system of a building or any number of buildings to be served. Water mains are supposed to remain in operating condition with full flow conditions under certain pressure. Water service pipes are to be connected with the main under full flowing conditions. So, special measures and techniques are to be followed for perfect connections, ensuring uninterrupted flow without contaminating water in the main.

The size of the service connections varies depending on the flow requirements for the water supply system in a building. The minimum size of a water service pipe shall not be less than 20 mm [12]. The section of the service connection pipe from the water main to near the property line of the site for building is installed and maintained by the respective water service providers, and at the end of the service pipe a water meter is installed. The piping downstream of the meter is installed and maintained by the property owner.

A curb stop, basically a gate valve, is installed in a stop box to house the valve, used to turn the water service on and off. It is usually located in the public right of way, outside the property line, on the water service connection pipe. The curb stop is generally operated by the water service provider by a long key, except in the case of an emergency when it may be operated by the concerned building or property owner.

To connect the service pipe on the main, tapping is done on the water main with the help of a special drilling machine. The machine is clamped onto the main over a washer with the help of an adjustable chain. Pressure is imposed through one spindle by the screw which forces the combined drill and taps into the main. When tapping is completed, another spindle fitted with the ferrule is turned on the tapping hole and the ferrule is fitted on the main with the help of a ratchet lever. Ferrules, as shown in Figure 7.1, are the fittings which allow tapping of the water main flowing full and allowing for a service pipe to be connected without interrupting the flow in the main. The upper part of the ferrule is made to swivel. To accommodate undue subsidence of underground soil, a flexible service pipe called a "goose neck" is sometimes connected with the ferrule. With this flexible pipe the service pipe is connected.

In another way, a saddle or strap is fitted to the main with bolts. On this saddle, the ferrule can be threaded. In PVC pipe, saddles are secured by solvent welding and the ferrules are usually moulded with the saddle.

7.7 Water reservoirs

In developing water supply systems in a building, a water storage tank may need to be incorporated, depending upon the potentiality of the source and the supply system to be developed inside the building. Water reservoirs play an important role in developing efficient and economic supply systems. They also play a role in maintaining the quality of water. Tanks for storing water are also needed for following purposes.

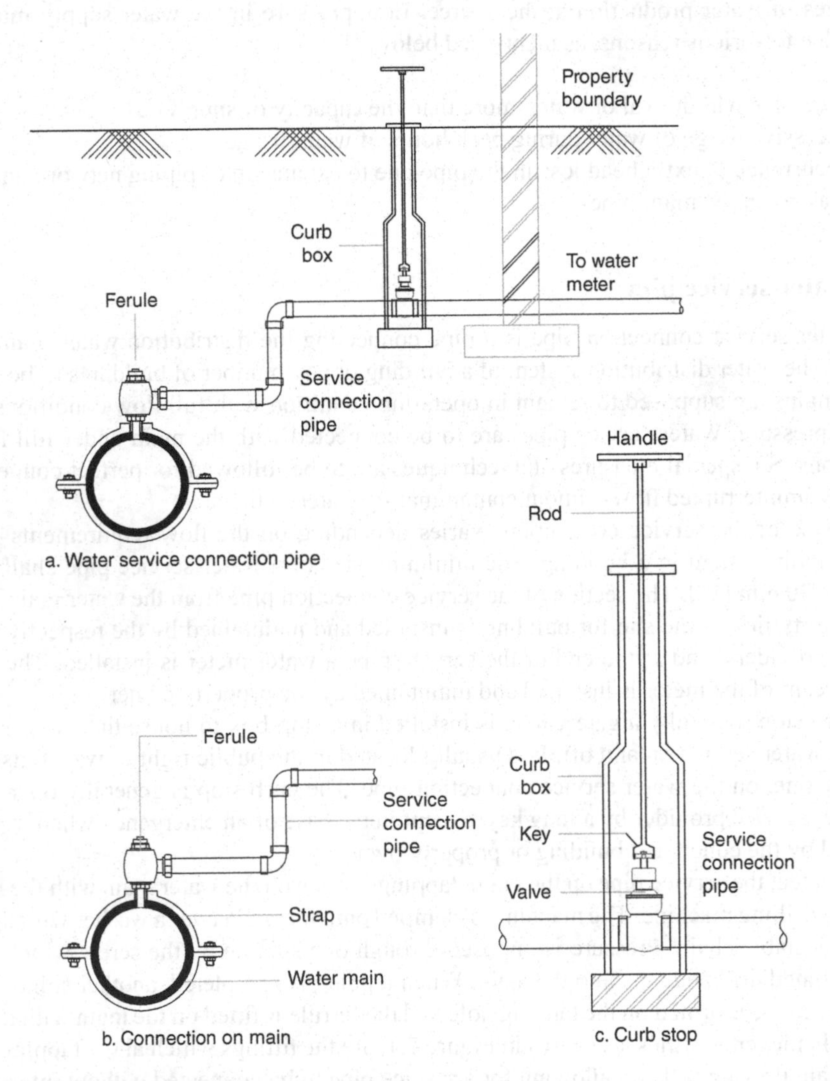

Figure 7.1 Building service connection to water main by ferrule.

1 To provide uninterrupted water supply when the public water supply mains are shut off for any reason
2 To meet the peak demand within a building when it cannot be met from a public main
3 To provide water for various occasional uses, like firefighting, and
4 Public water supplies when found not continuous.

Two types of water reservoirs are used depending on the type of supply system and its location in the building. These are as follows.

1 Underground reservoir and
2 Overhead tank.

Sizing storage capacity: The size of a water reservoir depends mainly on the water demand pattern of the building. There are various others factors on which the size of a reservoir depends. These factors are as follows.

1 Time of supply in the main
2 Flow and pressure in the main and
3 Water storage for fire suppression.

7.7.1 Underground water reservoir

When the water supply authorities fail to supply water round the clock, at the desired pressure, as they are supposed to, then an underground water reservoir is needed in a building to store water for use in the non-supplying periods.

The capacity of an underground reservoir should be the net difference between the peak demand and the flow during hours of supplies for direct water supply systems. For underground-overhead tank and pumping systems, an underground reservoir shall be sized considering the one-day demand of the building. If water is to be stored for firefighting, then the amount of water required for fire is to be added to the volume of water needed for fulfilling the one-day demand of the building.

Underground reservoirs shall be constructed with utmost care. The following measures should be accounted for during the development of an underground reservoir.

1 The tank shall be made watertight and not leak when empty or full.
2 There shall be two manhole covers, placed far apart from each other, on the top slab. At least one shall be of 550 mm diameter for ease of access to the reservoir. Manhole cover fixation shall be ensured watertight.
3 The structural strength of all walls and floors shall be capable of withstanding active earth pressure, buoyant pressure and surcharge loads, if any.
4 Tanks may be provided with catch rings of non-corrosive material.
5 A short piece of pipe for connecting the water service pipe, 150 mm longer than the width of the wall, shall be fitted into position while casting the wall of the reservoir.
6 The internal finish shall be very smooth and brightly coloured by non-toxic paint; a depression may be created by making a slope of the finished surface on the floor of the reservoir, under a manhole.
7 For very large reservoirs, to avoid interruption in operation during cleaning or repair works, compartment can be made by providing separation walls with interconnection pipes supported by gate valves.

Connection of pipes with underground reservoirs: For underground reservoirs, various service pipes need to be connected to develop the water supply system. For any reservoir, there is an inlet pipe to fill the tank with water. Inlet pipes are generally, depending upon the size of the tank, connected about 150 to 225 mm below the top slab of the reservoir. The other pipe installed in an underground reservoir is a suction pipe for the pump. If the pump is located on the reservoir, then the suction pipe is installed by inserting the top slab of reservoir, as shown in Figure 7.2. When a pump is placed by the side of the reservoir, then a suction pipe is installed by inserting the wall in front, as shown in the previous chapter on pump. For installing suction pipes on side walls, a short piece of pipe, 150 mm longer than

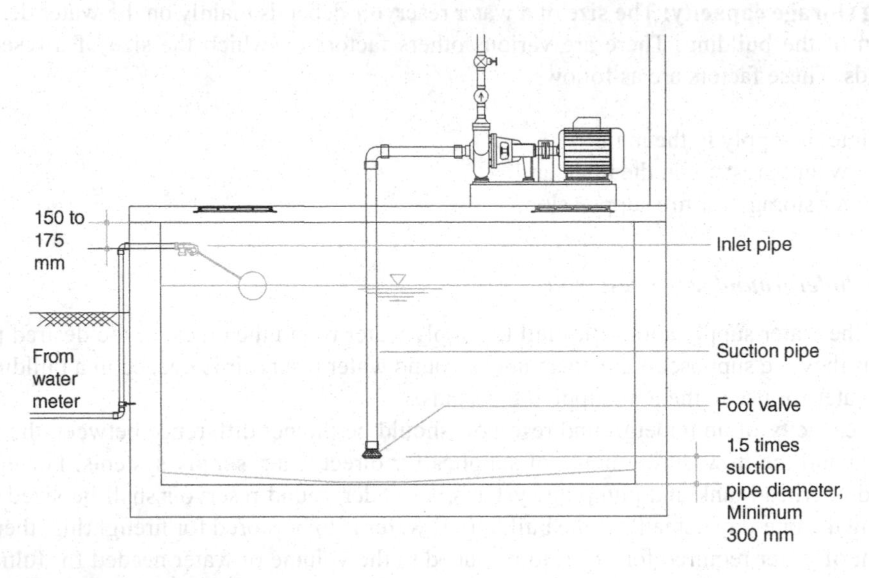

Figure 7.2 Underground reservoir pipe connections.

the wall thickness, shall be placed while casting the wall to ensure watertightness, or a pipe sleeve shall be provided for the same purpose.

7.7.2 Rooftop or overhead water tank

When a water supply system in a building is to be developed following an underground-overhead tank system, then an overhead tank of appropriate size shall be constructed at the proper location. The tank shall be placed at such a location that it can produce sufficient pressure for the topmost and high pressure-dependent faucet in the building.

The size of overhead tank can be found from formula 7.1.

Volume of water to be reserved (litres)

$$V = kPD - HF \qquad 7.1$$

Where

k = a constant, 1.0 for one time filling; 0.5 for two times filling; and 0.34 for 3 times filling in a day

P = number of population

D = average daily demand of water in litres/person/day

H = pumping hours in number and

F = water consumption rate during pumping, litres per hour.

Measures for overhead tank: The following measures should be accounted for during the development of an overhead tank.

1 Tank structural members should be designed to withstand both the load of the tank and the pressure exerted by water.

2 Tank bottoms shall be made clear off the roof surface to avoid damage due to leakage and overloading the roof slab.
3 A lightening arrestor, guard rail or wall shall be provided on the top slab of the reservoir.
4 Short pieces of all connection pipes (inlet, supply, overflow etc.) to the tank shall be placed in position while casting the elements of the tank.
5 Tanks shall be located near the centroid of the supply points with respect to rate of flow to the points.
6 The height of the tank bottom shall be at such a level that it ensures a minimum water pressure of 60 kPa in the topmost faucet inlet.
7 Instructions regarding manholes, watertightness, catch rings, sump, finishing and color mentioned for underground reservoir will be equally applicable for overhead tank.

Connection of pipes with overhead tanks: For overhead tanks, various service pipes need to be connected to develop the water supply system. For this tank, there is an inlet pipe to fill the tank with water. Inlet pipes are generally, depending upon the size of the tank, connected about 150 to 225 mm below the top slab of the reservoir. The other pipes installed on the tanks are as follows.

1 **Supply pipe:** Installed about 150 to 300 mm above the floor of the tank to avoid sediments entering into the supply pipe. The supply pipe inlet should be fitted with a strainer. To ensure watertightness, a short piece should be kept during casting of the masonry wall.
2 **Vent pipe:** A vent pipe of comparatively smaller diameter than the supply pipe should be installed on top of the supply pipe where it bends down, as shown in Figure 7.3.
3 **Overflow pipe:** Overflow pipes are provided about 100 mm below the inlet pipe to ensure the water level rises up to a certain fixed level. The recommended size of overflow pipe shall be chosen according to Table 7.5.
4 **Draining pipe:** A drain pipe of minimum 50 mm shall be put on the bottom slab preferably; a reducer of 100–50 mm should be placed at the inlet on which a grating is to be provided. The drain pipe shall be supported by a gate valve. The size of the drain pipe should be fixed according to the recommendation furnished in Table 7.6.

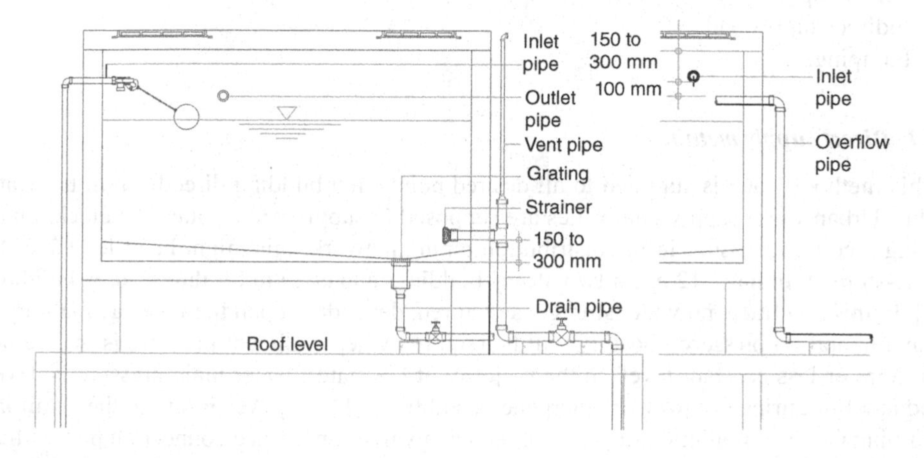

Figure 7.3 Pipe connections with overhead tank.

Table 7.5 Recommended sizes for overflow pipes for water storage tanks (converted to SI unit from source [13].

Sl. No	*Maximum capacity of water supply line to tank lpm*	*Diameter of overflow pipe (mm ID)*
1	0–190	50
2	191–379	63
3	380–760	75
4	761–1514	100
5	1515–2650	125
6	2651–3785	150
7	Over 3785	200

Table 7.6 Size of drain pipes for overhead water tanks (Converted to SI unit from source [13].

Sl. No	*Tank capacity litre*	*Drain pipe mm*
1	Up to 2840	25
2	2841 to 5680	40
3	5681 to 11360	50
4	11361 to 18930	63
5	18931 to 28390	75
6	Over 28390	100

7.8 Water supply methods

Water supply methods from underground sources or water mains to various locations of a building can be developed in different ways depending upon the condition of the sources. Each method of water supply has its own merits and demerits, more or less. These supply methods can be broadly defined in three ways, as follows.

1 Direct supply
2 Indirect supply and
3 Pumping.

7.8.1 Direct supply method

In this method, water is supplied to all desired points in a building directly from the water mains. Urban water supply authorities are supposed to supply water round the clock, maintaining a certain pressure in the distribution piping network; this might be at least 7 m for single-storey buildings, 12 m for two-storey buildings and or 17 m for three-storey buildings [14]. In this condition, no water storage is required, depending upon the water availability in the main, and no pressure boosting will be required when sufficient pressure is maintained at a more or less constant level. In the majority of US states, water main pressure is developed at 14 m during fire flow or emergency conditions [11, 15]. According to the "Uniform Plumbing Code", installation of pressure reducing valves on service connection pipes where the mains water pressure exceeds 550 kPa is mandatory [12]. Pressure reducing valves are to be installed on the pipe after the water meter. Figure 7.4 illustrates a direct supply method developed for a four-storey building.

When the water mains cannot provide water round the clock, but pressure in the main is sufficient, then incorporation of a rooftop water tank on smaller-height buildings, subject to the pressure in the main, will be required to store water for the non-supply periods. Water from the rooftop tank is then distributed to various points of use under gravity flow. To distribute water in various locations in a building by drawing water directly from the main, incorporation of such overhead water tanks on the roof due to discontinuous water

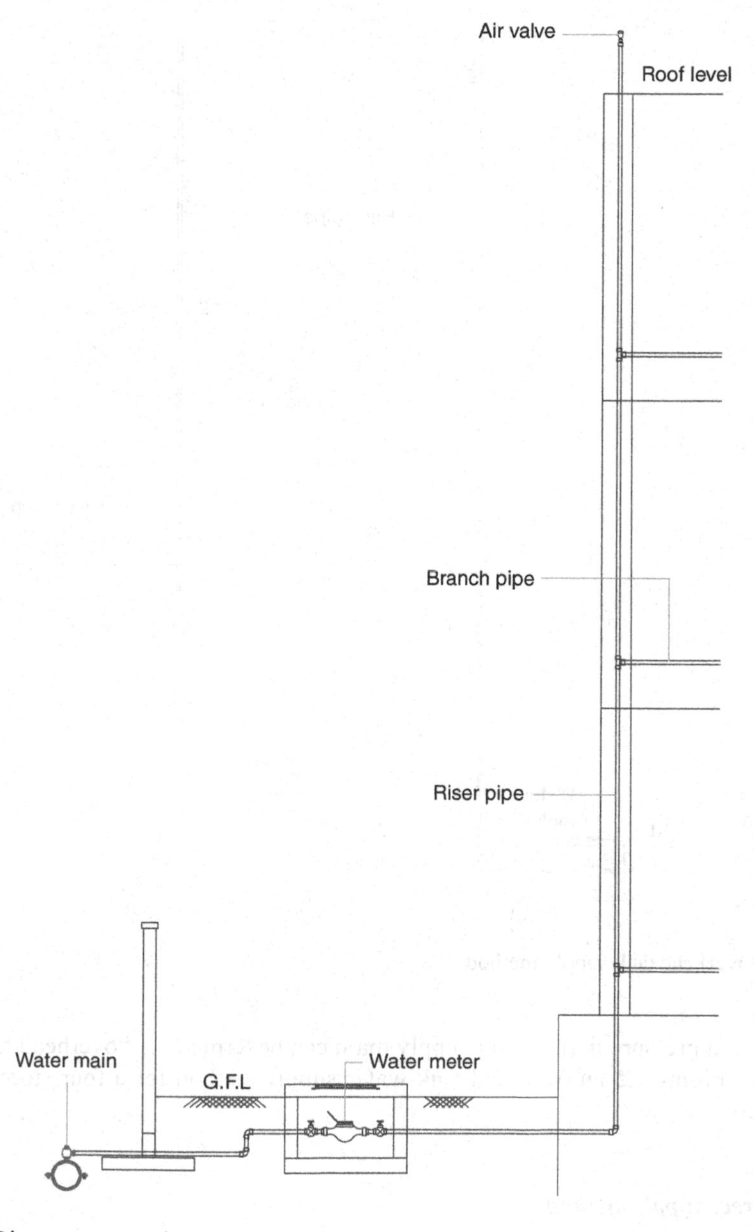

Figure 7.4 Direct water supply system.

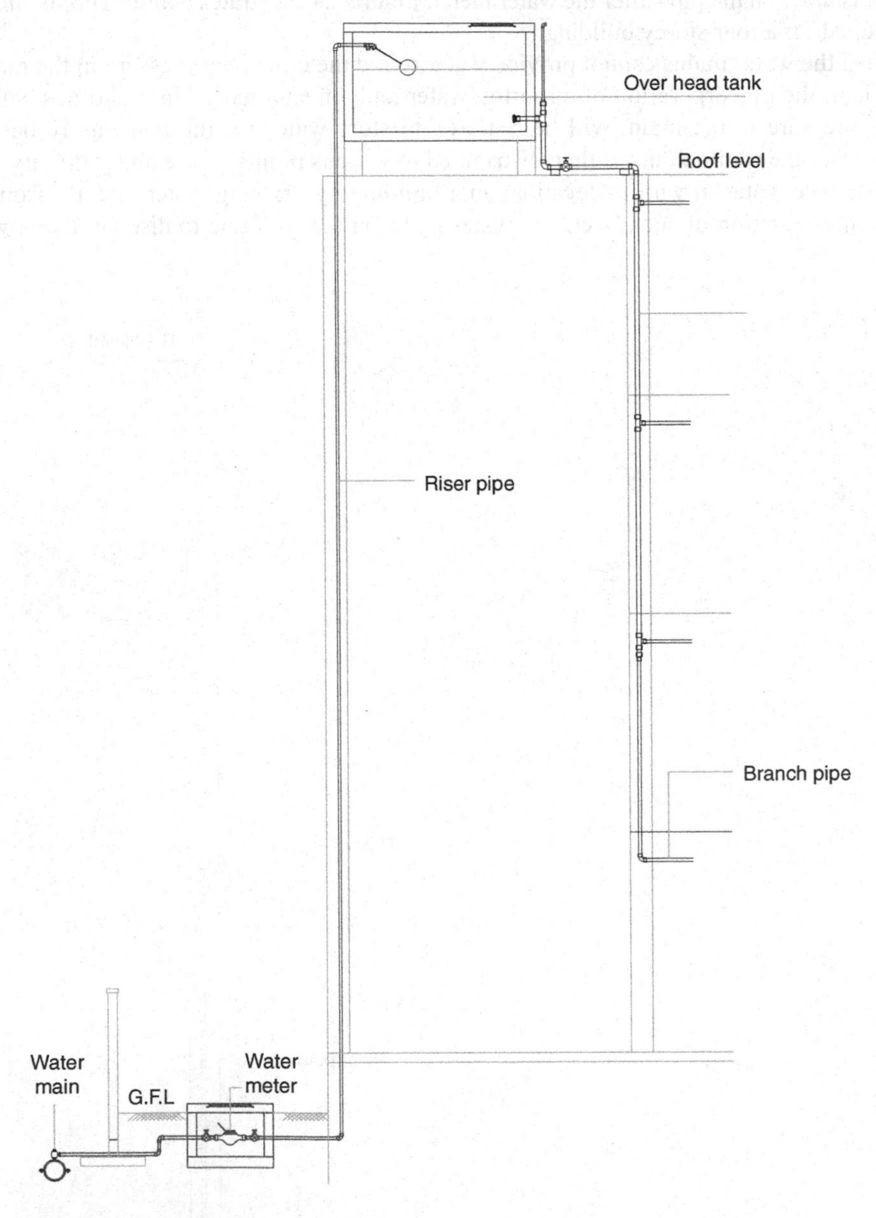

Figure 7.5 Overhead tank supply method.

supply at rated pressure in the water supply main can be termed the "overhead tank supply method". In Figure 7.5 an overhead tank water supply method for a four-storey building is illustrated.

7.8.2 Indirect supply method

When the water supply main cannot maintain the supply of water round the clock at the rated pressure, then storage of water at a low level, either on the ground or underground, is required, depending upon the pressure conditions. Generally, storing water in an underground reservoir is preferred for various advantageous reasons in comparison to above-ground storage. Water distribution from these ground or underground storage tanks to various locations in the building can be developed in two ways: 1. employing a rooftop overhead tank called an underground-overhead tank system and 2. pressurized supply by pumping.

7.8.3 Underground-overhead tank system

In this system, both underground and overhead tanks are employed in the water distribution system in a building, as shown in Figure 7.6. Water from the main or from any independent sources is stored in the underground water reservoir. This stored water is then pumped up to keep in the rooftop overhead tank. From the rooftop overhead tank, water is distributed to the desired locations under gravity through the piping network. Water distribution under gravity is also termed a down-feed distribution system.

7.8.4 Pumping system

Water can be supplied from any water source or reservoir to various locations of water usage by imparting pressure energy, employing one of the following two systems.

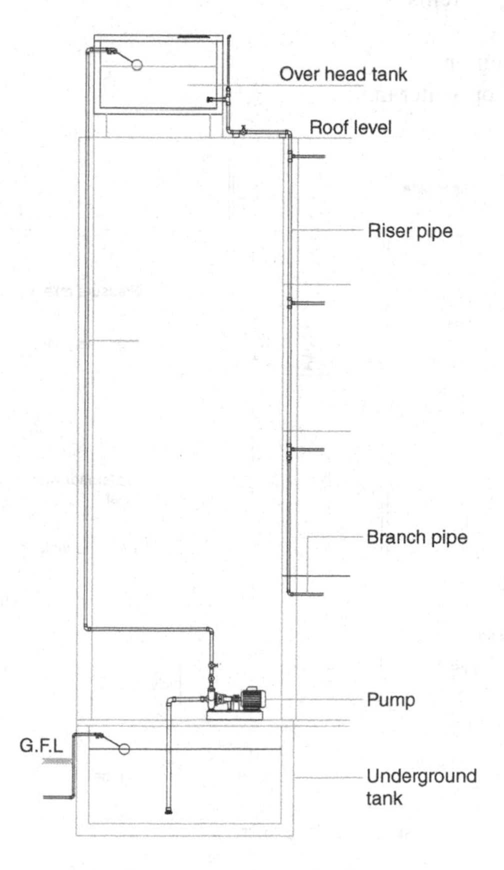

Figure 7.6 Underground-overhead tank water supply method.

1 Hydro-pneumatic tank system and
2 Booster pumping system.

Water distribution from lower-level sources to upper-level points of usage by direct pumping is also termed a up-feed water distribution system.

7.9 Hydro-pneumatic tank system

A hydro-pneumatic tank system includes a pressurized hydro-pneumatic tank, containing water under pressurized air, which works in conjunction with a water supplying pump called a fill pump, as illustrated in Figure 7.7. The operation of the pump is regulated to stop at a certain maximum pressure, at which the tank will contain some reserved water ready to furnish smaller, intermittent water demand. Additionally, hydro-pneumatic tanks are employed to achieve the following objectives.

1 To deliver water at predetermined pressure range so that continuous pumping can be avoided
2 To prevent a pump from starting up at minor supply requirement of water and
3 To minimize development of pressure surges (water hammer) in the distribution piping.

Pressure boosting by a hydro-pneumatic tank system has the following advantages compared to other boosting systems.

1 Prevents water hammer
2 Avoids use of rooftop water tank

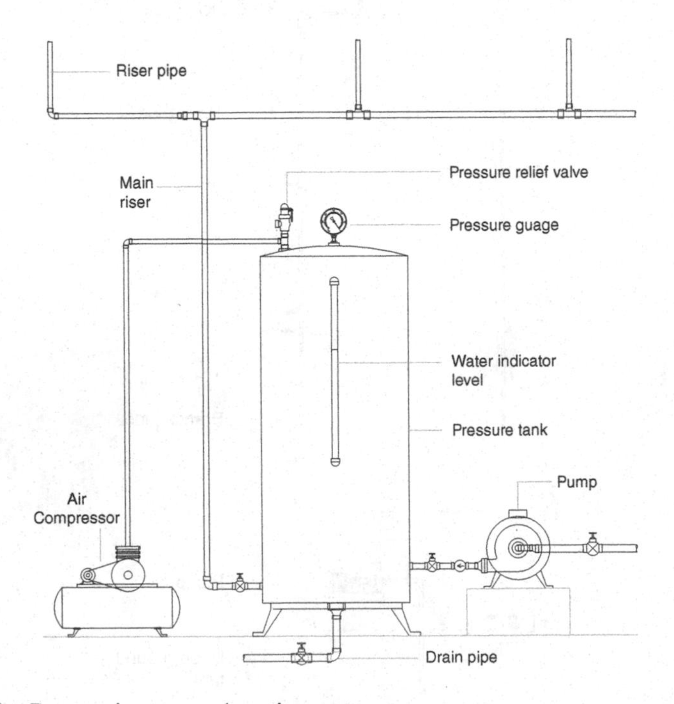

Figure 7.7 Hydro-Pneumatic pressure boosting system.

3 Pump run-time saving reduces cost of energy and
4 Water supply system can be designed for any desired pressure.

So, where there is high fluctuation of water demand during a day, such as in schools and apartments, incorporation of a hydro-pneumatic tank system will be very effective.

In a hydro-pneumatic pressure boosting system, there are some disadvantages as well, as mentioned below.

1 High initial cost
2 Tanks are heavy, need strong support and require maintenance
3 Additional floor space is required
4 Air volume needs to be regulated daily, which needs regular monitoring
5 A standby tank is seldom provided, so, during maintenance and repair of the tank, the running of a fill-pump makes the system very inefficient and
6 If there is no electricity, there will be no flow of water.

Hydro-pneumatic systems elements: There are two types of hydro-pneumatic tank system: 1. plain steel tank and 2. diaphragm-type or replaceable bladder pre-pressurized hydro-pneumatic tanks.

In plain steel tank hydro-pneumatic tanks, there are three major elements installed; these are to be properly sized, installed and maintained for economic and efficient operation of the system. The elements are as follows.

1 Hydro-pneumatic tank
2 Air compressor and
3 Fill-pump.

7.9.1 Hydro-pneumatic tank

The hydro-pneumatic tank is a closed vessel generally made of steel. The tank is partly filled with air and the remainder is filled with water. Air remains pressurized in direct contact with the water. Compressed air works as a cushion and absorbs pressure to exert on the water. This air cushion decreases in volume as air is absorbed by water causing loss of pressure in the tank thereby losing the ability to pressurize the water supply or distribution system. Hydro-pneumatic tanks perform efficiently under an air cushion of ¼ to ½ of the tank capacity [16]. Smaller tanks are generally a vertical type but larger hydro-pneumatic tanks of 7570 litre or more are usually a horizontal type.

In diaphragm type or replaceable bladder pre-pressurized hydro-pneumatic tanks, there is a flexible diaphragm of rubber between the air and the water in the tank. So, these tanks are also called bladder tanks, captive air tanks, etc. The tank is charged with air at initial start-up. No air compressor is needed in this type of tank, as because no air is lost through water due to remaining separated by the diaphragm. The on–off cycling causes the diaphragm to go up and down continuously. The flexing back and forth will cause the rubber diaphragm to break eventually. Once this diaphragm is busted, the tank must be repaired. Rapid cycling on and off is the first sign of damage to the diaphragm.

The tank usually possesses two pipes, one inlet pipe comes from the fill-pump and one outlet pipe that supplies water to the desired points. The inlet pipe size can never be smaller than the outlet pipe size.

The captive pressurized air inside the tank keeps the water supply system under pressure while the pump remains off. When drawing of water from the system starts, it causes a drop of pressure in the tank. When the pressure dropping in the tank reaches the "on" pressure setting, the fill-pump starts running and raises the volume and pressure of the water under the diaphragm in the tank. Smaller hydro-pneumatic tanks can therefore be used to control pressure booster pumps to be cycled on and off, like a pressure switch.

The hydro-pneumatic pressure tank maintains the pumping-cycle rate at an acceptable limit. If the tank becomes waterlogged anyhow, the on and off motor-cycles may be more than six times an hour. This will lead to higher energy costs, inefficiency of the system and premature motor failure.

Accessories: Hydro-pneumatic tanks shall be supported by various accessories for its effective operation, monitoring and maintenance. Following are some essential accessories.

1 Pressure gauge
2 Water sight glass
3 Air blow-off (automatic or manual)
4 Pressure relief valve and
5 Access manhole for larger tanks.

Sizing hydro-pneumatic tank: Correctly sizing a hydro-pneumatic pressure tank is very important, as it greatly affects the life of the pump. Choosing the right size for a pressure tank is a bit critical. Selecting a smaller pressure tank for particular water demand may result in overuse of the pump. So, it is safer to choose a slightly larger tank. Sizing methodology for both types of hydro-pneumatic pressure tank is discussed below.

Sizing of plain hydro-pneumatic tank: Hydro-pneumatic pressure tanks must be sized considering the capacity of the pump with respect to the peak water flow demand, so that the pump "off" cycle remains for a longer period than that of the "on" cycle, and the pump does not cycle too frequently.

The volume of tank V_t (litre) can be determined from the formula given below.

$$Vt = Va + Vv + Vw \quad 7.2$$

Where
Va = volume of air under high pressure (litre)
Vv = variable volume which may be filled by air or water (litre) and
Vw = volume of water remaining in the tank for use (litre).

The volume Vw, which is arbitrarily chosen, is kept about 10 percent of the water volume. In a tank supported by air compressor, this volume should merely be sufficient to extend over the inlet or outlet of the tank, so that there is no danger of air reaching up to those points.

The volume Vv can be determined from the following formula.

$$Vv = kC \quad 7.3$$

Where
k = time factor in minutes, generally considered 15 sec (0.25 min) to 30 sec (0.5 min) and
C = maximum water flow demand or pump capacity in lpm.

The volume *Va* is dependent on the volume *Vv*. The volume *Va* can be determined from the following formula.

$$Va = \frac{Vv \times P2}{P1 - P2} \quad 7.4$$

Where

P_1 = maximum absolute air pressure when air volume is at its minimum occupying space *Va* and

P_2 = minimum absolute air pressure when air volume is at its maximum occupying space *Va* + *Vv* = the minimum pressure to be maintained for the system.

Sizing of diaphragm type hydro-pneumatic tank: The volume of the tank shall be larger than the volume of water that the pressure tank will be holding. The amount of water held in a pressure tank is called drawdown, which is usable water. This amount of water is drawn from the tank between the time the pressure switch turning off and turning on the pump. The drawdown capacity is dependent on a variety of factors, but most important is the pump flow rate. An acceptance factor is used to determine the acceptance volume of the tank which is the maximum amount of water in the tank. The acceptance factor is the multiplier against the tank volume to determine the acceptance volume of tank.

Total tank volume = Total drawdown (D) / Acceptance factor (F)

Drawdown D can be determined from the formula

$$D = t(Q \times 60)/4 \quad 7.5$$

Where

t = time cycle (minute) and

Q = tank flow rate (litre per second).

And the acceptance factor F= 1 – (P_1 pressure tank precharge + 101) (P_2 cut-out + 101)

So, the pressure tank volume $Q = \dfrac{D}{(1 - P_1 / P_2)}$ 7.6

Where

D = pressure tank storage capacity (litre)

P_1 = minimum allowable operating gauge pressure (kPa) and

P_2 = maximum allowable operating gauge pressure (kPa).

The additional 101 kPa added to maximum and minimum operating pressures of fixtures is atmospheric pressure.

Hydro-pneumatic tank capacity can also be determined based upon withdrawal, in litres, of 2.5 times the capacity of the pump lps, and a low water level of not less than 10 percent of the total tank capacity or 75 mm above the top of the tank outlet, whichever is greater. Table 7.7 provides the values of high water levels and withdrawals for efficient operation of tanks with bottom outlets and a 10 percent residual, using which the tank capacity can be determined per the following expression.

Table 7.7 Hydro-pneumatic tank high water levels and withdrawal (based on bottom outlet tanks and a 10 percent residual) [17].

Sl. No	*Pressure range (kPa)*	*High water level withdrawal (% of total tank capacity)*	*Withdrawals (96% of total tank capacity)*
1	140–275	43	33
2	205–345	38	28
3	275–415	34	24
4	345–480	32	22
5	415–550	28	18
6	140–310	48	38
7	205–380	42	32
8	275–450	37	27
9	345–520	35	25
10	415–590	32	22

$$\text{Total tank capacity} = 2.50 \times \text{flow (lps)/withdrawal, in percent of tank capacity} \quad 7.7$$

Example: Determine the hydro-pneumatic tank capacity when pump capacity is 600 lps, operating under a pressure range of 275–415 kPa.

Referring to Table 7.7, the withdrawal from the tank is 24 percent of the tank capacity.

So, the total tank capacity = 2.50 × 600 lps/0.24 percent = 6250 litres.

There is a general rule of thumb for determining the drawdown capacity of a pressure tank based on the flow rate. Following are the rules of determining pressure tank capacity in litres for a range of pump flow rates.

1 For flow rates between 0 and 38 lpm, multiply flow rate by 1 to find capacity in litres i.e for 25 lpm, flow rate pressure tank drawdown capacity is 25 litres.
2 For flow rates between 38.1 and 76 lpm, multiply flow rate by 1.5 to find capacity in litres i.e for 50 litres, flow rate pressure tank drawdown capacity is 75 litres.
3 For flow rates larger than 76 lpm, multiply flow rate by 2 to find capacity in litres i.e for 100 lpm, flow rate pressure tank drawdown capacity is 200 litres.

7.9.2 Air compressor

In a hydro-pneumatic tank system, an air compressor is employed to fill air and compress it to build up pressure in the tank whenever it is drained and needed to be refilled. During operation, replenishment of absorbed air by water is in some cases also done by this air compressor.

Sizing air compressor: The size of the air compressor is determined considering the size of the hydo-pneumatic tank. Satisfactory operation has been attained by providing 0.0425 cum/min (2.55 cum/h) for tank capacities up to 1,893 litres and 0.0566 cum/min (3.4 cum/h) for capacities from 1,893 to 11,355 litres. For each additional 11,355 litres or fraction thereof, 0.0566 cum/min shall have to be added. (Quantities are expressed in cubic meter per minute free air at pressure equal to the high pressure maintained within the hydro-pneumatic tank) [17].

Table 7.8 Factors for sizing flow capacity of fill pump [17].

Administration building		*Apartments*		*Hospitals*	
Number of fixtures	*Minimum pump capacity (L/s)*	*Number of fixtures*	*Minimum pump capacity (L/s)*	*Number of fixtures*	*Minimum pump capacity (L/s)*
1–25	1.5	1–25	0.6	1–50	1.6
26–50	2.2	26–50	0.9	51–100	3.5
51–100	3.2	51–100	1.9	101–200	5.4
101–150	4.7	101–200	2.5	201–400	7.9
151–250	6.3	201–400	4.1	401–up	13.0
251–500	7.8	401–800	7.6		
501–750	15.0	801–up	13.0		
751–1000	17.0				
1001-up	20.0				

Industrial buildings		*Quarters and barracks*		*Schools*	
Number of fixtures	*Minimum pump capacity (L/s)*	*Number of fixtures*	*Minimum pump capacity (L/s)*	*Number of fixtures*	*Minimum pump capacity (L/s)*
1–25	1.6	1–50	1.6	1–10	0.06
26–50	2.5	51–100	2.2	11–25	0.9
51–100	3.8	101–200	3.8	26–50	1.9
101–150	5.0	201–400	6.3	51–100	2.8
151–250	7.0	401–800	9.5	101–200	4.1
251-up	10.5	801–1200	14.5	201–up	7.0
		1201–up	19.0		

7.9.3 Fill pump

In a hydro-pneumatic pressure tank system, a pump is needed to run when the hydro-pneumatic tank needs to be refilled by water and to supply the distribution system. So, this pump is also termed a fill-pump. This is basically a centrifugal pump which draws water from a water reservoir or from other sources. This pump starts running when the pressure in the tank reaches the minimum level and stops when the pressure in the tank rises to its maximum level. Usually, a pressure differential of 138 kPa is made between the pump start and the pump stop [18]. The pump capacity is to be large enough to refill the tank very fast.

Sizing of fill-pump: The fill-pump is named such when it is used for supplying water by employing a hydro-pneumatic tank pressure boosting system. The pump used, basically a centrifugal pump, is mostly sized to meet the peak demand of water flow in the distribution piping. The flow capacity is determined by multiplying the number of fixtures, not fixture unit, to be served in the building by the factors given in Table 7.8.

7.10 Booster pumping

Booster pumping is generally used to increase head in water supply systems in buildings and may be termed pressure boosting. In some cases, boosting of flow rate is also done, which is termed flow boosting. Booster pumping may be a pump or a set of pumps that draw water directly from the water supply main or from a water storage tank and then supply through the piping network, exerting pressure on water. When a booster pump is to be installed to draw water directly from the water supply mains, the water supply in the main must be ensured as adequate to meet the peak demand of the concerned building. Otherwise, there would be high risk of backflow and subsequent contamination of the main's water from buildings, which are not equipped with proper backflow preventers.

Booster pump arrangement: In booster pumping, one or more booster pumps are used, depending upon the volume of water being handled and the extent of the pressure boosting. It is advisable to use at least a set of two booster pumps for economy in operation. In larger projects, a set of three booster pumps is also used. The arrangement of booster pumps differs for different causes of boosting, as discussed below.

7.10.1 Flow boosting

In buildings, flow boosting might be needed when there would be an increase in demand for water for various obvious reasons and where there is wide variation in demand, such as in residential buildings, assembly buildings etc. For flow boosting, i.e. increasing the flow, multiple booster pumps possessing varying or same flow capacity at the same head are installed in parallel, as shown in Figure 7.8. In a parallel arrangement of pumps, the delivery pipes of the pumps are connected to get the cumulative flow capacity of the pump set.

In case of widely varied demand of water flow requirements, a parallel setting of pumps of varying flow capacity will be very effective. In operation, one of two different scenarios can be adopted; either one pump will run at its full capacity while the other adds flow when needed, or both pumps will run at varying speeds and adjust according to the flow requirement.

Flow boosting pumps setting: In the majority of cases, for economy and efficient pumping, a set of multiple booster pumps are installed instead of one booster pump. In general, sets of three booster pumps are more desirable than sets of two booster pumps.

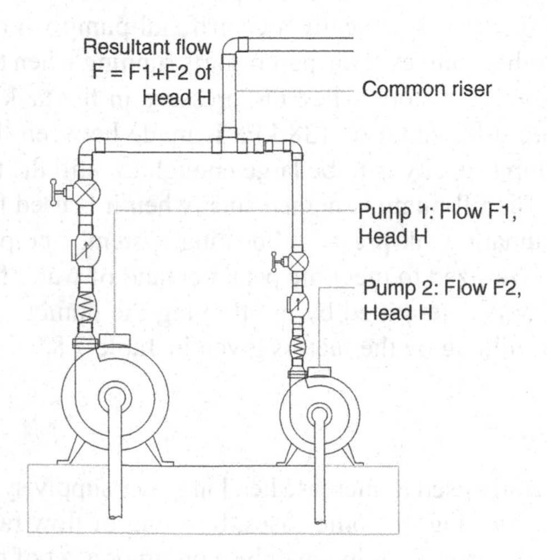

Figure 7.8 Parallel pumps setting for flow boosting.

For water supply at a constant flow rate, at least two booster pumps can be used for economy. Each pump shall be capable of supplying at least 50 percent of the peak water flow demand. But, for reliability, it is wise to install two pumps, each capable of supplying about 60 percent of the peak water flow demand [19]. To ensure uninterrupted supply of water, even during maintenance and repair of pumps, installation of three pumps set is essential, each of which shall be capable of supplying 50 percent of the peak water flow demand, of which two pumps will run in peak demand of flow while one will remain as a standby pump.

For water supply at variable flow rates to meet variable water demand, a set of multiple booster pumps of varied flow capacity is preferred.

7.10.2 Pressure boosting

Pressure boosting in buildings might be needed particularly when the building size is increased either vertically or horizontally. For pressure boosting i.e. increasing head of pumping, two or more pumps having the same flow capacity at a different or same head are installed in series, as shown in Figure 7.9. In a serial arrangement, pumps of the same flow capacity are installed in series to get the cumulative head of the all pumps installed.

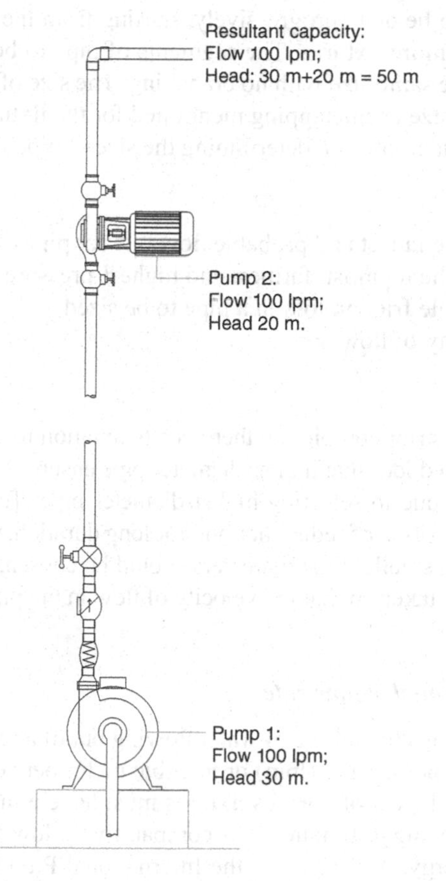

Figure 7.9 Pressure boosting by installing pump in series.

7.11 Water supply pipe sizing

The main objective of developing a water supply system in a building is to ensure uninterrupted availability of desired quality of water at the optimum flow rate where there is need of it. To achieve the objective of getting optimum flow rate, the basic job is to plan a water distribution piping layout starting from the source to every point of use and then to determine the appropriate size of piping for conveying water. The other important objectives are also achieved through designing the water supply piping, as follows.

1 Eliminating over-design, thereby saving resource and energy
2 Avoiding accumulation of sediments due to reduced velocity for over-pipe sizing
3 Avoiding extra load of oversized pipe on supporting structure
4 Saving cost, time of installation and space for accommodation by not oversizing of pipe and
5 Eliminating erosion, sound and water-hammering due to high velocity for under-pipe sizing.

The layout of water distribution piping is basically a piping network arrangement of horizontal and vertical pipes of limited height and length. Design of all segments of horizontal and vertical piping are to be done progressively, starting from the branch pipes supplying a group of at least two or more fixtures. The segments of pipe to be sized are that portion of the pipe network with the same size with no branching. The size of a pipe supplying a fixture is provided as the same size as the tapping mentioned for the fixture to be served. There are five factors that are considered when determining the size of pipe for water supplying. These are as follows.

1 Flow rate for a single faucet and probable flow rate for pipes supplying multiple faucets
2 Pressure needed in the topmost, farthest and highest-pressure needing faucet
3 Friction and allowable friction loss in a pipe to be sized
4 Limitation of velocity of flow and
5 Available pressure.

While sizing the water distribution piping, there is a temptation to choose the higher diameter pipes due to a preconceived idea that higher-diameter pipe ensures better flow. On the contrary, flow velocity is reduced due to selecting higher-diameter pipe. Too much reduction of flow velocity might cause deposition of sediments, and for long duration, water quality may degrade. In selecting the pipe size, smaller pipe diameters should be chosen, as they are comparatively cheaper, but care must be taken so that the velocity of flow in the pipe does not cross the limit.

7.11.1 Flow rate and probable flow rate

The flow rate of a faucet is the volume of water flowing out of a faucet in a given time. Flow rate is expressed in litre per minute (lpm) or in cubic meter per second (cusec). For the satisfaction of users, every faucet of various fixtures must have a minimum flow rate. In high water-scarce areas, code suggests faucets of comparatively low flow rates for the purpose of saving water and energy. According to the International Plumbing Code, the acceptable maximum flow rate of faucets in various fixtures is furnished in Table 7.9.

Table 7.9 Acceptable maximum flow rate of faucets in various fixtures. Converted to SI unit from source [20].

Sl. No	*Plumbing fixture*	*Maximum flow rates*
1	Lavatory faucet, private	5.68 lpm at 414 kPa
2	Lavatory faucets, public (metering)	1 litre per metering cycle
3	Lavatory faucets, public (other than metering)	1.9 litre at 414 kPa
4	Shower head (includes handheld)	7.6 lpm at 550 kPa
5	Sink faucet	6.8 lpm at 414 kPa
6	Urinal	1.9 litre per flush
7	Water closet	4.8 litre per flushing cycle, with minimum MaP (Maximum Performance of solid-waste removal) of 350 grams
8	Pre-rinse spray valves (food service industry)	Must meet federal requirements – Department of Energy (DOE) Energy Policy Act
9	Bar sinks (food service industry)	8.3 lpm 414 kPa

Water supply pipes may supply more than one faucet. The likelihood of running all the faucets installed on a section of pipe or piping system of a building at a time is quite unusual. It is observed that with the increase in number of faucets to be supplied, the number of faucets running simultaneously at a time decreases i.e. flow decreases. This estimated decreased flow rate is termed the probable flow rate for the total number of faucets concerned. Considering these facts, in calculating the probable flow rate for a given number of fixtures, the water supply fixture unit (WSFU) is devised.

The water supply fixture unit (WSFU) is a factor so chosen that load-producing effects of different kinds of plumbing fixtures and their conditions of service can be expressed approximately as a multiple of that factor. WSFU are dimensionless numbers assigned for specific fixtures relating to water flow rate needed, duration of discharge while using the fixture and the time interval between usage. The water supply fixture unit for any plumbing fixture is also a measure of the probable demand of water by that particular type of plumbing fixture, which must be ensured. Table 7.10 gives the value of these units for different fixtures. For fixtures having both hot and cold water supplies, the values for separate hot and cold water demands may be taken as three-fourths of the total value assigned to the fixtures in each case. These units have no direct mathematical relation to the rate of flow and depend on the following factors.

The maximum probable flow rate is considered to be the flow that will occur in the piping under peak conditions. It is not the total combined flow with all fixtures running fully open, at the same time, but an expected flow which is supposed to be proportional to the number of fixtures that may be expected to be running simultaneously. It is also termed as peak demand or maximum expected flow. The water supply piping is designed considering maximum probable flow rate in the pipe concerned. The estimated demand load or probable flow rate in litres per minute (lpm) for fixtures running intermittently on any water supply pipe, corresponding to the total number of water supply fixture units (WSFU), is given in Figure 7.10.

Table 7.10 Value of water supply fixture units for different fixtures [7, 21].

Sl. No	*Fixtures*		*Fixture unit*					
			Private individual dwelling)			*Public (general use)*		
			Cold	*Hot*	*Total*	*Cold*	*Hot*	*Total*
1	Ablution tap		0.75	0.75	1	1.5	1.5	2
2	Wash basin		0.75	0.75	1	1.5	1.5	2
3	Shower stall		1.5	1.5	2	3	3	4
4	Bathtubs or combination bath/shower		1.5	1.5	2	3	3	4
5	Sinks	a Kitchen	1.5	1.5	2	–	–	–
		b Service	–	–	–	2.25	2.25	3
6	Water closet	a Flush tank	3	–	3	5	–	5
	a	b Flush valve	6	–	6	10	–	10
7	Urinals	a Flush tank	–	–	–	3	–	3
	a	b 20 mm flush valve	–	–	–	5	–	5
8	Laundry machine.	a 3.6 kg	1.5	1.5	2	–	–	–
		b 7.3 kg	–	–	–	3	3	4
9	Drinking fountain		0.25	–	0.25	0.25	–	0.25
10	Dishwashing machine		–	1	1	–	–	–

7.11.2 Pressure and pressure loss

Pressure exhibits energy at one end of pipe to cause the flow of water along a pipe, towards the other end of the pipe. While flowing, water encounters resistance from the inner surface of the pipe in contact with water. Thereby, energy is lost as the water flows along the pipe, resulting in a fall in pressure. There occurs a pressure difference between the start and end points of the pipe. In a longer pipe flow more energy is lost, so, a greater pressure drop occurs. The rate of pressure drop i.e. the pressure drop per unit length of pipe depends upon the diameter of pipe and the velocity of flow.

The pressure difference ΔP in a flow between the ends of a pipe can be obtained using the Darcy-Weisbach equation as given below.

$$\Delta P = f \times v^2 \times \frac{\rho}{2} \times \frac{L}{D} \qquad 7.8$$

Where
v = flow velocity; v depends on the followings
L = length of the pipe
D = diameter of the pipe
ρ = density of the water; about 1,000 kg/cum for cold water and
f = friction factor

This equation is often expressed in another form addressing head or specific energy, as in equation 7.8.

Available pressure: Available pressure is the pressure found at a particular section of a pipe. Available pressure is found comparatively less than the pressure available at the start point of the flow of water of the concerned pipe.

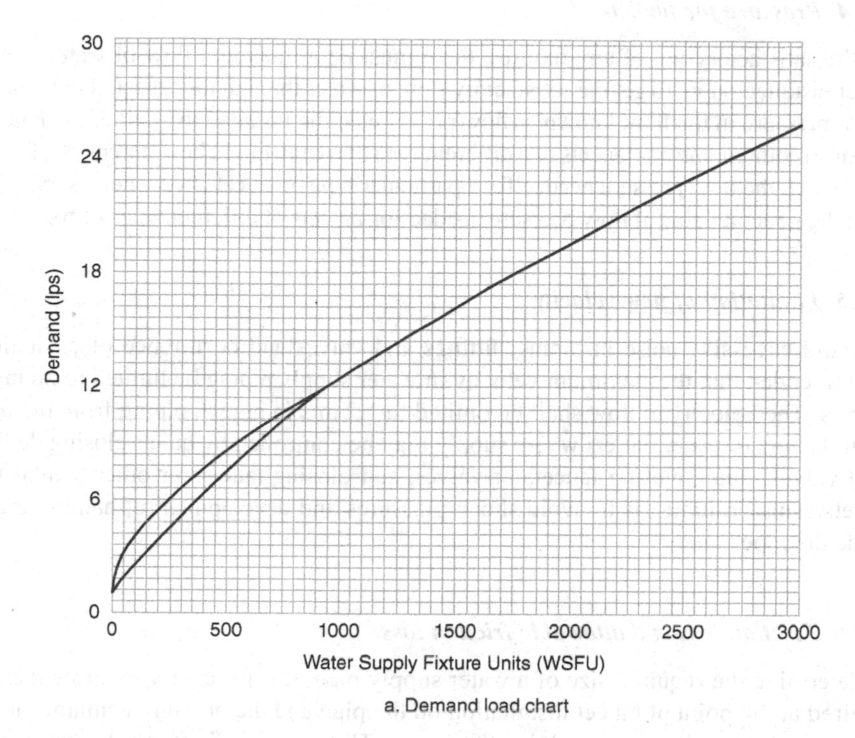

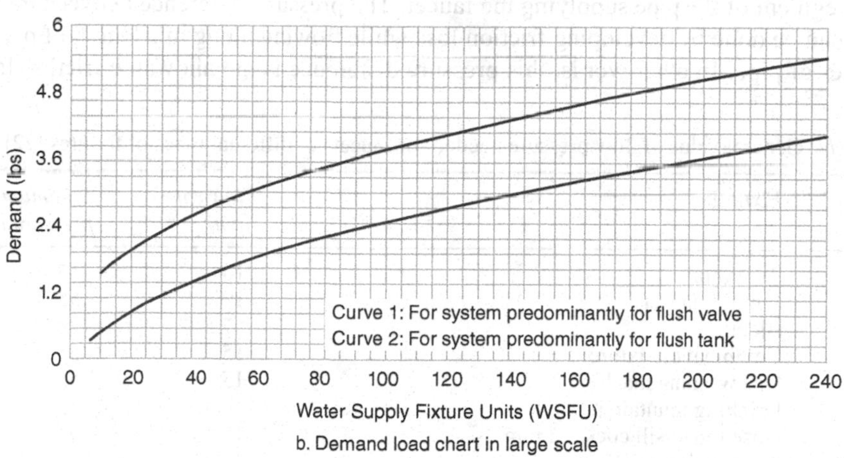

Figure 7.10 Charts for converting fixture units to water supply demand in litres per minute (lpm) for either a flush tank water closet system or flushometer valve system.

7.11.3 Pressure for distribution piping

Water pressure in distribution piping should be comparatively high in comparison to the faucet pressure needed at the end of distribution piping. Generally, the normal working pressure to be maintained in distribution piping is between 240 and 414 kPa. In no case should the minimum pressure in the distribution piping, under all flow conditions, be less than 140 kPa.

7.11.4 Pressure for faucets

For the satisfactory use of any fixtures, there must be a minimum flow of water through the faucet installed on it. To get the satisfactory flow of water through the faucet, there must be sufficient pressure in the flow, known as flow pressure, at the installation point of the faucet. Flow pressure requirements for faucets of a different type of fixture might be different. So, for accurate pipe sizing, the flow pressure needed for a particular type of faucet is very necessary. Table 7.11 gives the general value of flow pressure needed for faucets of different types of fixtures.

7.11.5 Limitation of flow velocity

To avoid excessive noise in piping, fittings and valves and occurrence of cavitation, it is recommended that the maximum velocity in water supply piping be limited to no more than 2.4 m/s. The velocity of flow shall be limited to 1.2 m/s in branch piping from mains, headers and risers to water outlets where supply may be controlled by quick-closing devices like flush valves, quick closing faucets or valves, self-closing faucets or other similar types of faucets. This limitation is to avoid shock pressures and development of hammering effects inside the pipe.

7.11.6 Friction loss and allowable friction loss

To determine the required size of a water supply pipe, it is necessary to know the pressure required at the point of faucet installation on the pipe and the pressure available at the start of the segment of the pipe supplying the faucet. The pressure difference between these two points can be done by developing friction loss while flowing along this length of particular diameter of pipe. In other words, this pressure difference is the allowable friction loss for

Table 7.11 General value of flow pressure needed for faucets of different types of fixtures [22].

Sl. No	*Fixture*	*Flow rate (lpm)*	*Minimum supply pressure (kPa)*
1	Aspirator	10	55
2	Bathtub faucet	19	55
3	Bidet	7.5	28
4	Combination fixture	15	55
5	Dishwashing machine	15	55
6	Drinking fountain jet	3	55
7	Hose bib or sill cock, 12 mm	19	55
8	Laundry faucet, 12 mm	19	55
9	Laundry machine	15	55
10	Lavatory faucet, ordinary	7.5	55
11	Lavatory faucet, self-closing	10	55
12	Shower head	19	55
13	Shower, temperature controlled	10	138
14	Sink faucet, 10 mm, 12 mm	17	55
15	Sink faucet, 20 mm	23	55
16	Urinal flush valve	56	110
17	Water closet with flush valve	132	170
18	Water closet with gravity tank	10	55
19	Water closet with close coupled tank, ball cock	11	55

the flow of water through this length of a particular diameter of pipe, so that the available pressure at the start of the concerned segment of pipe can be reduced to the pressure needed at the point of fixture installation for the concerned faucet, by allowing the friction loss equivalent to the existing pressure difference. The more the pressure difference, the more friction loss can be allowed by choosing a more reduced diameter of pipe. But care must be taken in doing so that velocity of flow does not exceed the maximum limiting velocity of 2.4 m/s.

Friction loss in pipes is a function of roughness of the inside surface, length and diameter of pipe and flow velocity. The friction loss due to flow of water through pipe, which is also expressed as head loss, is expressed by the formula known as the Hazen–Williams formula.

$$\Delta h = f \times \frac{L}{D} \times \frac{v^2}{2g} \qquad 7.9$$

Where
Δh = head loss in m
f = friction factor related to the inside roughness of the pipe
L = length of the pipe
D = diameter of the pipe
v = average flow velocity in the pipe and
g = acceleration due to gravity 9.81 m/sec^2.

A Hazen–Williams table has been developed based on the head loss in ten-year-old steel pipes. For different pipe ages and materials, the values need to be adjusted. The adjustment is done through the use of a correction factor, "C". The "C" value of a ten-year-old steel pipe is considered 100, or a multiplier of 1.0 is the value on which the table is based. New and smooth steel pipes have a "C" value above 100 or a multiplier below 1.0, which indicates lower head loss. Older and rough steel pipes have a "C" value below 100 and a multiplier above 1.0, so, the head loss would be higher. Different tables have been made based on testing ten-year-old steel pipes of different pipe sizes. Each of these tables has values for velocity in m/sec, velocity head and head loss per 100 m segment of pipe.

7.11.7 Friction loss for fittings and valves

Pipe fittings and valves cause extra resistance to the flow of water, resulting in additional head loss. There are two methods of determining head loss due to flowing through fittings and valves etc. They are known as the "K factor" and the "equivalent length of pipe" methods.

In the "K" factor method, fittings such as elbows, tees, strainers, valves etc. are assigned a "K" factor for particular diameters of pipe, based on the head loss measured through them. The velocity head is adjusted by multiplying by "K" factor

In the "equivalent length of pipe" method, the pipe fittings and valves of various diameters are assigned an "equivalent length of pipe in linear feet" value. The values indicate that a particular fitting of particular size will cause a head loss equal to head loss caused by that length of straight pipe of same size. For all the fittings and valves the equivalent length values are added to the actual length of pipe to get the total effective length of pipe to determine the total head loss. In Table 7.12 equivalent lengths of pipe in meters for various fittings and valves are given.

Table 7.12 Allowance in equivalent feet of pipe for friction loss in threaded fittings and valves. Converted in SI unit from source [23].

Diameter of fitting (mm)	Equivalent length of pipe in m for various fittings and valves						
	90°elbow	45°elbow	90°tee	Straight run/coupling	Gate valve	Globe valve	Angle valve
10	0.305	0.183	0.457	0.091	0.061	2.44	1.22
12	0.610	0.366	0.914	0.183	0.122	4.57	2.44
20	0.762	0.457	1.22	0.244	0.152	6.10	3.66
25	0.914	0.549	1.52	0.274	0.183	7.62	4.57
32	1.219	0.732	1.83	0.366	0.244	10.67	5.49
40	1.52	0.914	2.13	0.457	0.305	13.72	6.71
50	2.13	1.22	3.05	0.61	0.396	16.76	8.53
63	2.44	1.52	3.66	0.762	0.488	19.81	10.36
75	3.05	1.83	4.572	0.914	0.61	24.38	12.19
100	4.27	2.44	6.40	1.219	0.823	38.1	16.76
150	6.10	3.66	9.14	1.83	1.22	50.29	24.38

Table 7.13 Minimum sizes of fixture water supply pipes [24].

Sl. No	Fixture	Minimum pipe size (mm)
1	Bathtubs (1500 × 800 and smaller)	12
2	Bathtubs (larger than 1500 × 800)	12
3	Bidet	10
4	Combination sink and tray	12
5	Dishwasher, domestic	12
6	Drinking fountain	10
7	Hose bibbs	12
8	Kitchen sink	12
9	Laundry, 1, 2 or 3 compartments	12
10	Lavatory	10
11	Shower, single head	12
12	Sinks, flushing rim	20
13	Sinks, service	12
14	Urinal, flush tank	12
15	Urinal, flushometer valve	20
16	Wall hydrant	12
17	Water closet, flush tank	10
18	Water closet, flushometer tank	10
19	Water closet, flushometer valve	25
20	Water closet, one piece	12

7.12 Water supply pipe sizing methods

The minimum size of a fixture supply pipe is generally fixed by various plumbing codes. According to the International Plumbing Code 2015, the minimum size of individual fixture distribution lines are as shown in Table 7.13.

Pipe sizing for water supply distribution piping can be approached in three methods as mentioned below.

1 Empirical design method
2 Velocity design method and
3 Pressure loss method.

7.12.1 Empirical design method

In this method, piping is sized using rules of thumb based on equivalent cross-section of pipe served. This method may be applied in finding sizes of pipes of very simple and small water distribution networks very quickly. The main pipe, which serves a number of branch pipes of particular size – 12 mm in general – can be sized from the values furnished in Table 7.14.

From Table 7.10, it is further found that a main pipe that serves a number of branch pipes of particular size can be sized as follows.

1 Two 12 mm branch pipes should be served by a 20 mm main
2 Maximum six 12 mm or three 20 mm branch pipes can be served by a 25 mm main and
3 Maximum ten 12 mm or five 20 mm branch pipes can be served by a 32 mm main.

7.12.2 Velocity design method

Velocity of water flow in a pipe is a function of pipe size for a particular flow rate in the pipe. This method is the simplest way of determining the pipe size, particularly for preliminary design. In low-rise buildings of two or three storeys where sufficient pressure is not available, particularly in gravity flow systems, this method is suitably applied.

In this method, the total water supply fixture units (WSFU) to be served by a pipe to be designed, is to be determined. The total WSFU is converted into corresponding design flow load in lpm. The corresponding pipe diameter is found based on velocity falling within the range of acceptable flow velocity. The relation of pipe size with respect to flow rate based on velocity can be expressed as follows.

$$D = \sqrt{1.274Q/v} \quad (m) \tag{7.10}$$

Where
D = the internal diameter of pipe (m)
Q = the flow rate (cum/s) and
v = the velocity of flow (m/s)

Table 7.14 The number of 12 mm pipe discharging as much as a single pipe of particular size for the same pressure loss.

Sl. No	*Size of pipe (mm)*	*Number of 12 mm pipe with same capacity*
1	12	1
2	16	1.7
3	20	2.9
4	25	6.2
5	32	10.9
6	40	17.4
7	50	37.8
8	63	65.5
9	75	110.5
10	100	189
11	150	527
12	200	1200
13	250	2090

Table 7.15 Pipe sizing table based on velocity limitation; Galvanized Iron and steel pipe, standard pipe size [21].

Nominal size (mm)	*For velocity 1.2 m/sec.*				*For velocity 2.4 m/sec.*			
	Flow (lps)	*Load (wsfu)* ≠	*Load (wsfu)* *	*Friction (Pa/m)*	*Flow (lps)*	*Load (wsfu)* ≠	*Load (wsfu)* *	*Friction (Pa/m)*
12.70	0.23	1.5	–	172.3	0.47	3.7	–	651.5
19.00	0.42	3.0	–	126.1	0.84	8.4	–	472.8
25.40	0.68	6.1	–	96.7 9	1.36	25.3	7.7	361.5
31.80	1.17	17.5	6.0	71.5	2.34	77.3	23.7	269.0
38.10	1.60	37.0	9.3	60.9	3.20	132.3	52.0	227.0
50.80	2.63	93.0	29.8	46.2	5.27	293.0	171.6	176.5
63.50	3.77	174.0	75.6	37.8	7.54	477.0	361.0	142.9
76.20	5.80	335.0	209.0	29.4	11.60	842.0	806.0	113.5
102.0	10	688.0	615.0	23.1	20.01	1930.0	1930.0	86.2

(≠) Applying to piping which does not supply flush valve.

(*) Applying to piping which supplies flush valve.

(‡) Friction loss P corresponding to the flow rate Q, for piping having fairly smooth surface condition after extended service, from the formula.

$$q = 4.57\,p^{0.546}\,d^{2.64} \quad 7.11$$

Table 7.15 is furnished to determine the pipe size based on limiting velocity of flow. The velocity of flow in branch water supply piping shall be no more than 1.8 mps.

Pump delivery and suction pipe sizing

Example:

Let us find the size (diameter) of delivery pipe having a flow requirement of 10 lps at a minimum velocity of 1.2 m/s.

Flow Q = AV
So, area of pipe A = Q/V = $\pi \times D^2/4$
Then, diameter of pipe D = $\sqrt{4Q/\pi V}$
The area of pipe needed = 10/1200 = 0.00833 sqm. = 8333.33 sqmm.
So, diameter of pipe will be $\sqrt{8333.33 \times 4/\pi} = \sqrt{10{,}610.33} = 103$ mm.

It is to be noted that the commercially available pipe size is closest to the required pipe diameter, which is 100 mm. So, if a 100 mm delivery pipe is used, the flow will be a bit less than 10 lps at a velocity of 1.2 m/sec, as shown in Table 7.15.

The size of the suction pipe should be one size bigger than that of the delivery pipe size or at least the same as the size of the delivery pipe. In this example, a 100 mm diameter suction pipe will be a good selection.

Branch pipe sizing

Branch water supply pipes can be sized easily and quickly by considering the particular velocity of flow in the pipe. In determining the size of a branch pipe supplying multiple fixtures and faucets based on velocity limitation of water flow, it is necessary to know the flow load in the corresponding pipe, from node to node of the branch pipes, expressed in terms of the number of water supply fixture units (WSFU) to be served. The minimum size of pipes to supply faucets of various fixtures is more or less fixed, so, these do not need to be designed.

Example: Consider a toilet with fixtures: one basin, one water closet supported by a bib tap and one shower head. For supplying water to these fixtures and faucets, piping is laid down, as shown in Figure 7.11. All the branch pipes have to be sized.

Pipes AA', AA", and BB" are end pipes supplying the respective fixtures or faucets, so these pipes are to be sized according to the minimum pipe size required for the respective fixtures or faucets.

The cumulative fixture unit for segment BA is 3 (FU of AA': 1 + FU of AA": 2).

From Table 7.11, for velocity 1.2 m/sec, a 20 mm pipe can bear a load of 3 FU, so a 20 mm diameter of BA can be chosen.

The cumulative fixture unit for segment BB' is 4 (FU of BB": 1 + FU of water closet: 3).

From Table 7.11, for velocity 1.2 m/sec, a 20 mm pipe can bear a load of 3 FU, and if velocity increases to 2.4 m/sec, a 20 mm pipe can accommodate 8.4 FU. So, a 20 mm diameter of BB' can be chosen, for which a bit more velocity than 1.2 m/sec is needed for adequate flow.

The cumulative fixture unit for segment CB is theoretically 7 (FU of BA is 3 + FU of BB' is 4); but for a toilet battery, the maximum FU can be considered 6. From Table 7.11, for velocity 1.2 m/sec, a 25 mm pipe can bear a load of 6.1 FU, so, a 25 mm diameter of BA can be chosen; a 20 mm can be chosen if velocity is raised to 2.4 m/sec or less (about 1.8 m/sec) to accommodate 6 FU.

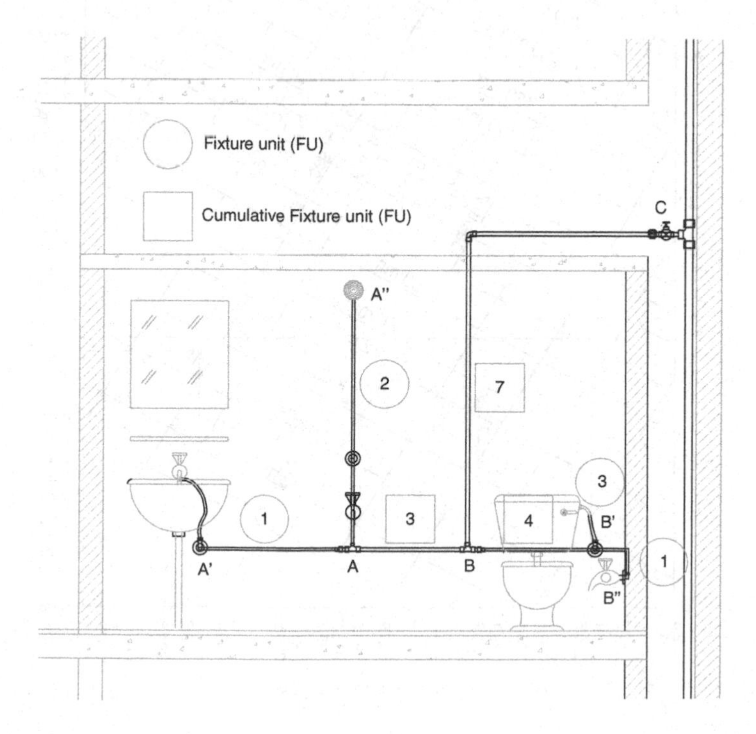

Figure 7.11 Water supply piping in a toilet.

7.12.3 Pressure loss method

The pressure loss method can be applied in determining pipe sizes in both down-feed gravitational systems and up-feed pressurized systems. In this method, pipe size is determined based on pressure available at the start of a pipe segment to be sized and pressure needed at the end (generally faucet pressure or pressure at outlet of a branch pipe). The difference of pressure can be lost by choosing a section for the equivalent length of that segment of pipe. As a rule of thumb, the equivalent length of the pipe segment to be sized is 50 percent more than the actual length i.e. 1.5 times the actual length. Corresponding to the water flow in lpm in the segment of pipe, the pipe size is determined by applying the allowable friction loss per 100 m of equivalent length of pipe to the rate of flow, as shown in Figure 7.12.

Down-feed system: In this system, an overhead water tank is used. The elevation of the tank produces the pressure to flow. The required pressure of the topmost faucet on upper storeys is the governing factor in determining the pipe size, particularly for the riser pipe of the top floor. When the required pressure for the topmost faucet is more than the pressure developed for the height of the tank, then the size can be determined by considering minimum flow velocity, as discussed earlier. However, the size cannot be less than the size of the riser found for the immediate lower floor. In this situation, faucets activating at low pressure are to be chosen for the top floors needed.

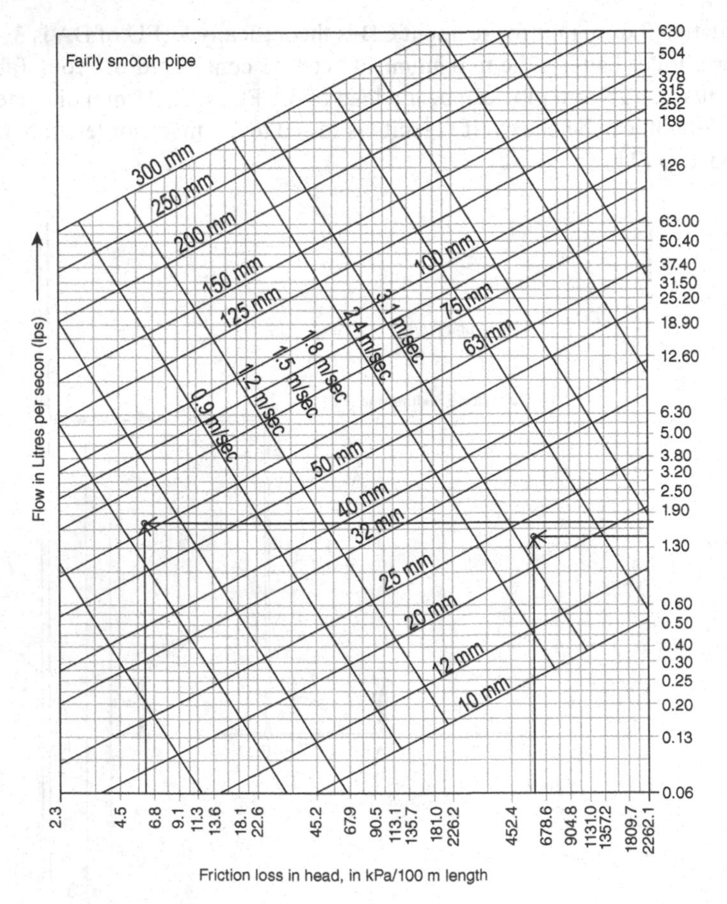

Figure 7.12 Flow chart for fairly rough pipe.

Example: Consider a six-storey residential building in which a water supply riser pipe is to be designed for supplying branch pipes for a toilet, as shown in Figure 7.13. Floor to floor height is 3 m. An overhead tank is placed 10 m away from the riser location, 4.2 m above the topmost fixture, a shower head, and 6.13 m above the flush tank for the water closet. Table 7.16 furnishes all the data obtained and the required values determined.

On the fifth floor, the permissible pressure drop in the pipe from the tank to the topmost faucet requiring pressure 34.5 kPa is $p = 4.2 \times 9.6 - 34.5 - 5 = 0.82$ kPa, considering pressure loss in the branch pipe is 5 kPa. If the faucet pressure is 55 kPa, then there will no permissible pressure drop, as the value for p will be negative. So, a fixture with such a pressure requirement should not be used. In this case of negative value, an arbitrary value of pressure loss of, say, 220 kPa pressure loss per 100 m pipe, can be chosen [25], and accordingly pipe size can be determined.

From Figure 7.12, for allowable pressure drop per 100 m length of pipe $= 0.82 \times 100 / 13.72 = 5.98$ kPa and flow rate 1.74 lps, the corresponding size is a bit higher than 63 mm, say 63 mm, which will cause velocity of flow less than 0.6 m/sec. It is to be noted that for velocity 1.2 m/sec, it is only 40 mm, and for 2.4 m/sec the corresponding pipe diameter is 32 mm. So, pipe size of 40 mm is considered to be within the velocity range. For choosing 40 mm pipe diameters, more friction loss will occur which may cause lower available pressure for the faucet. So, in this situation there will be a lesser flow rate from the faucet.

When the available pressure exceeds the required pressure, the additional gain in static pressure can be used to overcome the friction loss. Thus, in each lower floor, pressure gain is $3.048 \times 9.6 = 29.26$ kPa. So, the friction that can be lost in a 100 m pipe is $29.26 \times 100 / 4.57 = 640$ kPa. For this allowable pressure loss in the riser pipe, corresponding to a required flow rate of 1.52 lps,

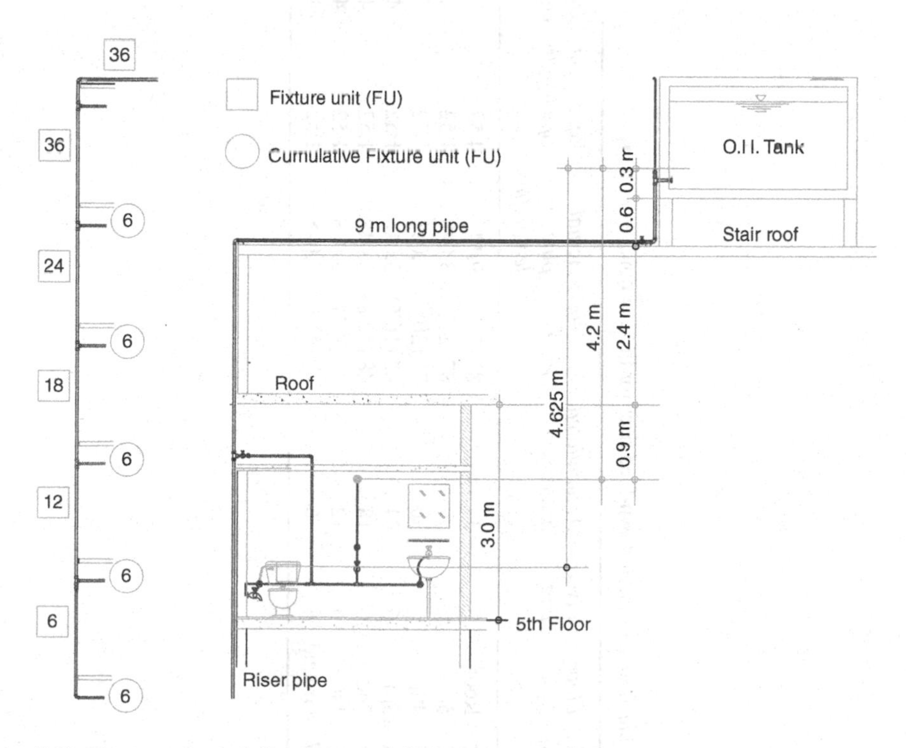

Figure 7.13 Fixture units and dimensions of riser pipe on top floor.

Table 7.16 Design of down-feed riser for the example.

Sl. No	*Floor*	*Fixture unit*	*Accumulated fixture unit*	*Flow lps (lpm)*	*Actual pipe length (m)*	*Equivalent pipe length (m)*	*Total equivalent length (m)*	*Pressure drop, kPa/100 m*	*Pipe size (mm)*	*Pipe size (mm) based on velocity = 2.4 m/sec*	*Pipe size (mm) based on velocity = 1.2 m/sec*
01	Roof		36	87 (23)	9.14	4.57	13.72				
02	5th	6	36	87 (23)	3.048	1.52	4.57	5.98	63	32	40
03	4th	6	30	76 (20)	3.048	1.52	4.57	29.26	25	25	40
04	3rd	6	24	62 (16.5)	3.048	1.52	4.57	29.26	25	25	32
05	2nd	6	18	48 (12.8)	3.048	1.52	4.57	29.26	25	25	32
06	1st	6	12	35 (9.2)	3.048	1.52	4.57	29.26	20	20	25
07	Ground	6	6	20 (5)	3.048	1.52	4.57	29.26	20	20	20

the required pipe size found is larger than 20 mm from the chart, so 25 mm is chosen. In this way, the other pipe sizes are chosen, as done before; the values are shown in Table 7.16.

Up-feed (pressurized) system: In an up-feed pressurized water supply system, the pressure required to overcome the gravity head in the riser pipe and the pressure desired at the highest fixture or the fixture requiring the highest pressure in any floor, whichever is greater, shall be subtracted from the pressure available at the pump delivery point. The difference in pressure will be the available pressure that can be lost through friction in the riser pipe section corresponding to the required flow rate in that riser section, found by consulting Figure 7.12, as done before. The allowable friction loss in pipe fittings, valves etc. is to be considered as the equivalent pipe length of those installed in the segment of the riser pipe to be designed, which is 50 percent more than the riser pipe.

7.13 Pressure management in sky scrapers

Pressure in a water supply system is developed due to the height of water column in pipes for conveying water from top to bottom levels under gravity or from bottom to top levels or for sending far away. It is already known that the development of pressure depends upon the water column and the distance to be served. So, pressure development is directly proportional to the height of the building and, so, critical in super or mega-tall buildings.

In a water supply system, development of maximum pressure is to be kept limited. The development of maximum pressure is limited to plumbing appliance and fittings, which is about 550 kPa. In some codes, maximum pressure is limited to 350 kPa [4]. The high pressure in a water supply system is managed by adopting one of the following methodologies.

1. Hydraulic pressure zoning of floors incorporating intermediate water tank or booster pump and
2. Installing pressure reducing valve (PRV).

A hydraulic pressure zone is defined as a pipe system for water supply, where at the connected faucet the required minimum pressure and maximum pressure or maximum static pressure are maintained.

By the above methods, a large pressure drop can be made, but for small or minor pressure drops, the following two strategies can be adopted.

1. Install globe valve to increase friction loss, thereby reducing pressure and
2. Use smaller diameter pipe to increase friction loss; but care must be taken so that velocity of flow does not cross the limit.

When the pressure in the distribution piping is lower than the desired pressure, it can be increased only by incorporating a booster pump on the pipe. The other way to avoid boosting pressure is to use faucets of low pressure. In gravity or down-feed water supply systems, low-pressure faucets of maximum 34.5 kPa can be used on the top floor just under the roof.

7.13.1 Hydraulic zoning

In skyscrapers or taller buildings, water supply by incorporating one giant pump would be dangerous, as might be done in high-rise buildings, considering the pressure it would generate. In skyscrapers and taller buildings, pumping of water is accomplished by incorporating

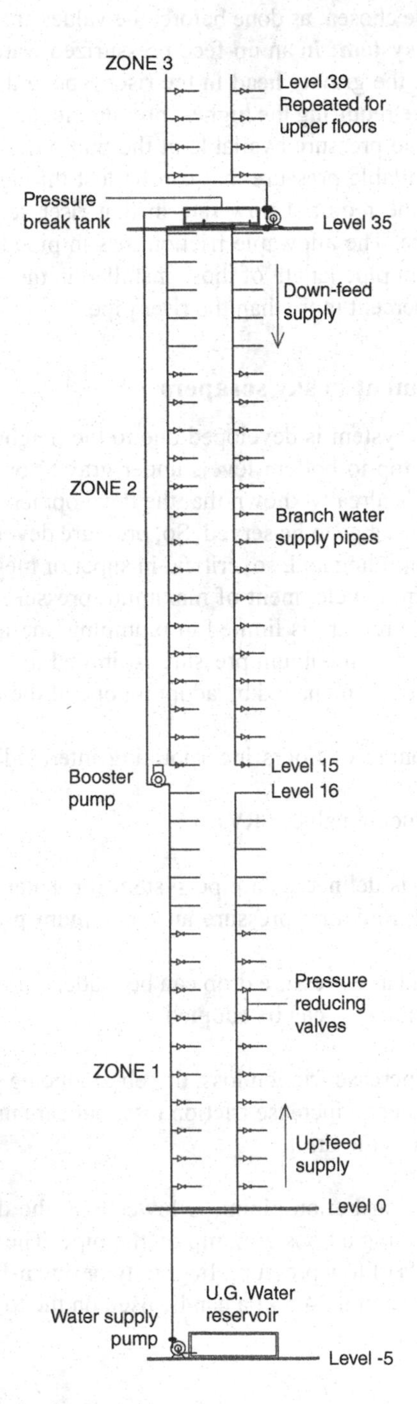

Figure 7.14 Water supply zoning in skyscraper.

a series of tanks at different levels throughout the building. These tanks are generally housed on mechanical floors in multiple storeys, staggered throughout the structure. The mechanical floors generally house electro-mechanical units for various other services as well. Water is lifted from an underground water reservoir at the lowest basement floor to a holding tank placed generally around 40 storeys above, from where water is supplied to lower floors under gravity. The first 20 storeys above the underground reservoir can be served by adopting the up-feed pressurized supply system with the pump used for lifting water to the holding tank above, as illustrated in Figure 7.14.

When the height of the building gets much higher, the pressure rating of the pipe, valves and fittings should be of higher pressure-rating capacity. These buildings might need to utilize 2068 kPa pressure rating fittings, equipments and accessories and multiple booster pumps could be used to lift water to the water tank.

Pressure management by zoning has some disadvantages as well. Advantages and disadvantages of zoning system for water pressure management are tabulated below.

Advantages

1 The required water supply pressure can be maintained
2 No space required for booster pumps on upper levels and
3 Less vulnerable in the event of pump failure.

Disadvantages

1 High pressure-graded pipes and booster sets are required and
2 Sensitive to electricity failure.

7.13.2 Pressure reducing valve

Pressure reducing valves (PRV) are primarily employed on the main pipe distributing water to the branches that lead to the faucets or fixtures to reduce excess pressure in the system, down the pressure needed and ensure faucet pressure at the pipe ends. Sometimes two or more PRVs are used in series to cause large pressure drops.

Pressure relief valves are also used to thwart (prevent) pressure surges and water hammer caused by stops and starts of large pumps in high-rise buildings. PRVs are equipped with anti-cavitation trim to prevent noise.

References

[1] World Health Organization (2011) *Guidelines for Drinking-Water Quality*, 4th Ed., WHO, Geneva.

[2] Wikipedia, "Drinking Water Quality Standards" Retrieved from https://en.wikipedia.org/wiki/Drinking_water_quality_standards.

[3] Housing and Building Research Institute (HBRI) (1993) *Bangladesh National Building Code 1993*, HBRI, Dhaka.

[4] Jain, V.K. (1985) *Handbook of Designing and Installation of Services in High Rise Building Complexes*, Jain Book Agency, New Delhi.

[5] Poultry Hub, "Water Consumption Rates for Chickens" *Poultry CRC (Chris R Clarkson)*, Retrieved from www.poultryhub.org/nutrition/nutrient-requirements/water-consumption-rates-for-chickens/ on 26 December 2020.

[6] International Carwash Association (ICA) (2002) "Water Use in the Professional Car Wash Industry" Report, p. 34, Retrieved from https://www.carwash.org/docs/default-document-library/Water-Use-in-the-Professional-Car-Wash-Industry.pdf on 22 May, 2021.

[7] Deolalikar, S.G. (1994) *Plumbing: Design and Practice*, Tata McGraw-Hill Publishing Company Ltd, New Delhi.

[8] Rahman, M., Quayyum, S., and Quayyum, S. (Undated) *Sustainable Water Supply in Dhaka City: Present and Future*, Bangladesh University of Engineering and Technology BUET, Dhaka.

[9] Haq, Syed Azizul (2005) "Rainwater Harvesting: Next Option as Source of Water" Proceeding of 3rd Annual Paper Meet and International Conference on Civil Engineering, IEB, Dhaka.

[10] Worm, J., and van Hattum, T. (2006) *Rainwater Harvesting for Domestic Use*, Agromisa Foundation and CTA, Wageningen, The Netherlands.

[11] Vali Ghorbanian, S.M., Karney, B., and Guo, Y. (2016) "Pressure Standards in Water Distribution Systems: Reflection on Current Practice with Consideration of Some Unresolved Issues" DOI: 10.1061/(ASCE)WR.1943–5452.0000665. © American Society of Civil Engineers.

[12] Uniform Plumbing Code (2015) "6 Water Supply and Distribution" 603 Water Service, 603.1 Size of Water Service Pipe, Retrieved from https://up.codes/s/size-of-water-service-pipe on 26 December 2020.

[13] International Plumbing Code (2012) "Chapter 6 Water Supply and Distribution" Retrieved from https://codes.iccsafe.org/content/IPC2012/chapter-6-water-supply-and-distribution on 26 December 2020.

[14] CEPT University (2012) "Basics of Water Supply System" *Training Module-1 for Local Water and Sanitation Management*, p. 9, Maharashtra Jeevan Pradhikaran (MJP), Gujarat, India, Retrieved from https://pas.org.in/Portal/document/ResourcesFiles/pdfs/Module_1%20Basics%20of%20water%20supply%20system.pdf on 26 December 2020.

[15] 10 State Standards (2007) "Recommended Standards for Water Works" ⟨http://10statesstandards.com/index.html⟩ Found at Ghorbanian et al. (2016) "Pressure Standards in Water Distribution Systems: Reflection on Current Practice with Consideration of Some Unresolved Issues" *American Society of Civil Engineers*, DOI: 10.1061/(ASCE)WR.1943–5452.0000665, Retrieved from http://hydratek.com/wp-content/uploads/2017/07/Ghorbanian-Karney-Guo-Pressure-Standards-in-Water-Distribution-Syste . . . . pdf on 26 December 2020.

[16] Department of Health (DoH) (2011) "Hydropneumatic Tank Control Systems" DOH 331–380, Washington State, Retrieved from www.doh.wa.gov/portals/1/Documents/pubs/331-380.pdf on 26 December 2020.

[17] United States Department of Defence (USDoD) (2019) "Unified Facilities Criteria, *Plumbing Systems*, Retrieved from www.wbdg.org/FFC/DOD/UFC/ufc_3_420_01_2004_c11.pdf on 1 January 2021.

[18] Cycle Stop Valves (2019) "Pressure Tank Maintenance" Retrieved from https://cyclestopvalves.com/pages/pressure-tank-maintenance.

[19] Guyer, J. Paul (2012) *Introduction to Pumping Stations for Water Supply Systems*, Continuing Education and Development, Inc, p. 24, Retrieved from https://sswm.info/sites/default/files/reference_attachments/GUYER%202012%20Introduction%20to%20Pumping%20Stations%20for%20Water%20Supply%20Systems.pdf on 26 December 2020.

[20] International Plumbing Code (IPC), "Sec. 604.4" Retrieved from www.fcgov.com/building/pdf/green-plumbing.pdf on 26 December 2020.

[21] Cyril, M. Harris (1991) "Practical Plumbing Engineering" McGraw-Hill company.

[22] Engineering Tool Box, (2004). Fixture Water Requirements. [online] Available at: https://www.engineeringtoolbox.com/fixture-water-capacity-d_755.html on 26 December, 2020

[23] National Bureau of Standards (1940) "Plumbing Manual" p. 41, and National Plumbing Code, Sec D.7.8.

[24] Utah Plumbing Code (2015) 604.5 "Size of Fixture Supply" Retrieved from https://up.codes/s/size-of-fixture-supply on 26 December 2020.

[25] McGuinness, William J., and Stein, Benjamin (1964) *Mechanical and Electrical Equipment for Buildings*, 4th Ed, John Willy and Sons Inc, New York, London, Sydney, p. 37.

8 Hot water supply systems

8.1 Introduction

Hot water is a condition of potable water, considered to be at a temperature equal to or greater than 43.33°C [1]. Hot water use is mostly preferred and desired when and where the atmospheric temperature prevails at about 24°C or below. So, in areas where atmospheric temperature around this range predominates throughout the year, the water supply system is generally developed exclusively as a hot water supply system. But in areas where temperatures above 25°C predominates during most of the year and for a short period temperature goes below 24°C, both hot and normal water supply systems are developed side by side, in conjunction with another. A hot water supply system involves some extra appliances and equipment, needs extra measures to be taken for safety and sanitation and, most importantly, needs a different approach in planning, design and installation in relation to normal water supply development systems. Again, in exclusive hot water supply and dual hot-and-normal water supply development, the approaches would be different. In this chapter, hot water supply system development in conjunction with normal water supply systems is elaborately discussed.

8.1.1 Hot water production

In hot water supply, hot water is produced in water heaters by heating the water available of normal temperature. The heating of water in heaters occurs mostly by the process of conduction and convection of heat transfer; however, by using a radiation process of heat transfer, water can also be heated.

Conduction is the process of heat transfer from the source of the heat to the water to be heated, which occurs due to vibration of the molecules of water that is heated and transferred through materials of the heating element or container. The transfer of heat by convection arises due to the physical movement of heated molecules within water. By radiation, heat is transferred through both air and space.

To produce hot water, several principles are employed.

1 Hot water production by direct heating in individual heaters near points of use
2 Indirect heating using heat exchanger from a distant location, connected to the heating system of heater
3 Central hot water is produced and distributed to all points of usage.

Due to heating, the properties of water change with the intensity of heat, which is measured by its temperature in either Centigrade or Fahrenheit. Due to increase in temperature, water expands, and at 100°C it becomes steam.

DOI: 10.1201/9781003172239-8

8.1.2 Hot water temperature for personal requirements

Hot water temperature requirements for various purposes of personal use differ widely. For personal hygiene and body cleansing, the temperature requirement may slightly vary from person to person. Hot water temperature requirements for various purposes of personal use are tabulated in Table 8.1.

8.1.3 Importance of hot water supply

The major factors governing the use of hot water are comfort, necessity and habit. Hot water is generally and mostly used for bathing, cleansing and washing purposes. High temperature causes quick melting of oil and grease and removes dirt easily from human skin, fabrics and utensils as well.

8.1.4 Challenges of hot water supply

Hot water supply systems are developed for health benefits and comfort of users, but its development approach needs utmost care due to facing various challenges. With a view to developing an efficient and safe hot water supply system, the following issues are to be well addressed.

1 Risk of hot water supply
2 Energy conservation and
3 Water efficiency.

8.2 Risk of hot water supply

Though hot water supply systems are developed for the health benefits and comfort of users, but its improper design and development may turn it into a hazard that risks three areas of human health and life, as mentioned below.

1 Scalding
2 Legionella infection and
3 Explosion of heater.

Both these areas of risk arise from the hazard of incorrect storage and delivery of hot water for use. Only by taking some effective measures these risks can be avoided or minimized. However, it is to be brought to mind that in many cases, control measures taken to reduce either one of these risks causes the potential increase of risk in the other.

Table 8.1 Hot water temperature requirement for different purposes of personal uses [2].

Sl. No	*Purpose of use*	*Temperature (°C)*
1	Hand-washing	40
2	Shaving	45
3	Showers	43
4	Therapeutic baths	35
5	Bidet applications	38

8.2.1 Scalding

Scalding is a form of burning or painful affecting of skin caused by heated fluid, generally hot water or steam. The temperatures at which various conditions of scalding can occur are more or less constant, but the degree of potential scalding depends on the flow rate of hot water and the contact time. Skin sensitivities also play a role in the effect of scalding. The risk of scalding increases rapidly with the increase of hot water temperature above 45°C. Table 8.2 shows the exposure time for human skin to cause major scalding from hot water of different temperature ranges.

The mixture of hot and cold water from a faucet may suddenly vary in temperature and rate of flow. This situation may arise due to undersized piping when water is turned on at another connected outlet. Failure to remedy such undersized piping may cause injury by scalding.

8.2.2 Legionella

This is one kind of naturally occurring organism that falls into the group of bacteria named *Legionella pneumophilia*, which may grow in water supply systems with favourable temperature conditions. The favourable temperature range for proliferation of Legionella bacteria is between 20°C and 45°C. At temperatures below 20°C and above 45°C, Legionella bacteria will not reproduce but will remain alive as a potential threat, growing subject to favourable temperatures, and will die at temperatures above 60°C. Legionella bacteria are present at low levels in almost all surface water sources. So, these bacteria is likelihood to be available in potable water supplies and grow well when temperature reaches the warm condition in nutrient rich hot water environment in heaters, pipe faucets fixtures etc. Scale, iron rust, biofilm, and the presence of other microorganisms provide nutrients for the growth of Legionella bacteria.

The risk associated with the exposure to *Legionella pneumophilia* growing in hot water supply systems is Legionnaire's disease, due to inhaling contaminated hot water during bathing or face washing, etc. The disease may be even caused by inhaling aerosolized hot water mist contaminated with the bacteria. Legionnaire's disease is a type of respiratory infection affecting the lungs. The symptom of this disease is like those for pneumonia.

To prevent growth of Legionella bacteria, various measures are to be taken, particularly in maintaining a system environment not conducive to growth. The measures recommended to control Legionella bacteria are as follows.

1 All the elements of hot water distribution system must be maintained in a hygienic condition

Table 8.2 Exposure time for human skin to cause major scalding from hot water of different temperature ranges [3].

Sl. No	*Water (°C)*	*Major burn in*
1	49	5 minutes
2	52	1.5–2 minutes
3	54	30 seconds
4	57	10 seconds
5	60	less than 5 seconds
6	63	less than 3 seconds
7	66	1.5 seconds
8	68	1 second

2 The temperature of water at the point of mixing must be maintained either below 20°C for cold supplies or above 55°C for hot supplies and
3 Hot water must be stored at or above 60°C.

8.2.3 Explosion of heater

Water is a non-compressible liquid but expands when heated. For each 260 K increase in temperature, water expands about 0.2 percent by volume. In a closed system, pressure build-up when heating is very rapid. It is found that in a 151-litre water heater installed on 550 kPa system pressure, pressure of 1000 kPa could be built up due to temperature increase of 258 K [4], which is roughly 27.5 kPa increase of pressure for every 256 K of temperature rise.

So, when water is heated in a closed heater, pressure builds up rapidly. A heater might turn into a closed container when the valves or pipes of both inlet and outlet get closed anyhow. As the temperature rises, the pressure can exceed the allowable working pressure of about 1034 kPa for which the heater is generally designed.

Water heater tanks are typically designed with the shape of the top heads convex and the bottoms concave. Excessive pressure will deform the concave bottom and make it flatten first and then turn into a bulged configuration. Ultimately, the heater might burst out suddenly.

In gas water heaters, if the products of combustion are restricted from ventilating to the chimney through the flue passage, then life-threatening circumstances might occur due to the generation of carbon monoxide gas.

8.3 Hot water supply requirements

8.3.1 Hot water temperature for various requirements

Hot water temperature requirements for various fixtures' and equipments' use differ widely. For hygienic wash and cleansing, the temperature requirements may vary for different fixtures and equipments. Hot water temperature requirements for various fixtures' and equipments' use are tabulated in Table 8.3.

8.3.2 Hot water quantity requirements

While determining the hot water requirements of an establishment, the following critical factors shall be taken into consideration for correct estimation.

1 Ambient temperature
2 Population to be served
3 Purposes of hot water use for particular occupancy buildings

Table 8.3 Hot water temperature requirement for various purposes of uses.

Sl. No	*Purpose of use*	*Temperature*
1	Drinking water	25°C
2	For kitchen sink	46–48°C [5]
3	Sterilization	30 minutes at 121°C
4	Linen washing	71°C for at least 24 minutes
5	Dish utensils sanitizing	60–88°C

Table 8.4 Typical residential usage of hot water per task. Converted in SI unit from source [6].

Sl. No	*Use*	*High flow (litre)*	*Low flow when water saver used (litre)*
1	Food preparation	18.93	11.36
2	Hand dishwashing	15.14	15.14
3	Automatic dishwasher	56.78	56.78
4	Clothes washer	121.13	79.49
5	Shower or bath	75.71	56.78
6	Face and hand washing	15.14	7.57

Table 8.5 Hot water consumption per occupant or person in typical buildings [7].

Type of building	*Consumption per occupant*	*Peak demand per occupant*	*Storage per occupant*
	litre/day	*litre/hr*	*litre*
Factories (no process)	22–45	9	5
Hospitals, general	160	30	27
Hospitals, mental	110	22	27
Hostels	90	45	30
Hotels	90–160	45	30
Houses and flats	90–160	45	30
Offices	22	9	5
Schools, boarding	115	20	25
Schools, day	15	9	5

4 Plumbing appliance and fixture count and
5 Use of special equipment, appliance or fixtures for unusual or metered water demands.

Typical residential usage of hot water per task is shown in Table 8.4, and hot water consumption per occupant or person in typical buildings is shown in Table 8.5.

8.3.3 Hot water supply components

Hot water supply systems are typically comprised of the following major components.

1 Heat transfer equipment or heater
2 A fuel source for heater
3 Cold water feed to the heater and
4 Hot water distribution piping.

8.4 Water heaters

Water heaters heat water up to the desired temperature. These units are generally cylindrical with diameters ranging from 300 to 1250 mm, depending upon their water storage capacity, size of heating elements etc. The tank is generally made of galvanized iron sheet with anti-corrosive lining and insulation. Water heaters are equipped with a thermostat to adjust and to maintain a certain temperature of water inside. Both the cold water inlet and the hot water outlet are fitted on the top of the tank. A drain valve is installed at the bottom of the tank to facilitate draining of waste or dirty water when needed.

Two water pipes are connected to the hot water heaters: the cold water inlet pipe and hot water outlet pipe. A shutoff valve is normally installed only on the cold water inlet pipe. The shutoff valve must not be installed on the hot water outlet pipe. If a shutoff valve is installed on hot water outlet pipe, it will be risking an explosion. A pressure and temperature relief valve is installed right on the water heater, coming off of the tank top or upper area of its side.

Types of water heaters: There are various types of hot water heaters based on various aspects. The various aspects according to which the water heaters are classified are as follows.

1 With respect to location of fuels for heating
2 With respect to storage capacity and
3 With respect to fuels used for the heaters.

With respect to the location of heating elements and the heaters' location, the heaters are broadly categorized in two ways as follows.

1 Direct fired water heaters and
2 Indirect fired water heaters.

With respect to storage capacity, there are three types of water heaters as follows.

1 Storage type
2 Semi instantaneous type and
3 Tankless or instantaneous type.

Based on type of fuels used, heaters are classified as follows.

1 Gas-fired water heaters
2 Electric water heaters
3 Water-jacketed tube heaters
4 Solar water heaters
5 Heat pumps and
6 Oil-fired water heaters.

Based on pressure conditions in the tank, water heaters may be classified as follows.

1 Non-pressure type water heater and
2 Pressure type water heater.

8.4.1 Direct fired water heaters

In direct fired water heaters, the source of heat, i.e. the fuel, is located where the water is to be heated. Direct fired water heaters are mostly used when comparatively small hot water generation is required. With the exception of electric water heaters, supply of sufficient air for combustion is required for various other heaters. At the same time, provision of flue management needs to be addressed for where the heaters are to be located.

8.4.2 Storage type water heaters

In storage type water heaters, the heated water is stored in an insulated tank, as shown in Figure 8.1, to keep it ready for use throughout the day. The tank may be of a horizontal or vertical type. As hot water starts to diminish due to use, it is replenished by incoming cold water, which is then heated. Storage type heaters heat water when the temperature falls below the desired level, even if there is no use of hot water. The temperature of hot water in the storage tank is usually maintained at 60°C, which is thermostatically controlled. Hot water heaters need some form of fuel to be used for producing heat. Storage type water heaters are preferred in the following conditions.

1 Large volume of hot water is needed at intervals or for limited periods
2 Demand of hot water highly fluctuates and
3 Availability of limited heating energy.

The advantages and disadvantages of storage type water heaters can be tabulated as below.

Advantages	*Disadvantages*
Smaller demand on fuel or power	Occupy more space.
Low maintenance cost	During peak demand it may go out of hot water
Simple installation and replacement	Limited hot water supply depending on size of storage capacity
Less risk of being without supply	Risk of freezing of water in winter

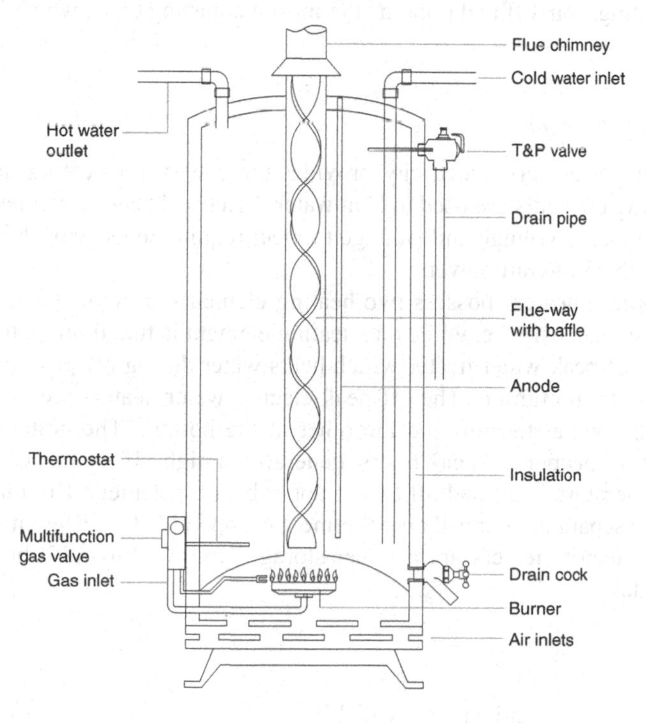

Figure 8.1 A storage type gas water heater

8.4.3 Gas-fired water heaters

In this type of water heater, natural gas, manufactured gas or propane gas is used as the source of fuel. Gas is burned to heat water using a burner inside the heater. Three types of burners are used: atmospheric burner, forced-draft burner and condensing burner. In this type of heater, venting and chimney that extend to outdoors are required if it is to be installed indoors. The entire tank is insulated to prevent the loss of heat. Gas water heaters are mostly storage types, but instantaneous gas heaters are also available.

Gas-fired water heaters are comparatively more efficient than electric water heaters in terms of energy consumption. These heaters are susceptible to loss of efficiency and malfunctioning subject to inadequacy of the air supply needed for combustion.

The burner is fired by triggering ignition using a pilot flame. The combustion chamber, housing the burner, is located at the lower part, which is vented by a baffled flue that extends through the tank center and ends up above the top of the tank, as shown in Figure 8.1. Sufficient fresh air supply for the oxygen contained in it is required for efficient burning. About 0.354 cum of air for 0.289 kW (i.e. 1.23 cum air for 1 kW) heat generation must be allowed to enter the combustion chamber from the base of the heater [8].

The burner is fired by triggering ignition using a pilot flame. The combustion chamber, housing the burner, is located at the lower part, which is vented by a baffled flue that extends through the tank center and ends up above the top of the tank, as shown in Figure 8.1. Sufficient fresh air supply for the oxygen contained in it is required for efficient burning. The burner assembly has several components. The pilot light is used to light the burner; a thermocouple is used to provide gas, which is connected to the main gas valve of the burner assembly. The pilot light is adjusted by either an air shutter or pilot light adjustment screw. The air shutter also adjusts the burner flame. The main valve controls the burner assembly. It consists of a control knob that has three settings: on, off and pilot; a "thermostat adjusting dial", which is used to set the water temperature.

8.4.4 Electric water heaters

In electric hot water heaters, electricity provides the energy for heating. In these heaters, one or two heating elements are used to heat water. Electrical heating elements are available in a variety of standard voltage and wattage to meet requirements. Most heaters operate on 240 volts and with 4500-watt power.

Generally, electric heaters possess two heating elements, as shown in Figure 8.2, functioning non-simultaneously, i.e. while one heating element is functioning, the other remains off. There is an off-peak water heater which heats water during off-peak hours when electrical demand is at a minimum. The off-peak electric water heaters are provided with two heating elements: one at the top and the other at the bottom. The bottom element is for heating during the cheaper off-peak hours, generally at night. If the heater runs low during the day, the temperature is topped off by the upper heating element. Both heating elements are controlled by separately controllable thermostats, operable for different temperature settings. Off-peak electric heaters are a larger storage type used to fulfil hot water demand throughout the day.

Advantages

1 Low investment cost and widely available
2 Off-peak models can be run at comparatively low cost

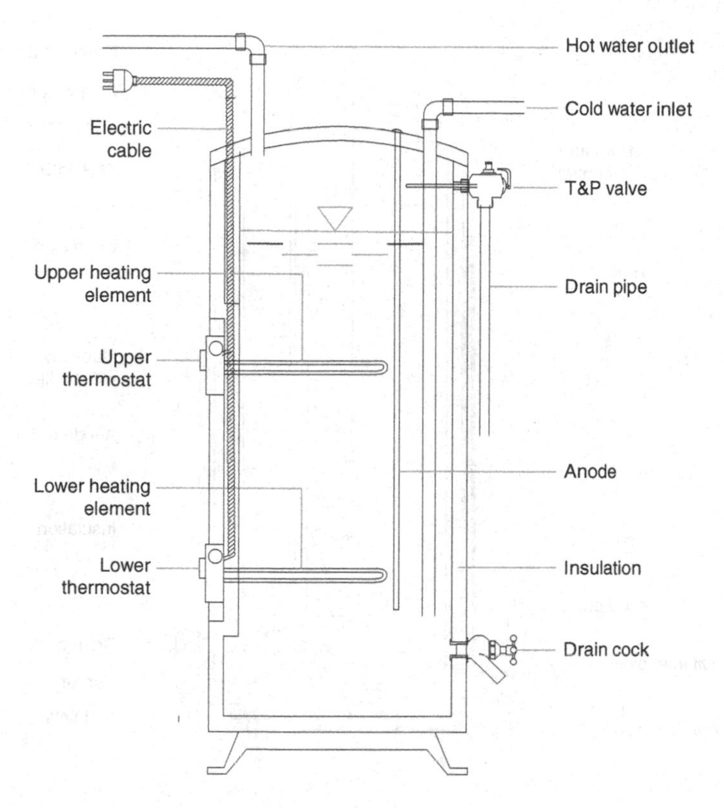

Figure 8.2 An electric water heater

3 Easy installation
4 No venting or chimney is required
5 No soot or grime is generated and
6 No chance of fuel leakage.

Disadvantages

1 Higher operating costs with rising electricity prices
2 Higher greenhouse gas emission rate
3 Use may be restricted by the concerned regulatory authorities and
4 No electricity, no operation.

8.4.5 Oil-fired water heaters

These heaters are more or less similar to gas-fired water heaters. These heaters use a forced-draft oil burner operated by an oil pump, driven by an electric motor that is usually mounted at the bottom of the hot water tank, fed by light-grade oil from a tank. Use of kerosene or ultra-low sulfur diesel is recommended as fuel. A flue vent stack is extended to the top of the water tank and continued on into a chimney leading to the outdoors. Figure 8.3 illustrates an oil-fired water heater.

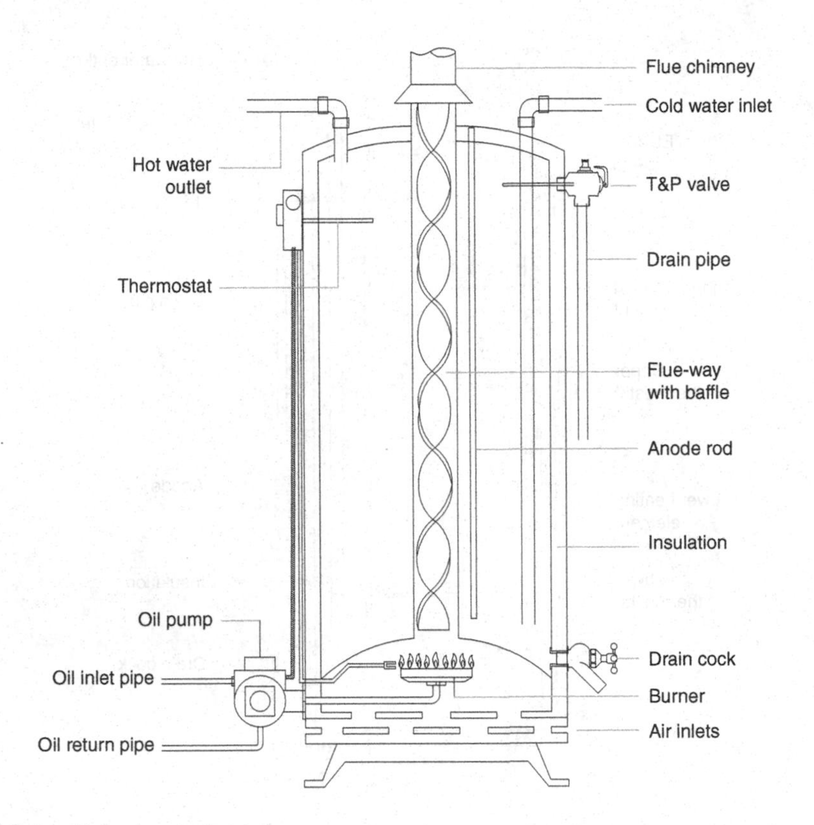

Figure 8.3 An oil-fired water heater.

Oil burners must have provision of flame retention and interrupted ignition and be equipped with a primary control system which enables safety shutdown within 15 seconds, in the event of flame failure.

Regular maintenance of heaters is required for their efficiency. Special attention to the oil pump is essential. Any soot and burn marks, oil leaks, noisy running, odour etc. are signs of improper operation of the oil-burning unit, which needs immediate action. If an oil leak is observed, the operation of the heaters must be shut down until sealed.

8.4.6 Water-jacketed tube water heaters

In this type of water heater, cold water, flowing through coiled tubes jacketed by hot water or steam in the heater tank supplied from a boiler, is thus heated and finally flows out of the heater through the outlet pipe, as shown in Figure 8.4. These heaters are an indirect, instantaneous type of water heaters. The coiled tube inside the heater performs as a heat exchanger.

8.4.7 Semi-instantaneous water heaters

Semi-instantaneous water heaters are essentially an instantaneous type of heater but are designed to contain small hot water storage of a capacity of approximately 38 to 76 litres,

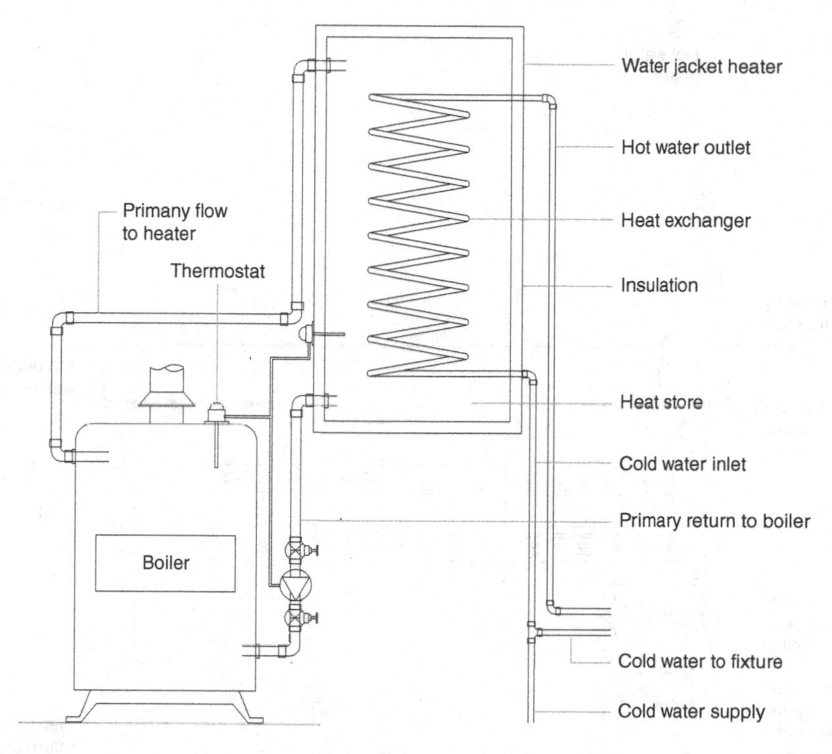

Figure 8.4 Water-jacketed tube water heater.

depending on the rated heating capacity. This stored quantity of water should be adequate to allow the temperature control system to respond to sudden fluctuations in water flow and to maintain the delivered water temperature within ±2.7°C. This type of heater is preferred where there is serious restriction of space for the installation of a storage type water heating system.

In this heater, the heat exchanger is partially enclosed within the hot water storage tank, as shown in Figure 8.5. The heat exchanger consists of several tubes through which the heating medium is flown. A circulation pump is installed to force the water being heated from the bottom of the storage tank to flow around the tubing of the heat exchanger, thereby getting heated and then entering the tank. When the temperature in the storage tank rises to a pre-set value, as measured by the temperature sensor, the control valve closes, shutting down the flow of the heating medium into the heat exchanger.

8.4.8 Tankless or instantaneous water heaters

Tankless hot water heaters do not have any space in the tank to store hot water. These heaters generate hot water on demand and can supply hot water continuously at the desired temperature. So, these are also termed instantaneous type water heaters, in which inflowing cold water is heated up to the desired temperature very quickly while flowing through a series of heating coils inside the unit before finally discharging at the desired temperature, as illustrated in Figure 8.6. Instant water heaters are usually installed at or near the fixtures of use.

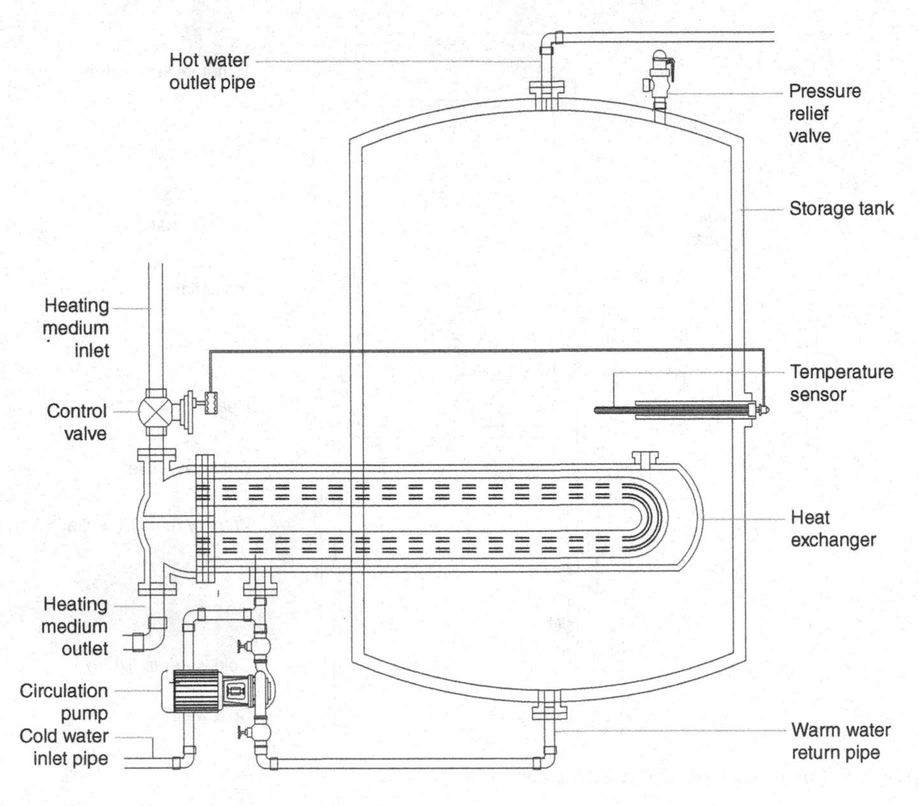

Figure 8.5 An indirect semi-instantaneous water heater

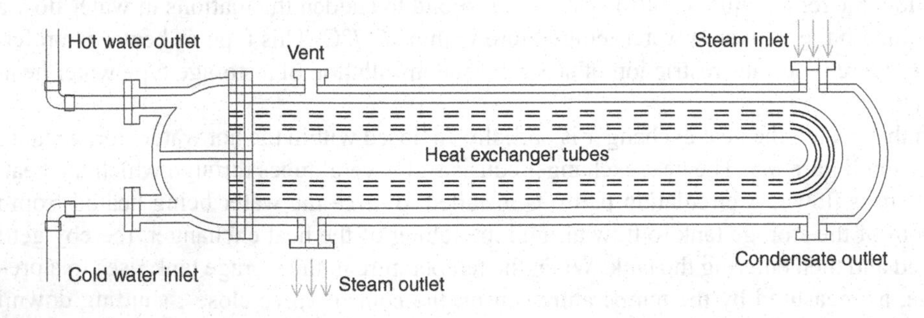

Figure 8.6 Conventional instantaneous water heater

Generally, tankless water heaters are designed to provide hot water flow at a rate of 11 to 22 lpm at 49°C. Tankless heaters can also be regulated to generate water at 50°C, so that the risk of scalding can be avoided without installing a temperature relief valve.

Tankless water heaters are energy efficient because they remain off when hot water is not in use, and heating rate is set based upon demand of flow rate. In these types of heaters, cold water passes slowly through the heat exchanger to assure adequate heat transfer. Therefore, the rate of hot water flow upon demand is comparatively less than can be provided by storage water heaters.

Mostly, one tankless water heater is used for one fixture; but more than one tankless water heater, connected in parallel, can be used, depending upon the number and type of fixtures or faucets to be served. Tankless water heaters shall be sized to provide adequate hot water flow of about 7.5 lpm for a fixture or faucet.

Power consumption for an instantaneous type water heater is too high in comparison with storage type water heaters. Usually, power consumption for storage type heaters is up to 6 kW (single-phase), whereas it is about 18kW (3 phase) for an instantaneous type [9].

The advantages and disadvantages of tankless instantaneous water heaters can be summarized as below.

Advantage	*Disadvantages*
Does not require too much space and can be placed in any location in a house	The heater must be sized carefully: undersized heaters will affect water temperature and flow
Unlimited hot water supply, as long as there is fuel or power	High energy demand and cost
It has long life and is economical	High initial cost: twice as much as the cost of conventional storage tanks
Reduce risk of water damage or loss	Cannot meet the high hot water demand
Quick generation of hot water	There is no backup of heat source during an emergency
No chance of bacterial growth	Only non-green source of fuel: electricity, gas and oil can be used

8.4.9 Indirect-fired water heaters

In an indirect fired water heating system, the heating source is remotely located from the water heating equipment. Generally, heating energy of hot water or steam from a boiler is conveyed through a coiled piping called a heat exchanger, which remains inside the heater or storage tank, by means of a separate supply and return piping, as shown in Figure 8.7. The use of other hydronic heating sources like heat pump or solar heat can also be grouped in an indirect fired water heating system.

The internal coil pipes of a heat exchanger are designed for low pressure drops and high heat transfer capabilities, to make the indirect water heaters most efficient. Indirect-fired water heaters are generally preferred where a large volume of hot water is needed and where there is excess steam or hot gases available to supply for this purpose.

8.4.10 Heat pumps

A heat pump is an effective device to produce hot water, drawing heat from surrounding warm air or even from the ground, as illustrated in Figure 8.8, and transfers it to the water in an enclosed tank to heat. The technique of heating is like the reverse of the normal refrigerating system.

Heat pumps are inefficient in a very cold and closed environment. It needs to be installed in an area of temperature 4.5 to 32°C. Sufficient clearance, as much as 2.1 m from floor to ceiling and space up to 28.32 cum around, is required to capture enough heat from the air. A nearby drain is to be provided to discharge the condensation.

Using heat pump water heaters, both energy and cost can be saved. About 30 to 60 percent of energy can be saved by using heat pumps, which also reduce humidity [10].

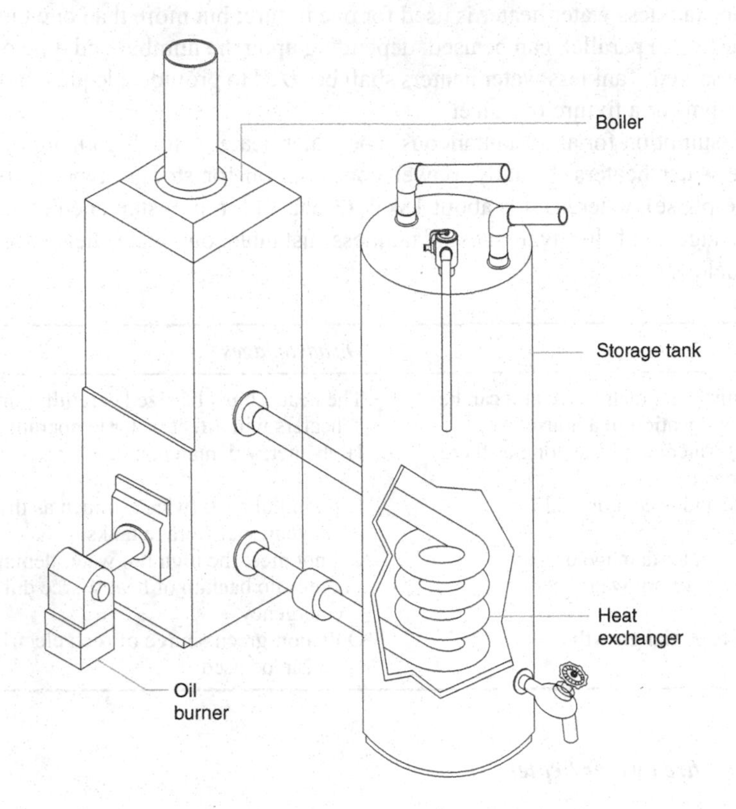

Figure 8.7 Indirect fired water heating system.

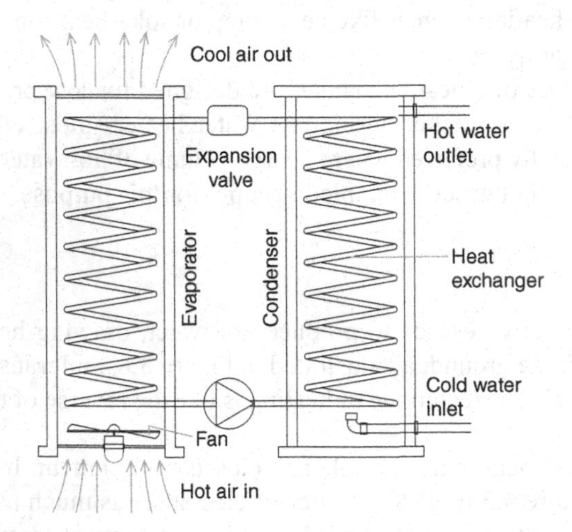

Figure 8.8 Water heating by heat-pump.

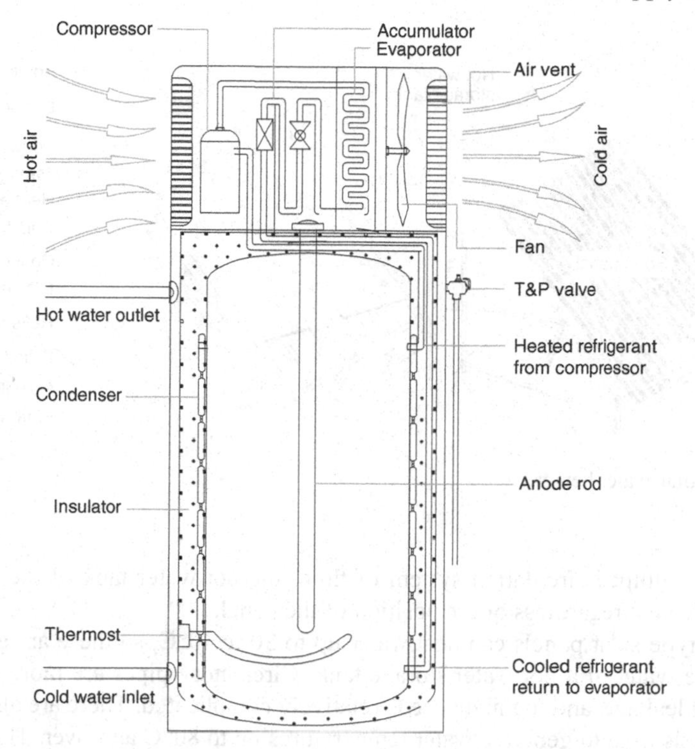

Figure 8.9 Heat pump water heater.

In hybrid-labeled hot water heaters, heat pump technology is used for preheating water of conventional hot water heaters, as shown in Figure 8.9. During off-peak periods, only the heat pump is run to heat water, thereby maximizing energy efficiency and saving costs. During peak periods of hot water demand, these heaters switch to standard electric resistance heating system automatically.

8.4.11 Solar water heater

In a solar water heater, the thermal energy of sunlight is utilized to heat water in a collector and storage tank, as shown in Figure 8.10. Solar water heaters have three major components, as mentioned below.

1 The solar panel
2 Heat circulator and
3 Solar storage tank.

In a solar panel, exposed to the sun, the cells absorb the sun's heat and transfer it to an antifreeze-like fluid in a closed-loop system that flows upward to the hot water storage tank. In thermo-siphonic circulation systems, the tank is located slightly above the panel and uses natural heat convection to circulate the fluid through the panel by flowing upward to the tank and returning to the panel. The heated fluid thus heats the water in the tank while flowing

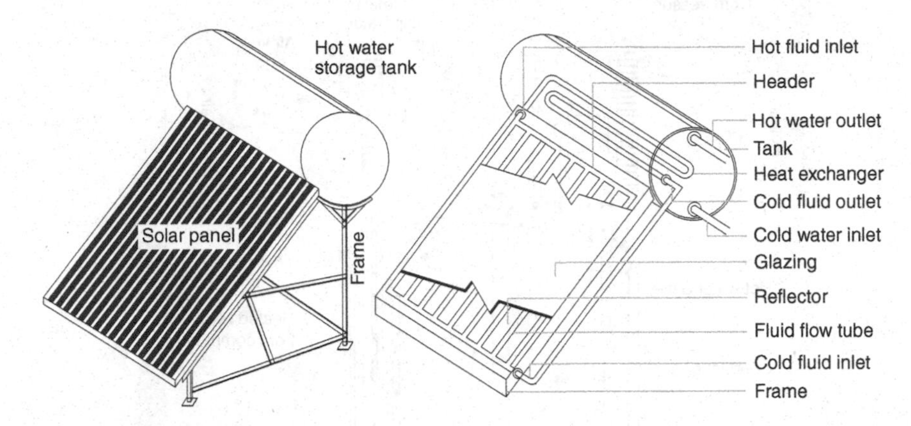

Figure 8.10 Solar water heater.

through. In a pumped circulation system of fluid, the hot water tank of the heater can be mounted anywhere regardless of the position of the panel.

Flat-plate type solar panels can heat water up to 30 to 40°C, so these are generally used to preheat the water of a hot water storage tank. Circulatory pipes are more susceptible to blockage and leakage, and maintenance is relatively complicated. There are also concentrating solar panels used to generate water temperatures up to 80°C and over. These panels are more efficient because of lower heat loss to the surrounding area. In this case, circulation of heated fluid is done by pumping and the system becomes costlier.

Advantages of solar water heater

1 A safer system, as there are no chances of any gas leakage, no cause of explosion and no electrical mishaps
2 Economical, as no payment required for electricity or gas and
3 Convenient in using any time of the day.

Disadvantages of solar water heater

1 Roof space is occupied by solar panels
2 Solar heaters' efficiency depends upon the abundance and availability of sun heat
3 Good insulation of the heater is needed to use hot water during night and
4 Less efficient in rainy, cloudy or foggy days.

8.4.12 Non-pressure type water heaters

In this type of heater, the pressure in the tank is atmospheric pressure. From the heater, water flows under gravity owing to pressure developed due to the height difference between the heater and the faucet to which it is connected. There is no chance of developing excess pressure or vacuum in these heaters. So, no air vent or vacuum breaker needs to be installed on this type of heater.

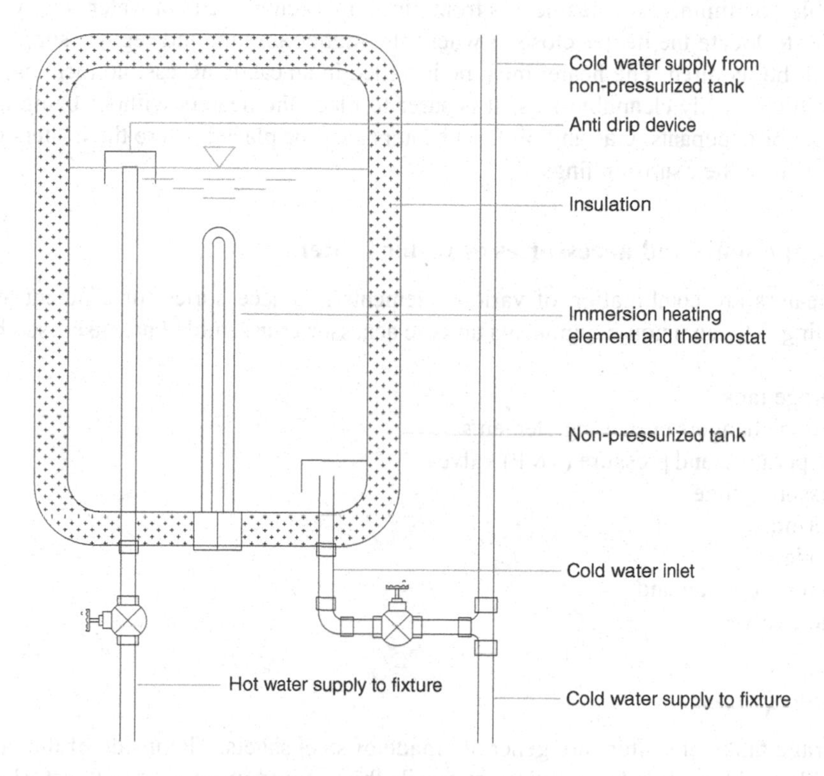

Figure 8.11 A non-pressurized instantaneous electric water heater

This type of heater is generally used to supply hot water directly to only one service faucet. So, these heaters are installed over the fixture where hot water is needed. The inlet connection of cold water to the heater is controlled by the gate valve and the supply is made under gravity, from the non-pressurized water tank remaining over the heater. No hose or other connection can be made on an outlet pipe. Non-pressurized water heaters may be of an electric, gas or solar heating system. Figure 8.11 illustrates a non-pressurized water heater.

8.4.13 Pressure type water heaters

These types of water heaters are kept under pressure greater than atmospheric pressure. Pressure type water heaters are mostly used in pressurized hot water distribution systems. According to ASME B31.1 hydrotest, the vessel shall be so constructed to withstand at least 1.5 times the system pressure [11]. These heaters remain always full of water, thus eliminating the danger of being emptied and thereby eliminating cut-off of electric heating coils. Pressure type heaters are susceptible to bursting or squeezing due to implosion caused by malfunctioning of connected accessories and incorrect installation.

8.4.14 Positioning water heaters

Positioning water heaters at ideal locations brings efficiency and energy saving in hot water supply. It is recommended to keep hot water pipe runs, from the heater to the faucets, as short as possible, to minimize the heat losses from pipes. In a centralized hot water supply system, it is good to locate the heater close to where the higher amount and regular supply of hot water will be required. The heater must be installed in an easily accessible location, and on 150 mm high, easily cleanable legs. It is safer to place the heaters without being exposed to the user or occupants. Gas and oil fired heaters shall be placed where the heaters get sufficient air from their surroundings.

8.5 Components and accessories of water heaters

Water heaters are combination of various elements and accessories for efficient and safe functioning. The common components and the accessories assembled are discussed below.

1 Storage tank
2 Heat exchangers or heating elements
3 Temperature and pressure (T&P) valve
4 Pressure gauge
5 Thermostat
6 Anode
7 Vacuum breaker and
8 Drain valve.

8.5.1 Storage tank

The storage tanks of heaters are generally made of steel sheets. The inside of the tank may be provided with a glass lining of about 5 mils thick, fused or bonded to the steel surface, to protect the tank from corrosion and to ensure hot water quality suitable for sanitizing purposes. There are also tanks with a cement lining of about 12 mm thick. The pressurized storage tanks are designed to sustain at least 1034 kPa pressure.

In addition to some common provisions mostly provided, there might be some special provisions needed for special purposes. For example, larger tanks are generally equipped with a hand-hole for cleaning and inspection. Tanks may have provision for connection with return pipes, which is needed in recirculating the hot water supply system. Connecting more tanks in parallel requires connection arrangements in the concerned tanks.

8.5.2 Heat exchanger

This is a device that transfers heat energy between two or more fluids, between a solid surface and a fluid at different temperatures and in thermal contact. Heat exchangers in water heaters are basically a coil of tube, though plate type heat exchangers are also available. There may be a single walled heat exchanger and a double-walled heat exchanger. In a double-wall heat exchanger, there is slight gap between the two tubes that constitute the heat exchanger coil. Single wall heat exchangers are limited to a maximum pressure of 207 kPa [12].

Care must be taken so that the tube of heat exchangers do not erode or corrode, so that no contamination of hot water can occur. It is observed that if water is hard in nature, it causes scaling or liming on the heat exchanger. The rate of scaling increases with the rise in temperature and period of usage due to losing solubility of calcium carbonate and other scaling compounds present in water. Therefore, the heat exchanger needs to be flushed periodically to prevent damage to it.

8.5.3 Thermostat

A thermostat is installed on water heaters to control the temperature of hot water in the heater automatically. There is a bi-metal switch located at the back of the thermostat. Inside the thermostat, there is a switch governed by a bi-metal disc made by fusing two different metals of different coefficients of thermal expansion. So, when heat rises, the two metals expand differently, causing a bend in the disc. The bending action is used in opening and closing the switch of the thermostat.

8.5.4 Anode

Anode, also known as sacrificial anode, is used in water heaters as a vital part of the protection of the storage tank body from getting rusted. The concept is that when two metals are physically connected underwater, one will corrode to protect the other. Here the anode corrodes earlier, thereby protecting the steel body of the tank from early and rapid corrosion. Anode is a rod-like element made of magnesium or aluminium, formed around a steel core wire. Anode is kept vertical inside the tank by screwing into the top cover of the tank. For a longer life for the heater, more than one anode is used.

8.5.5 Vacuum breaker or vacuum relief valve

Installation of a "vacuum breaker" prevents the heater tank from forming a vacuum inside, thereby protecting it from collapsing or being crumpled up. A vacuum breaker is installed on the cold water inlet pipe to the heater. When the cold water supply to the heater is turned off anyway and the inside water is drained of drawing hot water, a vacuum is created inside the tank. The amount of vacuum or negative pressure is directly proportional to the height of the tank above the level where the water is being drained out. Even a slight vacuum can exert tremendous pressure on the tank, causing collapse if not built strong enough. Access of air through a vacuum breaker on the cold water inlet pipe can eliminate such a problem.

8.5.6 Drain valve

Hot water heaters must be provided with a drain valve to drain out the tank content periodically. Periodical draining of tank contents includes removing sediment comprising debris or mineral deposits accumulated at the bottom of the water heater tank. When huge amounts of sediment accumulate in the heater, a knocking or popping noise is generated in the heaters, which need the flushing of the tank by draining its content.

Temperature and pressure valve: This was already discussed in Chapter 6, on T&P valves.

8.6 Two-temperature hot water service

In a building, there may be a requirement for hot water supply to serve appliances or fixtures at two temperature levels. In this case, the required temperature level for predominant usage is to be considered first. If the predominant usage is found at lower temperatures, it is a good approach to heat water for the hot water supply to the lower temperature level first and then use a separate booster heater to further heat the hot water of lower temperature for higher temperature service. In particular, occupancy building hot water supply to all toilets or baths is generally developed at 60°C, but in a kitchen, the hot water temperature may be required at 80°C. In this case, a central hot water supply system supplying hot water at 60°C can be developed, and for a kitchen, a separate booster heating system can be installed to raise hot water temperature from 60°C to 80°C, as represented by Figure 8.12. Where the bulk of the hot water is heated at the higher temperature level, the lower temperature hot water can be obtained by mixing cold water with hot water either centrally or at the point of use. Sometimes, an instantaneous water heater is intended to raise or boost the temperature of hot water to a higher temperature for specific purposes, such as for the sanitizing rinse on a high temperature automatic dishwashing machine etc.

8.6.1 Heat energy requirement for heating

When a mass is heated, heat is transferred due to temperature differences, referred to as heat flow. The SI unit for heat flow is Joules/sec or watt (W), the same as power. One watt is defined as 1 Joule/sec. Specific heat is the change in internal energy with respect to change in temperature at constant volume. Water has a specific heat of 4.19 kJ/kg°C.

The amount of heat energy is quantified by the British thermal unit (Btu), which is the amount of heat needed to raise 1 lb of water by 1°F; and the calorie, which is the amount of heat needed to raise 1 gram of water by 1°C (or 1 K).

The amount of heat required in watts to heat a mass for particular period in order to raise one temperature level to another, can be expressed as follows.

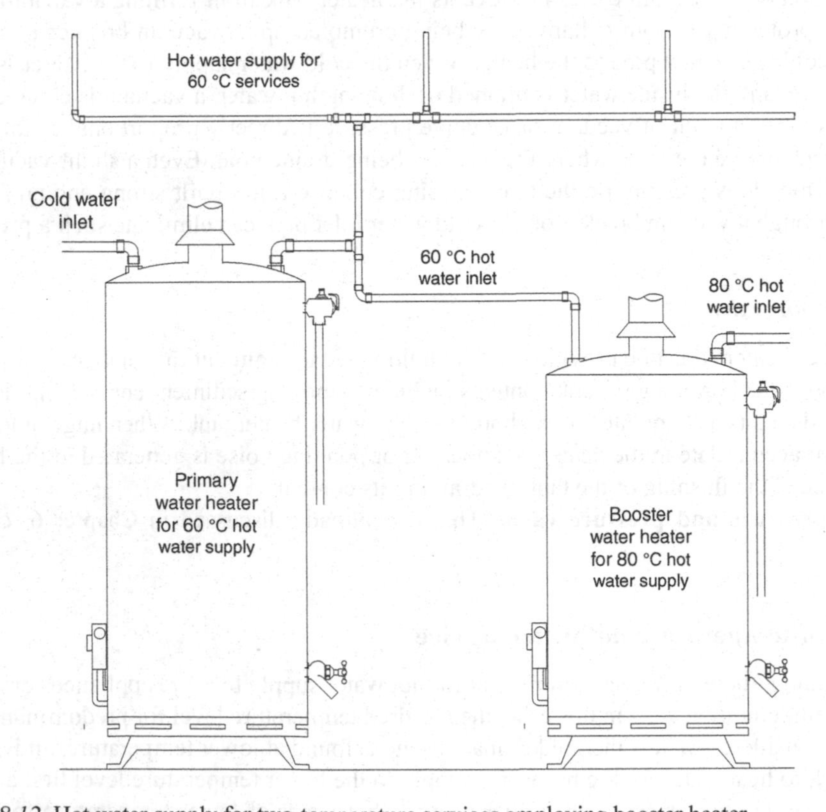

Figure 8.12 Hot water supply for two-temperature services employing booster heater.

$$Heat\,input\,rate = \frac{M}{D}\rho\,\Delta T \qquad 8.1$$

Where
M = mass (kg)
ρ = specific heat (kJ/kgK)
ΔT = temperature difference between hot and cold side (K) and
D = duration of heating (hour).

Example: Let a residential building require 2000 litres of hot water at a temperature of 60°C for its general use and 500 litres at 80°C for special purposes. If the primary water heater has a one-hour recovery period, the booster heater is chosen for a two-hour recovery period and the cold water temperature is 10°C, then the required heat input rate for the heaters could be calculated in the following ways.

1 Heat input rate for the primary heater = 2000 × 4.19 × (60 – 10) / (1 × 3600) = 116.39 kW; say 117 kW.
2 Heat input rate for the booster heater = 500 × 4.19 × (80 – 60) / (2 × 3600) = 5.82 kW; say 6 kW.

8.7 Hot water distribution systems

Hot water distribution from the point of generation or storage to points of use can be done in various ways. The hot water distribution system itself plays an important role in achieving efficiency. Efficient hot water distribution systems can be achieved by minimizing excessive loss of water, temperature and pressure. Hot water distribution systems can be basically divided into following two systems.

1 Direct or non-circulated system and
2 Circulation system.

8.7.1 Direct or non-circulated hot water supply systems

In a direct hot water supply system, hot water from a heater or hot water storage tank is supplied to the fixtures or appliances directly, as illustrated in Figure 8.13. This type of hot water distribution system is preferred for small installations such as one or two toilets or kitchens in residential or hostel building etc. This system is also preferred for buildings of various occupancies where hot water is occasionally used or needed for short periods. This system is comparatively economical when only the cost of development is considered. Direct or non-circulated hot water supply systems have many disadvantages as well.

In direct or non-circulated hot water supply systems, when water does not flow for a considerable period of time, particularly during the night when hot water may not be in use, the water that remains in the pipes and appliances cools down to room temperature. Afterwards, to get hot water flow from the faucet, the cool water from the hot water line is to be wasted out, meaning a considerable time of waiting, depending upon the length and size of the pipe from the heater to the faucet. The maximum time of waiting to get hot water that can be allowed is 30 seconds. This system is susceptible to loss of energy

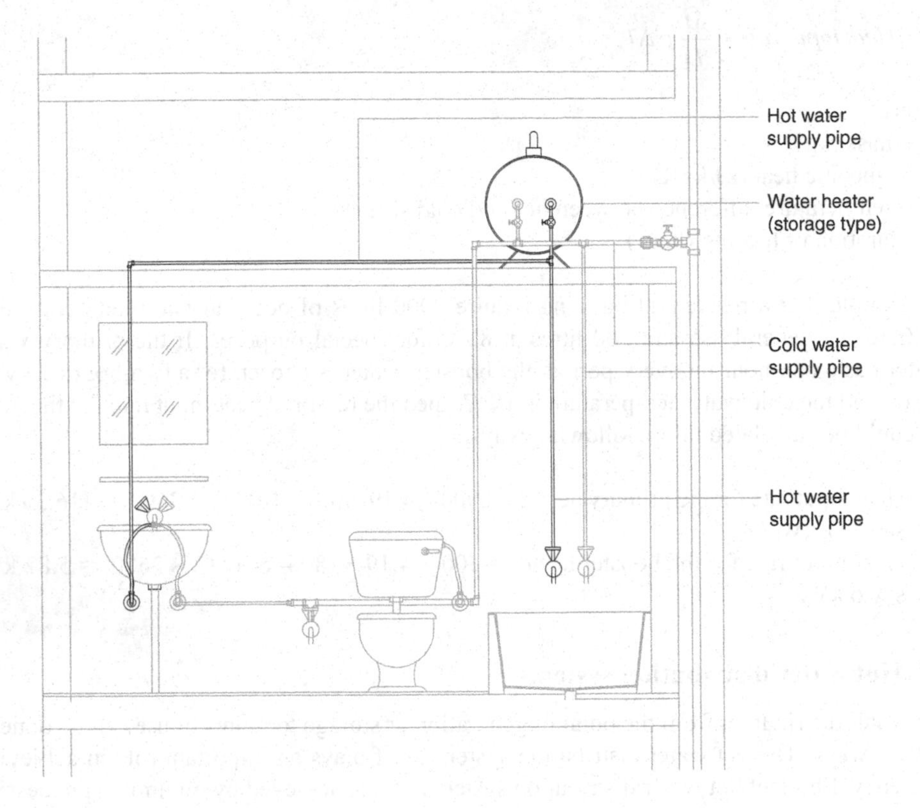

Figure 8.13 Direct hot water supply system in a toilet.

due to the cooling down of hot water and loss of water resources at the same time. These problems of direct hot water supply can be avoided by developing a circulated hot water distribution system.

Some codes allow a direct hot water supply system to be supplied for a 30 m pipe length from the water heater to the farthest faucet. The International Plumbing Code now allows only 15 m from the water heater to the farthest faucet, in order to ensure that water reaches the farthest fixture at a temperature of at least 55°C [13].

8.7.2 Circulated hot water supply systems

A circulation type hot water distribution system has been developed to reduce the wait time to get hot water, to minimize water and energy loss caused during the waiting period and to prevent cooling down of water temperature.

There are two methods of circulated hot water supply system as mentioned below.

1 Open-loop system and
2 Closed-loop system.

Open-loop system: In this system, hot water is supplied to all faucets from the heater. Under the farthest fixture from the heater, an on-demand circulation pump is installed by which hot

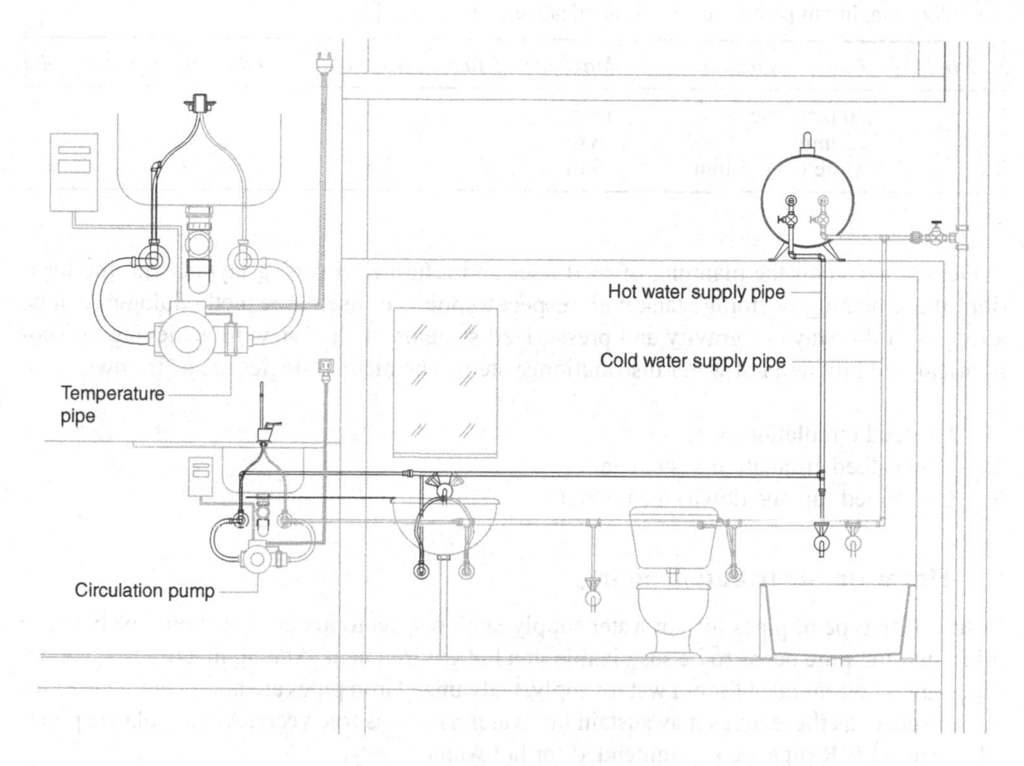

Figure 8.14 Hot water is circulated and returned after supplying the farthest fixture.

water is circulated to return after supplying the cold supply pipe, as shown in Figure 8.14. To operate the pump, there should be an electricity supply outlet nearby.

Closed-loop system: In this system, a dedicated hot water return pipe is installed at the end of the hot water supply pipe supplying to the farthest fixture or fixtures group, and is returned to the water heater. On the return pipe and near to the heater, an on-demand circulation pump is installed to push back or circulate the warm water into the heater. In cases with more than one hot water supply pipe, each one might have a return pipe connected to a common return header that leads to the water heater or storage tank. In such a case, a circulation pump is installed on the return header pipe.

A closed-loop hot water circulation system offers more expense in operation than a direct hot water supply system, but this system saves energy and resource. Furthermore, recirculation helps with maintaining required water temperatures.

In a circulation type hot water supply system, hot water is seldom returned from near the connection to the faucets. The return of hot water is made from a certain distance away from the faucet connection, depending upon the diameter of the supply pipe. The distance of returning hot water to the end of supply pipe is termed dead leg. It is likely that the hot water remaining in dead legs will cool down little when not in use, which is acceptable. The allowable dead leg for different diameters of supply pipe is given in Table 8.6.

Closed-loop circulation systems can again be planned and designed in two ways, as mentioned below, basing on the pressurizing methods used.

1 Gravity system and
2 Pressurized system.

Table 8.6 Maximum permissible hot water dead leg pipe length [9].

Sl. No	*Pipe size diameter*	*Maximum distance between faucet and hot water return point*
1	20 mm or less	12 m
2	25 mm	8 m
3	More than 25 mm	3 m

Depending upon the planning of feed riser and return riser piping layouts and the location of the heating or storage tank with respect to points of use, three methodologies can be adopted in the ways of gravity and pressurized systems, with a view to achieving an economical and efficient hot water distribution system. The methodologies are as follows.

1 Up-feed circulation system
2 Down-feed circulation system and
3 Combined (up and down) feed system.

8.8 Hot water distribution piping

In selecting type of pipes for hot water supply, with respect to material, it should be borne in mind that the pipe needs to be sustainable in a hot environment. Among metal pipes, copper pipes are recommended for hot water supply. Galvanized iron pipes are not recommended for this purpose, as these pipes may sustain hot water as few as five years. Among plastic pipes, CPVC and PP-R pipe are recommended for hot water supply.

All pipes installed in a hot water distribution system can be grouped as in the following, depending upon its service and location.

1 Branch and sub-branch supply pipe
2 Feeder risers and main feeder pipe and
3 Return risers and main return pipe.

8.8.1 Branch and sub-branch pipes

In hot water distribution piping networks, branch pipes are those pipes connected to the hot water feeder risers. The branch pipe may supply a faucet or a group of faucets fed by a network of pipes, which may be termed a sub-branch of the branch pipe. The branch pipes must be as short as possible and are designed based on the fixture units to be served, considering a flow velocity of 1.2 to 1.5 m/sec.

8.8.2 Feeder risers and main feeder pipes

Feeder risers for a main feeder pipe in hot water distribution systems are mostly vertical pipes supplying the branch hot water supply pipes.

8.8.3 Return risers and main return pipes

Return pipes are the segment of a circulating hot water distribution piping system that conveys the warm water, at a temperature that has dropped below the design temperature, back into the water heater to get reheated. No branch pipe or faucet is connected to the return pipe. The connection of the return pipe is made near to the end of the farthest branch pipe that supplies hot water to one or more fixtures requiring maintenance of desired temperature. On the return pipe, a balancing valve to throttle the returning flow and a check valve to prevent the reversal of returning flow, which may be caused due to discharging faucets, are to be installed.

Care should be taken in sizing the return pipe so that it is not undersized. The sizing of a hot water return pipe is to be based on the heat loss method. The heat loss in the circulation system must not exceed 10°C. The velocity of flow in hot water return piping shall be no more than 1.5 mps.

Recirculation system design procedures are as follows.

1 Determine required recirculation flow rate (based on heat loss of supply pipe)
2 Determine flow-friction head loss in recirculation line, heater supply pipe and etc.
3 Size recirculation line and
4 Select a pump based on flow requirement and head loss.

To maintain the hot water distribution system in equilibrium, the heat loss that occurs in piping shall be equal to the heat gained from the water heater.

8.9 Closed-loop circulation system layout

8.9.1 Gravitational closed-loop circulation systems

Hot water circulation can be done using gravitational force, where the source of hot water, i.e. heater, is placed above the highest part of hot water usage. The system will not function if the hot water storage tank is placed below any hot water supply lines. In this loop, the hot water return pipe is turned back to the heater after serving the farthest fixtures or fixtures groups by the hot water feed risers. In this system, a circulation pump is unavoidable when a non-pressurized hot water storage tank is used. Gravitational closed-loop hot water circulation systems can again be developed in the three following layout systems.

1 Gravitational down-feed circulation system
2 Gravitational up-feed circulation system and
3 Gravitational combined-feed circulation system.

Gravitational down-feed circulation system: In this system, a main hot water feeder pipe is extended from the source of hot water, from which the top ends of feeder risers supply all the faucets by gravity, down to the lowest part of points of usage, are connected, as shown in Figure 8.15. The return risers, starting from near the bottom end of the supply risers, are connected to the main return pipe to feed into the hot water storage tank. An automatic air relief valve and a vacuum breaker are to be installed above the highest water level of the system, on the pipe extending from the apex point of the feeder main.

Gravitational up-feed circulation system: In this system, a main hot water feeder pipe is extended down to the bottom of the feeder risers, up-feeding to the highest part of points of usage. The top end of feeder risers, serving as return risers, are connected to a main return pipe feeding the hot water storage tank, as shown in Figure 8.16. An automatic air relief valve and a vacuum breaker are to be installed above the highest water level of the system, on the pipe extending from the apex point of the feeder main.

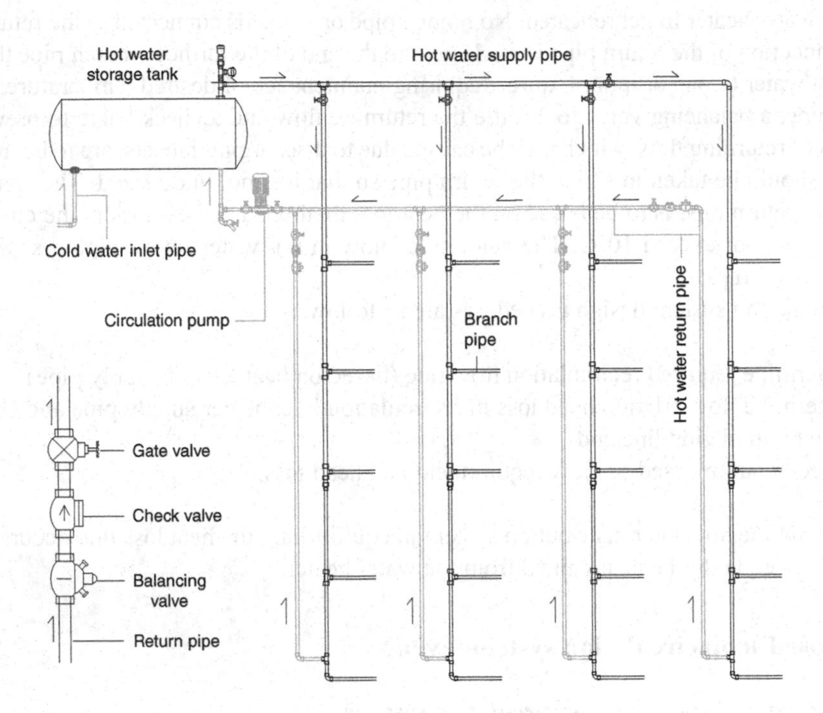

Figure 8.15 Gravitational down-feed hot water circulation system.

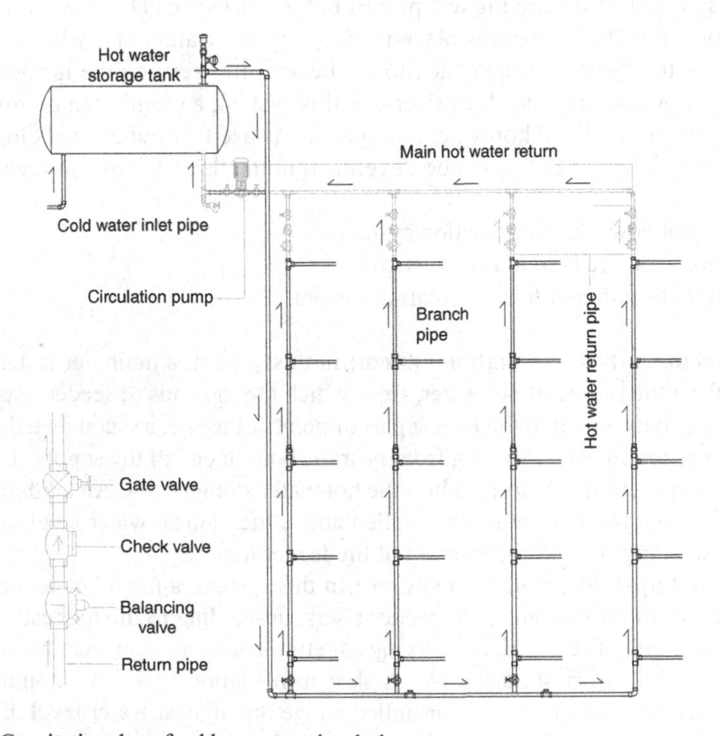

Figure 8.16 Gravitational up-feed hot water circulation system.

Gravitational combined-feed circulation system: In this system, a couple of feeder risers are used to circulate the hot water supply where one feeder is down-feeding and the other is up-feeding, as shown in Figure 8.17. The down-feed risers are extended from the main feeder pipe drawing from the hot water storage heater and remain above the top faucets. The up-feed risers are supplied from the bottom of the down-feed riser; the top end of the up-feed risers is connected to the main return pipe, through which warmed water is pumped back to the hot water storage tank by a circulation pump. An automatic air relief valve and a vacuum breaker are to be installed above the highest water level of the system, on the pipe extending from the apex point of the feeder main.

8.9.2 Pressurized closed-loop circulation systems

When the cold or normal water supply system is designed as a pressurized system, then the adjoining hot water distribution system is to be designed as a pressurized system to maintain nearly equal pressure in hot and cold water outlets at all points of usage of the distribution system. In a pressurized hot water distribution system, the supplying points of usage remain above the point of hot water generation. Pressurized closed-loop hot water circulation systems can be developed in the three following layout systems.

1 Pressurized up-feed circulation system
2 Pressurized down-feed circulation system and
3 Pressurized combined-feed circulation system.

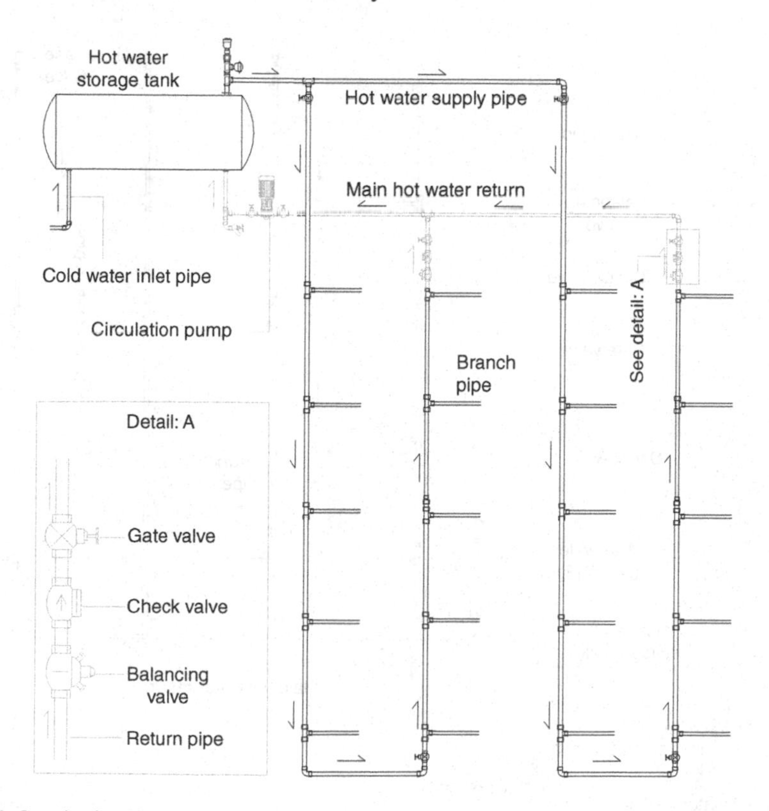

Figure 8.17 Gravitational combined-feed hot water circulation system

Pressurized up-feed circulation system: In this system, a main hot water feeder pipe is extended from the source of hot water, to which the bottom ends of the feeder risers supplying all the faucets above are connected. The flow is upward due to pressure exerted in the feeder risers that supply all faucets up to the highest part of points of usage. The return risers are either the extensions of feeder risers, with the top end connected to the main return pipe extended to feed into the hot water storage tank below, or separate return risers connected to the risers near the highest faucets, with the bottom end connected to the main return pipe extended to feed into the hot water storage, as shown in Figure 8.18. An automatic air relief valve is to be installed on the apex point of all feeder risers.

Pressurized down-feed circulation system: In this system, the main hot water feeder pipe is extended from the hot water heater or source upward to the highest level of points of water usage, as shown in Figure 8.19. Hot water is then supplied to all of the feeder risers by connecting the main feeder pipe with the feeder risers and supplying by gravity all of the faucets below. The extensions of down-feed risers are then connected to the main return pipe to circulate warm water to the storage tank or heater for reheating. An automatic air relief valve is to be installed on the apex point of the main feeder risers.

Pressurized combined-feed circulation system: In this system, a couple of feeder risers are used to circulate the hot water supply, where one feeder is up-feeding and the other is down-feeding, as shown in Figure 8.20. The up-feed risers are extended from the main feeder pipe drawing from the hot water storage and remain below all the faucets. The down-feed risers are supplied from the extension of the top of up-feed risers; the extension of the bottom end of the down-feed risers are connected to the main return pipe, through which warm

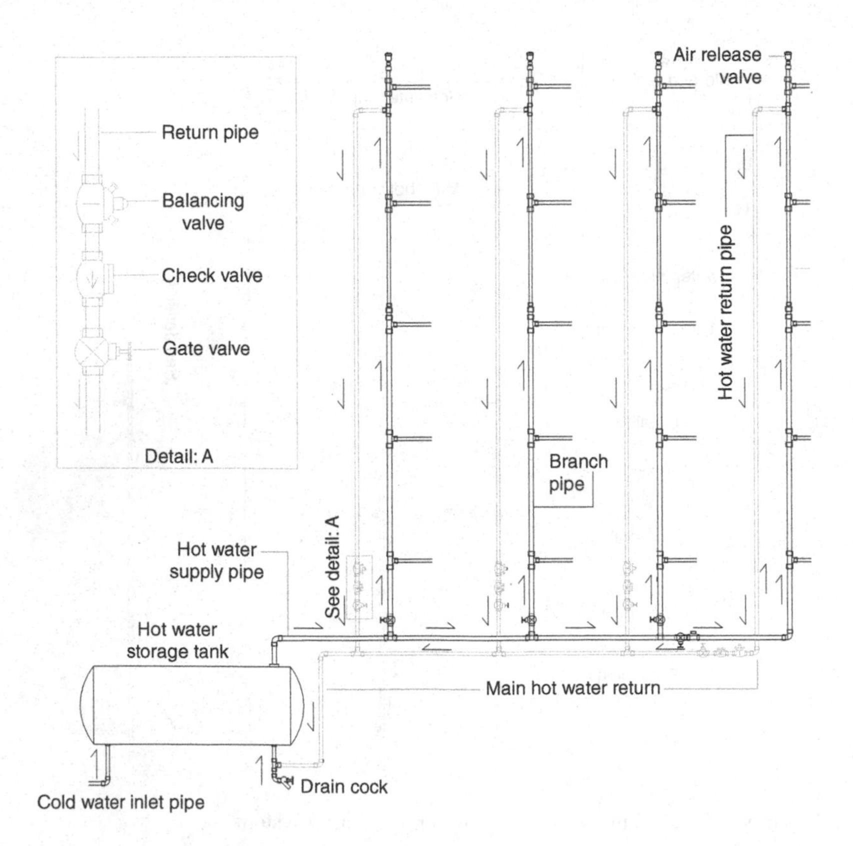

Figure 8.18 Pressurized up-feed hot water circulation system.

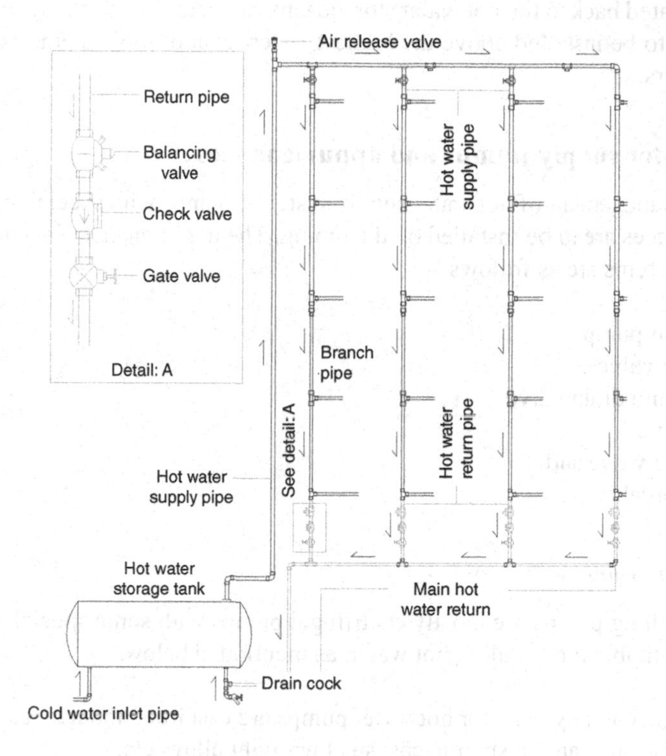

Figure 8.19 Pressurized down-feed hot water circulation system.

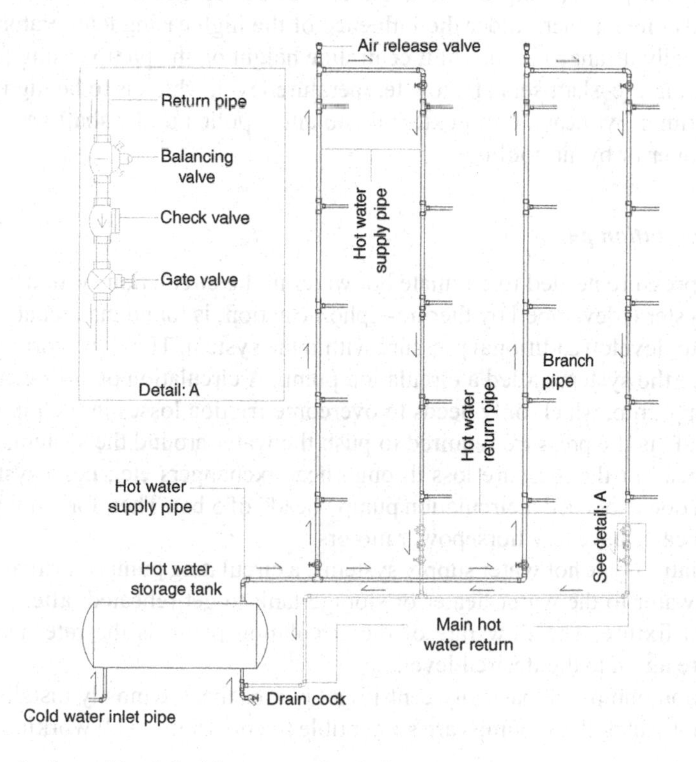

Figure 8.20 Pressurized combined-feed hot water circulation system.

water is circulated back to the hot water storage tank or heater for reheating. An automatic air relief valve is to be installed above the highest water level of the system, on the apex point of up-feed risers.

8.10 Hot water supply pumps and appurtenances

For efficient management of hot water supply systems, pumps with special features and various appurtenances are to be installed on the piping. The appurtenances installed in hot water distribution systems are as follows.

1 Circulation pump
2 Balancing valve
3 Isolating and drain valve
4 Check valve
5 Air release valve and
6 Vacuum breaker.

8.10.1 Hot water pumps

Hot water handling pumps are mostly centrifugal pumps with some special features and am arrangement suitable for handling hot water, as mentioned below.

1 Materials generally used for hot water pumps are cast iron, nodular cast iron, cast steel, cast chrome steel and austenitic cast steel wrought alloys etc.
2 To prevent any shifting of the alignment of centrelines of the pump and motor shaft towards one another, under the influence of the high casing temperature, the pump feet are usually arranged at the shaft centreline height on the pump casing [14].
3 Operating the shaft seal at a low temperature level, which is to be highly maintained, is done either by means of an external coolant supplied to the shaft seal chamber or heat exchanger or by air cooling.

8.10.2 Circulation pump

When the pressure needed to circulate hot water in the circulation system of a hot water distribution system, developed by thermo-siphonic action, is found inadequate, then it becomes necessary to develop additional pressure within the system. This is performed by introducing a pump into the system called a circulation pump. A circulation pump functions simply as a closed-loop pump, which only needs to overcome friction losses in the piping. A circulation pump "head" is the pressure required to push the water around the system, not to overcome the static head or the pressure loss through heat exchangers etc.: i.e. a system with a 50 m static head does not need a circulation pump "head" of 5 bar. Therefore, hot water circulation pumps typically have low horsepower motors.

In circulation type hot water supply systems, a circulating pump is primarily used to return the warm water to the water heater or storage tank to get reheated, after flowing to supply the farthest fixture. The flow rate of the circulating pump is the rate needed to raise the temperature again to the desired level.

Circulation pumps are basically centrifugal pumps that are mostly installed online. Due to handling hot water, these pumps are susceptible to corrosion, so all working parts exposed to hot water should be of brass, bronze, stainless steel or any other non-ferrous material, with special gaskets, seals and mountings.

The operation of circulating pump is controlled by incorporating two systems: 1. temperature controlled and 2. time controlled.

Temperature controlled: The operation of a circulating pump is governed by a temperature sensing circuit that will shut off the hot water circulation pump once the water temperature reaches a pre-set temperature and will start when the water temperature in the pipe drops below a second pre-set temperature. There are on-demand circulation pumps of smaller capacity that are generally used for a single fixture like a basin, bathtub or sink etc. These pumps are controlled by a temperature sensor fitted on the hot water supply line. When the hot water in the incoming supply line is detected, with a fall of temperature maximum 258k the electronic controller sets the pump to on and runs it until the temperature rises up to the desired level.

Time controlled: The operation of a circulating pump is governed by a timer that keeps the pump shut off during the period when hot water is not in use. The important advantage of the system is that the timer can be set to activate at specific time intervals depending on hot water demand.

8.10.3 Balancing valve

In circulation type hot water supply, to ensure efficiency and economy it is essential to balance the hot water piping loop so that at all circulation points, optimum flows are regulated to maintain the desired temperature. This is generally achieved by installing a temperature-balancing valve. Temperature-balancing valves measure the water temperature as it passes across its internal temperature-sensing element and adjust the flow of water accordingly. These valves are needed to be adjusted on site to achieve desired flow at controlled temperature.

Balancing valves are pressure independent and temperature-sensitive flow control valves that maintain the desired flow rate irrespective of pressure differential across the valve within the differential pressure range. When a hot water supply system is operating at less than design maximum load conditions, differential pressures tend to increase, resulting in overflow through the control valves. Balancing valves prevent overflow at conditions above design maximum flow. At any flow condition less than design maximum, balancing valves do not fulfill any purpose. Balancing valves are also known as commissioning or double regulating valves. In hot water supply systems, they are generally installed on return pipes to balance the flow between various parts of the system.

8.11 Hot water supply in skyscrapers

In skyscrapers, the hot water supply system is also developed to manage the system pressure within the limit, as in the cold water supply system. Following are some existing methods of hot water supply for super tall buildings.

1 Vertical zoning of hot water supply systems, in which the height of a zone is defined by the value of permissible hydraulic pressure for heaters and armature located at lower levels
2 Installation of pressure controllers for the heaters and
3 Installation of appliances, equipment, accessories and faucets capable to withstand higher pressure.

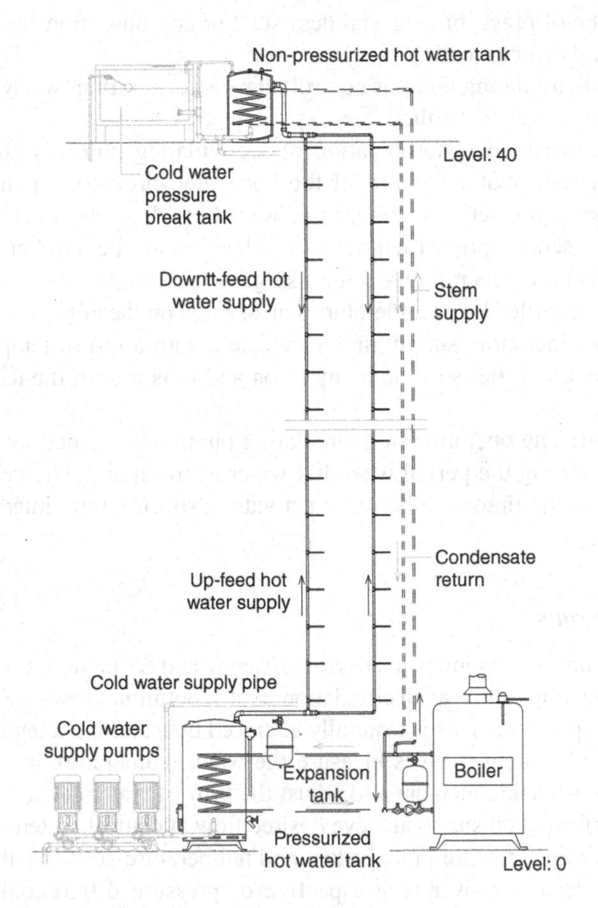

Figure 8.21 Hot water supply system in skyscrapers.

In high-rise buildings, multiple "pressure zones" might be required, depending on the height and vertical extension of the building. In a multiple zoning system, both pressurized up-feed and gravitational down-feed systems can be employed, as illustrated in Figure 8.21. Each zone has its own hot water tank, generally indirectly heated from its remotely located boilers at the bottom of each respective zone. Each zone has, at its base or top, its own water heater and a hot water circulation system. If space is not available to install storage heater tanks at the top of each zone, temperature management is done by installing instantaneous or semi-instantaneous heaters on the steam supply risers running from the boilers in order to supply hot water to an acceptable limit of temperature in each zone. The operating pressure of the hot water supply system must not exceed the maximum operating pressure of the heater's casing. There must be compatible expansion equipment for each pressure break or zone.

8.12 Insulation of hot water supply elements

The primary objective of providing insulation is to save energy by minimizing heat loss from hot water conveying pipes and storage tanks. Good insulation offers hot water at the point of use that is closer to the desired temperature than it would be without insulation. As a result, insulated hot water systems will save energy by avoiding frequent reheating of warm water and wastage of water. If insulation is not done properly, the operating cost of a hot water supply system may increase up 10 to 15 per cent due to rapid thermal loss and more energy required for reheating [15].

Generally, insulation of pipes and their appurtenances is are communicated in terms of thermal conductivity, k-value or λ value. A k-value, expressed in watts per meter Kelvin (W/(m K)) or λ value in W/m °C (watts per meter per degree Centigrade), is the rate of heat flow through a homogeneous insulating material. The lower the k or λ value, the greater the insulating value for a given set of conditions.

8.12.1 Insulating tank

In the past, external insulation blankets on water heater tanks were widely used. Now, water heater tanks are built with much higher levels of internal insulation. The hot water tank cylinder is generally insulated with 75 mm fibreglass insulation. In addition, some heaters may be provided with insulating outer jacket or "skin" that helps keeps heat and noise inside the heater and improves its operation.

The tank may be insulated with rigid polyurethane foam or flexible foam insulation with foil cladding that completely surrounds the tank. The tank insulation must be a value of R-16. Insulation thickness is about 50 mm. Heat loss of an insulated storage type water heater may be allowed less than 10°C/hr.

8.12.2 Insulating pipe

Hot water pipes are to be insulated wherever they are exposed and accessible, particularly when located in unheated areas. There are various types of insulation for pipes. Foil-faced fibreglass strip type insulation or foam insulation sleeves are generally used to insulate pipes.

Strip type insulation is made by rolling the strip around the pipe. Foam insulation sleeves are easy to install on pipes. The foam sleeve simply slips onto the pipe. In the case of fitting a sleeve insulator on an existing pipe, it needs to be cut along its length and then fit over the pipe, as shown in Figure 8.22. After fitting the insulator, the cut is made good with adhesive

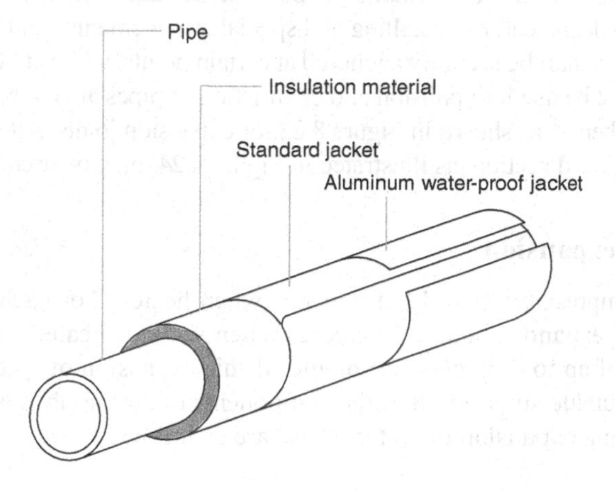

Figure 8.22 Pipe insulation

Table 8.7 Thickness of insulation for various diameters of pipe in hot water supply [16].

Nominal pipe size NPS (mm)	*Recommended minimum thickness of insulation (mm)**			
	Temperature range (°C)			
	50–90	*91–120*	*121–150*	*151–230*
	Hot water	*Low pressure steam*	*Medium pressure steam*	*High pressure steam*
Less than 25	25	40	50	63
32–50	25	40	63	63
63–100	40	50	63	75
125–150	40	50	75	88
200 and larger	40	50	75	88

Based on insulation with thermal resistivity in the range 7–8 W/mK (typical for mineral wool at room temperature).

tape. A good insulating material should be of adequate thickness. Hot water pipes are generally insulated with a minimum of 25 mm of insulation. For good performance, the insulating material shall possess the following characteristics.

1 It should not be flammable.
2 It should be vermin-proof.
3 It should be impervious to moisture and
4 It should be sufficiently robust for its purposes.

As a measure of energy conservation for hot water supply pipes, the insulation thickness should be adequate according to the values given in Table 8.7.

8.13 Pipe enlargement

Pipes conveying hot water are subjected to variation in length and form due to change in temperature of the hot water flowing inside. Due to a rise in temperature, a pipe elongates and expands. If this enlargement in pipe dimension is not well taken care of, the piping system will be subjected to undue stress and strain, resulting in possible damage to joints, fittings, valves etc. To overcome these problems, careful installing and special arrangements shall be made on the piping system. The pipes shall be securely anchored at certain points, while at other points sliding or flexible hangers must be used. Expansion and contraction of pipes are taken care of by the addition of large radius bends, as shown in Figure 8.23, or expansion joints. Alternatively, a U-bend, an offset or a change in direction, as illustrated in Figure 8.24, may be used for the purpose.

8.14 Hot water expansion

Water is a non-compressible liquid but expands when heated. For each 261 K increase in temperature, water expands about 0.2 percent. When water is heated from 4°C to 99°C, it causes expansion of up to 4 percent of volume. If this expansion of water is not well managed, it can create undue stress in all of the components of the supply system. There are two methods of managing expansion of water. These are as follows.

1 The open vent pipe system and
2 Expansion vessel or tank.

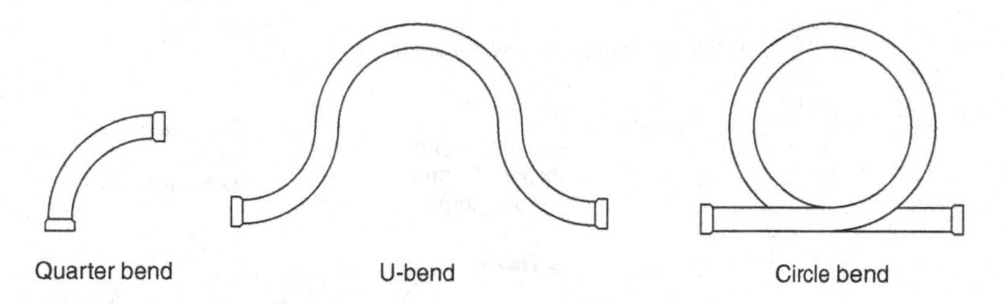

Figure 8.23 Factory-made pipe expansion bends.

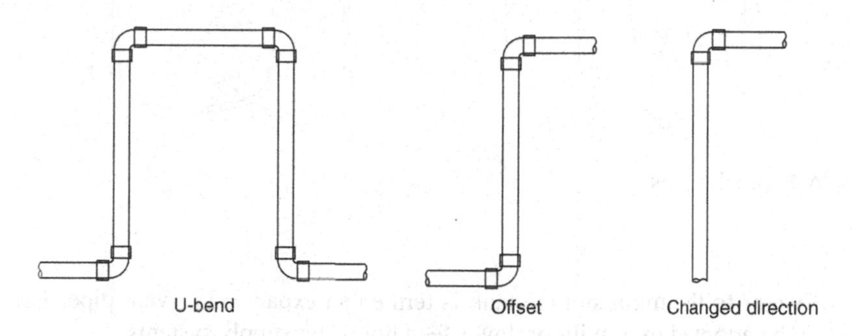

Figure 8.24 Various pipe configurations used to accommodate pipe expansion due to thermal change.

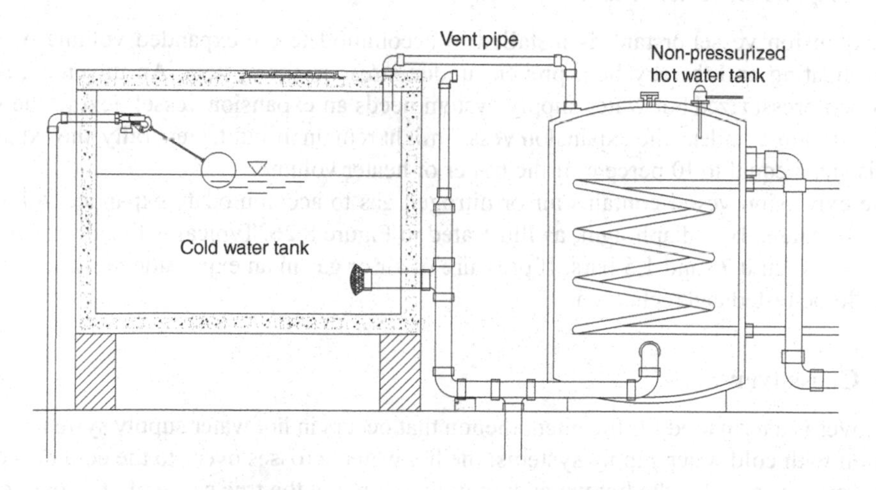

Figure 8.25 Vent pipe on water heaters.

8.14.1 The open vent pipe system

In this system, a vent pipe is installed on the water heater, which is generally extended to and terminated in the cold water storage tank, as shown in Figure 8.25. It is also a safety measure to address the expansion of hot water, as the open vent system does not cause an increase of pressure due to expansion of water volume. Vented hot water systems are desirable for heating up a large volume of water. The pipe used in open vented systems to convey the expanded

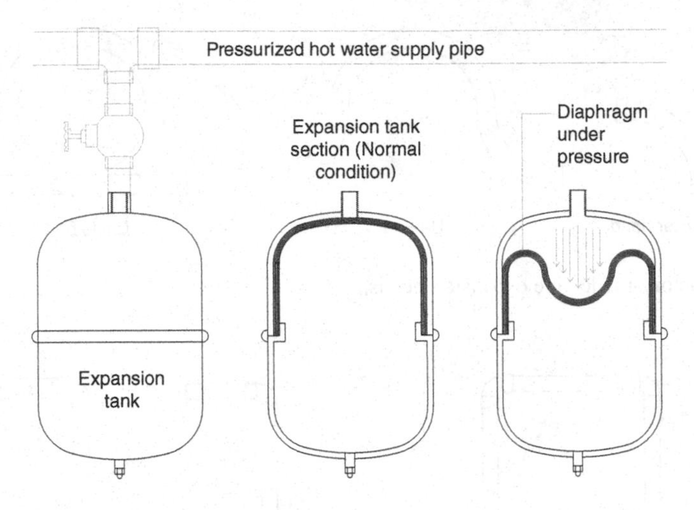

Figure 8.26 Expansion tank.

volume of water to the main storage tank is termed an expansion or vent pipe. The vented systems can be adopted in gravity or down-feed hot water supply systems.

8.14.2 Expansion vessel or tank

The expansion vessel or tank is installed to accommodate the expanded volume of water due to heating and thereby help prevent undue stress in the system. An unvented, sealed or looped pressurized hot water supply system needs an expansion vessel next to the water heaters. In large boilers, the expansion vessel might remain in-built. Generally, the expansion tank is sized equal to 10 percent of the boiler or heater volume.

The expansion vessel contains air or nitrogen gas to accommodate expanded volume of water, separated by a diaphragm, as illustrated in Figure 8.26. Typically, the air or gas pressure is between 0.75 and 1.5 bars. If pressure of air or gas in an expansion vessel decreases, it should be tested and recharged.

8.15 Crossover

Crossover is a term used for the phenomenon that occurs in hot water supply systems in conjunction with cold water supply systems: the hot water "crosses over" to the cold or the cold water "crosses over" to the hot water supply line. Finding the true cause of crossover can be very difficult, as there might be various reasons. Malfunctioning check valves on circulation systems might allow water to flow the wrong direction, while pressure balancing, flow control or temperature-sensing components of the respective valves used in the supply system can also malfunction, which will cause crossover.

The most common understanding is that when pressure in the hot and cold water lines differs widely, water will forcibly pass from high pressure to low. Though both hot and cold water piping systems are set separately, there is a point in the valve or faucet hot and cold water mixes where crossover occurs. So, in order to minimize the pressure difference in both hot and cold water supply pipes at a particular point, both hot and cold water supply, in particular a pressure zone or circuit, shall be developed following the same pressurized up-feed or gravitational down-feed system, drawing cold water from the same source of cold water storage tanks.

8.16 Sizing hot water supply components

All of the elements involved in a hot water supply system are to be rightly sized for efficient performance of the system. A well-designed hot water supply system provides hot water at the desired temperature to the outlet with little or no delay and saves energy. For this purpose, the following components of hot water supply systems in a building need to be designed.

1 Water heater sizing.
2 Piping
   a Hot water supply pipe (riser)
      i Branch supply pipes
      ii Vertical supply pipes (riser)
   b Hot water return pipes
3 Circulation pump.

8.16.1 Sizing water heaters

Storage type water heater sizing: Sizing storage type water heaters does not follow any definite procedure; rather, it is to be chosen, out of the following methods of determining size, based predominantly on experience. An oversized unit will incur higher costs and increased energy costs due to higher standby energy losses. On the contrary, an undersized heater will not meet satisfaction. So, to get the right size for a heater, the following major factors are to be considered.

1 Simultaneous use of hot water using appliances and faucets are the fundamental determining factors
2 The number of people combined with how and when hot water is used
3 Flow rate of shower (minimum 9 lpm or standard 18 to 25 lpm)
4 Capacity of bathtub, washing machine, dishwasher etc.
5 Climatic condition affecting cold water temperatures and
6 Storing hot water at a higher temperature increases the system's overall capacity but reduces the storage volume.

For storage tank heaters, two factors are very important for sizing: the amount of water to be held and the recovery rate. The recovery rate is expressed as "first hour rating" (FHR). Recovery rate is the amount of hot water the water heater is capable of providing in a given period of time, generally in one hour. The amount of hot water provided depends on the input rating of the water heater. The amount of fuel consumed by a gas heater in an hour is the input rating of a heater. Input for the electric water heater is expressed in kilowatts (kW).

One kW is equal to 3,413 BTUs of electricity. The storage capacity, C, in litres of water heater needed can be found from the following formula.

$$C = T(D - H) \quad 8.2$$

Where
D = dump load i.e. the maximum rate at which a facility can use hot water (lph)
T = time of peak demand (hours) and
H = heating capacity of water heater (lph).

Example: For a dump load of 600 lph, use of water heater with a heating rate of 300 lph during peak using time of 2.5 hours, storage capacity for the heater would be needed.

$2.5 \times (600 - 300) = 750$ litres

Basically, the amount of water to be held primarily depends upon the consumption rate through simultaneous use of various faucets and appliances. Heater tank capacity and flow rate must satisfy the peak usage rate. Among the fixtures, showers are the key determining elements. So, very simply, the minimum size of a storage water heater can be well estimated by multiplying the shower-head flow rate by the average length and number of simultaneous shower use in a given period of time.

Instantaneous water heater sizing: Instantaneous water heaters must be sized to provide hot water of at least 49°C and at a rate of at least 7.5 lpm to each faucet. The summation of the likelihood of faucet flow rate to be used simultaneously should be less than the flow rate of the heater. Consider the flow rate of lavatories and food preparation sinks (1.9 lpm), shower heads (7.5 lpm), three-compartment sinks (7.5 lpm) etc.

8.16.2 Sizing expansion tank

The size of an expansion tank V_f can be determined from the following formula, based on Boyle's Law.

$$V_f = \frac{e \times C}{1 - (P_i / P_f)} \quad 8.3$$

Where
e = expansion coefficient; calculated from Table 8.8, corresponding to the difference between the lowest water temperature (heating off) and the max working temperature
C = total capacity of water of the system, in litres; considered to be between 4 and 8 litres for every kW of boiler output
P_i = initial charge pressure (absolute) of vessel; pressure must not be lower than the hydrostatic pressure at the point where the tank is connected to the system and
P_f = maximum operating (absolute) pressure of the pressure relief valve.

8.16.3 Sizing the branch supply pipes

The basic principles of hot water hydrology are as same as normal water hydrology. So, water supply pipe sizing for normal water can be followed for hot water supply pipe sizing. The velocity of flow in hot water supply piping shall be no more than 1.5 meters per sec (mps) [17].

Table 8.8 Expansion coefficient corresponding to temperature [18].

°C	*Expansion coefficient*
0	0.00013
10	0.00025
20	0.00174
30	0.00426
40	0.00782
50	0.01207
60	0.0145
65	0.01704
70	0.0198
75	0.02269
80	0.0258
85	0.02899
90	0.0324
95	0.0396
100	0.04343

8.16.4 Sizing the return pipe

The sizing of hot water return pipes can be done based on the heat loss method. The heat loss in the circulation system must not exceed 10°C; 60°C to 50°C [19]. So, the maximum temperature drop from the heater to the farthest faucet shall be 5°C and, in the return pipe, rest 5°C can be lost. The velocity of flow in hot water return piping shall be no more than 1.2 meters per second (mps) [18].

The required circulated water flow in the return pipe can be calculated by the following formula

$$Q = q/(\rho C_p dt) \tag{8.4}$$

Where
Q = pump capacity (cum/s)
h = heat loss from the pipeline (kilo watt)
ρ = density of water (kg/cum) (988 kg/cum at 50°C)
Cp = specific heat of water (J/kg °C) (4182 J/kg °C at 50°C) and
dt = temperature drop (°C).

The simplified equation for flow rate in cubic meters per second (cum/s) can be found as below.

$$Q = h / 4200dt \text{ cum/s} \tag{8.5}$$

Where
Q = water flow in cubic meter per second (cum/s)
h = heat flow rate (kW or kJ/s) and
dt = temperature difference (°C).

Heat loss in insulated pipe is assumed to be about 30 watts per meter. Considering the average specific heat loss of about 30 wpm from a pipe 200 m long (100 m supply and 100 m return length), the flow needs to compensate for a 10°C temperature drop.

$$200(\text{m}) \times 30/1000\ (\text{kW}) / (4200 \times 10\ (^{\circ}\text{C})) \times 1000 = 0.143\ \text{liter/s}$$

From Table 7.15 in Chapter 7, for velocity at 1.2 mps (though recommended 1.5 mps) for flow of 0.143 lps, the corresponding pipe size would be a 12 mm pipe.

The ASHRAE applications handbook suggests a simplified rule of thumb for flow consideration in return pipes. It is estimated that 3.8 lpm return flow should be considered for every 20 hot water fixtures, or should allow 1.9 lpm for each 20 or 25 mm riser; 3.8 lpm for each 32 mm and 40 mm riser; and 7.5 lpm for 50 mm or larger hot water supply risers [20].

8.16.5 Sizing the circulation pump

The capacity of the circulation pump used in hot water distribution systems must be carefully chosen so that it is not oversized. If the pump is oversized, risk of corrosion increases due to high flow velocity eroding the pipe wall. Like all other pumps, it is also sized by its flow capacity and the head needed for the hot water circulation system.

Flow: The total flow of the circulation pump is determined by the heat loss occurred in the hot water supply pipe from the heater to the farthest fixture when the fixtures are not in use. The required flow through the hot water distribution system is a function of the heat loss over time and the allowable temperature drop to the balancing valves. In selecting and sizing the hot water circulating pump, a total of 10°C temperature loss – 5°C loss from the water heater out to the farthest fixture and 5°C loss from the farthest fixture back to the circulator near the water heater – is considered. In this case, formula 8.5 can be used to determine the flow required for the circulation pump, which can be modified as below.

$$\text{Flow (lps)} = \text{System heat loss (kj/h)} / (15077 \times \text{Temperature drop } (^{\circ}\text{C}))$$

Head: For the pump head requirement, the appropriate flow (lpm) is assigned to each section of pipe based on the BTUH loss requirements above, and from pipe friction loss charts, a total head (m) or pressure drop in kilo Pascals (kPa) can be determined. A quick and simple way to estimate heat loss in insulated pipe is to assume 43 to 52 w/mK, ignoring the hot water supply and return pipe size. Using the friction loss tables, the "head" requirements for the circulating flow through the longest run of pipe in the system can be calculated.

Power: The BHP required for the circulation pump is to be calculated in the same way as the BHP for general centrifugal pumps.

References

[1] Environmental Services Department (ESD) (2015) "Guideline to Sizing Water Heaters, Water Temperature and Potable Water Supply Requirements" p. 4, Retrieved from www.maricopa.gov/DocumentCenter/View/5888/Hot-Water-Supply-Requirements-PDF on 26 December 2020.

[2] American Society of Plumbing Engineers (ASPE) (2015) "Domestic Hot Water Systems" *Reprinted from Plumbing Engineering Design Handbook* 2, p. 7.

[3] State Government of Victoria (2014) *About . . . Hot Water Safety*, Victorian Building Authority (VBA), Victoria.

[4] Bradford White Corporation (2020) "Pressure Damaged Water Heaters (#115)" Retrieved from www.bradfordwhite.com/pressure-damaged-water-heaters-115/ on 26 December 2020.
[5] Thermostatic Mixing Valve Manufacturers Association (TMVA), "Recommended Code of Practice for Safe Water Temperatures" Retrieved from file:///C:/Users/pm/Downloads/Recommended%20Code%20of%20Practice%20for%20Safe%20Water%20Temperatures%20(3).pdf on 17 June, 2021.
[6] ASHRAE Application Handbook, Chapter 45, Table 4.
[7] Engineering Tool Box (ETB) (2003) "Hot Water Consumption per Occupant" Retrieved from www.engineeringtoolbox.com/hot-water-consumption-person-d_91.html on 27 December 2020.
[8] Smith Corporation (2006) "Residential Gas and Electric Water Heater" *Service Handbook*, p. 10, Retrieved from www.hotwater.com/lit/training/320991-000.pdf.
[9] Ho, Benjamin P.L. (2010) "Hot Water Supply" Presentation Slides, Department of Mechanical Engineering, The University of Hong Kong, Slide 28, Retrieved from http://ibse.hk/MEBS6000/mebs6000_1011_02_hot_water_supply.pdf on 27 December 2020.
[10] U.S Department of Energy, "Heat Pump System" Retrieved from www.energy.gov/energysaver/heat-and-cool/heat-pump-systems on 27 December 2020.
[11] American Society of Mechanical Engineers (ASME) (2013) "ASME Code for Pressure Piping" B31 ASME B31.1–2001, p. 101, Retrieved from www.nrc.gov/docs/ML0314/ML031470592.pdf on 27 December 2020.
[12] Minnesota Office of the Revisor of Statutes (2016) "Minnesota Administrative Rules" 4714.0603 Cross-Connection Control, Section 603.5.4, Retrieved from www.revisor.mn.gov/rules/4714.0603/on 27 December 2020.
[13] International Plumbing Code, 2015 Section 607.2.
[14] KSB, "Hot Water Pump" Retrieved from www.ksb.com/centrifugal-pump-lexicon/hot-water-pump/192280/#:~:text=In%20order%20to%20prevent%20any,pumps%20in%20ring%2Dsection%20design on 27 December 2020.
[15] Cyril, M. Harris (1991) "Practical Plumbing Engineering" McGraw-Hill, New York.
[16] Engineering ToolBox, (2003). Piping – Recommended Insulation Thickness. [online] Retrieved from https://www.engineeringtoolbox.com/pipes-insulation-thickness-d_16.html on 27 December, 2020
[17] BNP Media (2015) "Central Hot Water in High-Rise Buildings" Issue: 4/05, Retrieved from www.pmengineer.com/articles/85190-central-hot-water-in-high-rise-buildings on 27 December 2020.
[18] Automatic Heating Global Pty Ltd, "Expansion Tank Sizing and Commissioning" Retrieved from www.automaticheating.com.au/solutions/expansion-tanks-pressurisation-systems/expansion-tank-sizing-commissioning-maintenance/ on 27 December 2020.
[19] Engineering Tool Box (ETB) (2016) "Hot Water-Return Pipe" Retrieved from www.engineeringtoolbox.com/hot-water-circulation-return-pipe-d_1918.html.
[20] Continuing Education and Development, Inc, "Design Considerations for Hot Water Plumbing" Course No: M06–029 by A. Bhatia, pp66, NY 10980, Part 6, p. 66. Greyridge Farm Court Stony Point, NY, Retrieved from www.cedengineering.com/userfiles/Design%20Considerations%20for%20Hot%20Water%20Plumbing.pdf on 27 December 2020.

9 Plumbing fixtures and appliances

9.1 Introduction

Plumbing fixtures are the first element of a drainage system for a building, which receives and generates wastewater and eventually disposes of it into the drainage pipe. The majority of the fixtures are supported by water supply systems to get water to the user, and a discharge outlet pipe for disposing of wastewater in the drainage piping.

The major portion of used water is released as wastewater, which is received or collected by the fixture and then discharged through the discharge outlet. As the fixtures are close and exposed to the user while working, their size, shape, number, position material etc. directly affect users. In this chapter, all aspects and issues related to these concerns of fixtures are discussed.

9.2 Fixtures

Plumbing fixtures are a kind of receptacle for human body-related wastes which are ultimately discharged into the drainage system, through its discharge outlet. Plumbing appliances are a special class of plumbing items intended to perform special functions. Various types of fixtures and appliances are used in the plumbing system which must meet various requirements of the users.

9.2.1 Basic requirement for fixtures and appliances

To achieve the objectives of a plumbing system in a building, the fixtures must fulfil some requirements to satisfy the health, ergonomic and aesthetic need of users. The following are those basic requirements.

1 Fixtures should be made of non-absorbent, non-toxic material
2 Fixtures shall be so designed that they are free of foul spaces and sharp edges
3 Fixtures must be supported by non-corroding drainage fittings
4 Fixtures requiring retention of water must have plugging and overflow arrangements
5 Fixtures' surface finishing shall be made very smooth and shiny and
6 Fixtures' size and shape should be user-friendly.

9.2.2 Common fixtures and appliances

There are various types of plumbing fixtures used in buildings for various human needs, particularly their health and sanitation. Common fixtures and appliances are mentioned below.

DOI: 10.1201/9781003172239-9

Fixtures:

1 Lavatories
2 Bathtub and trays
3 Kitchen sinks
4 Water closets
5 Urinals and
6 Bidets.

Appliances:

1 Flushing devices
2 Drinking fountains
3 Washing machines and
4 Interceptors.

Fixture trims: These are the parts that mostly remain attached to fixtures or used for their installation, operation and use but do not include any fittings. Almost all of the fixtures more or less have trims of different features.

9.2.3 Materials for fixtures

Plumbing fixtures are manufactured of different materials, but the most common materials are as follows.

1 Ceramic
2 Porcelain
3 Stainless steel and
4 Plastic.

Ceramic: Theoretically, ceramics are non-metallic and inorganic solids prepared through a process of heating and cooling. Through heating, ceramics are basically made dense. It is, by property, a non-reactive and refractory material, capable of withstanding extremes of temperature, general wear and tear and attacks from acids and alkalis. By definition, graphite and diamond are also ceramic.

Porcelain: Porcelain is the hardest and strongest clay material. Porcelain is tough in durability, and translucent in visual appearance. The ceramic material is heated in a kiln at an extremely high temperature, causing the clay to become less dense and porous. It is the most common material that is used to make the majority of plumbing fixtures because of this. Porcelain is also one type of ceramic.

Vitreous china: The coating of vitreous china acts as a protective armour for porcelain and ceramic items due to possessing durable sanitary and stain resistant properties. The vitreous china coating is applied later on in the process of heating the element on which it is to be coated.

Stainless steel: Stainless steel fixtures made of heavy gauge sheets are sleek in design, very strong, durable, heat-resistant and comparatively cheap. Stainless steel does not tarnish and does not require any special cleaning products. Stainless steel surfaces might exhibit smudges and scratches from daily use. Generation of noise is associated with the use of stainless steel fixtures and must be taken care of where necessary.

Plastic: Plastic fixtures might be popular only on economic grounds, as plastic fixtures are the cheapest of materials used for making plumbing fixtures. Plastic fixture are light and so easy to install. Plastic fixtures are not durable. In plumbing fixtures, there will develop environmental stress, cracking and surface cracking which might provide a niche for pathogens.

9.2.4 Materials for faucets and accessories

1 **Bronze:** More expensive than other materials, bronze is durable. Its darker color provides an antiquated feel.
2 **Brass:** Brass is an alloy of copper and zinc with small amounts of other metals. Brass as a material is highly recommended for faucets, though it is expensive. Brass is dense and sturdy, corrosion less, and so long-lasting. Brass faucets with a brass finish tend to be durable and easy to clean. There is chance of breaking down of some types of brass, over time.
3 **Zinc alloy:** Zinc alloy is made by adding aluminium, magnesium and copper to zinc. Zinc is less costly and less durable than other metal faucets. Zinc and zinc-alloy faucets, in contact with water, tend to corrode.
4 **Stainless steel:** Stainless steel is slightly less expensive than brass. Steel generally rusts well before other materials when exposed to water.
5 **Plastic:** Plastic is the least expensive but is not durable.

9.2.5 Finishing of metals

Finishes are not only a matter of choice; the right finish keeps metal well protected. Various types of finished surface of metals are available.

1 **Chrome:** Chrome is popular and versatile as a finished surface. It is tough to scratch or damage. There is possibility of water spots on a chrome finish, which is difficult to clean.
2 **Bronze:** Satin bronze and oil-rubbed bronze-finished faucets are made. Bronze finishes adopt either physical vapor deposition (PVD) or clear epoxy coating processes. PVD coating is done by charged metal atoms that bond chemically to the surface material, which helps resist corrosion and scratches. Bronze-finished faucets are durable and easy to maintain but more expensive.
3 **Nickel:** Nickel finishes are the most long-lasting, but they are also the most expensive.

9.2.6 Allocation of fixtures

A number of plumbing fixtures in toilets for buildings other than residential buildings shall be provided with respect to the number of users of the toilet. So, a sufficient number of fixtures must be provided for the convenience and comfort of users. Table 9.1 furnishes the recommended number of fixtures to be provided with respect to the number of users.

9.3 Lavatories

Lavatories, the generic name for basins or bowls, are the fixtures, basically used to receive wastewater generated while abluting, washing, cleansing etc. and then discharged into the drainage system through its discharge outlet. The bowl capacity varies between 4 and 5 litres. In the bowl, there is an overflow outlet about 25 mm below the rim level, up to which

Table 9.1 Recommended numbers of fixtures provided with respect to the number of users [1].

Occupancy		*Water closets*		*Lavatories*	*Bathtubs/ Showers*	*Drinking fountains*	*Others*
		Male	*Female*				
Assembly	Nightclubs	1 per 40	1 per 40	1 per 75	–	1 per 500	1 service sink
	Restaurants [g]	1 per 75	1 per 75	1 per 200	–	1 per 500	1 service sink
	Theatres, hall, museum, etc.[g]	1 per 125	1 per 65	1 per 200	–	1 per 500	1 service sink
	Coliseums, arenas (less than 3000 seats)	1 per 75	1 per 40	1 per 150	–	1 per 1000	1 service sink
	Coliseums, arenas (3000 seats or greater)	1 per 120	1 per 60	Male: 1 per 200 Female: 1 per 150	–	1 per 1000	1 service sink
	Churches[b,g]	1 per 150	1 per 75	1 per 200	–	1 per 1000	1 service sink
	Stadiums (less than 3000 seats), pools, etc.[g,h]	1 per 100	1 per 50	1 per 150	–	1 per 1000	1 service sink
	Stadiums (3000 seats or greater)[g]	1 per 150	1 per 75	Male: 1 per 200 Female: 1 per 150	–	1 per 1000	1 service sink
	Business	1 per 50		1 per 80	–	1 per 100	1 service sink
	Educational	1 per 50		1 per 50	–	1 per 100	1 service sink
	Factory and industrial	1 per 100		1 per 100		1 per 400	1 service sink
	Passenger terminals and transportation facilities	1 per 500		1 per 750	–	1 per 1000	1 service sink
Institutional	Residential care	1 per 10		1 per 10	1 per 8	1 per 100	1 service sink
	Hospitals, ambulatory nursing home patients[c]	1 per room[d]		1 per room[d]	1 per 15	1 per 100	1 service sink floor
	Day nurseries, sanatoriums, non-ambulatory nursing home patients etc.[c]	1 per 15		1 per 15	1 per 15[e]	1 per 100	1 service sink
	Employees other than residential care[c]	1 per 25		1 per 35	–	1 per 100	–
	Visitors other than residential care	1 per 75		1 per 100	–	1 per 500	–
	Prisons[c]	1 per cell		1 per cell	1 per 15	1 per 100	1 service sink
	Asylums, reformatories etc.[c]	1 per 15		1 per 15	1 per 15	1 per 100	1 service sink

(*Continued*)

Table 9.1 (Continued)

Occupancy		*Water closets*		*Lavatories*	*Bathtubs/ Showers*	*Drinking fountains*	*Others*
		Male	*Female*				
Mercantile		1 per 500		1 per 750	–	1 per 1000	–
Residential	Hotels, motels	1 per guestroom		1 per guestroom	1 per guestroom	–	1 service sink
	Lodges	1 per 10		1 per 10	1 per 8	1 per 100	1 service sink
	Multiple families	1 per dwelling unit		1 per dwelling unit	1 per dwelling unit	–	1 kitchen sink per dwelling unit; 1 automatic clothes washer connection per 20 dwelling units
	Dormitories	1 per 10		1 per 10	1 per 8	1 per 100	1 service sink
	One- and two-family dwellings	1 per dwelling unit		1 per dwelling unit	1 per dwelling unit	–	1 kitchen sink per dwelling unit; 1 automatic clothes washer connection per dwelling units[f]
Storage		1 per 100		1 per 100		1 per 1000	1 service sink

a The fixtures shown are based on one fixture being the minimum required for the number of persons indicated or any fraction of the number of persons indicated. The number of occupants shall be determined by the International Building Code.

b Fixtures located in adjacent buildings under the ownership or control of the church shall be made available during periods the church is occupied.

c Toilet facilities for employees shall be separate from facilities of inmates or patients.

d A single-occupant toilet room with one water closet and one lavatory serving no more than two adjacent patient rooms shall be permitted where such room is provided with direct access to the toilet from each patient room and with provisions for privacy.

e For day nurseries, a maximum of one bathtub shall be required.

f For attached one- and two-family dwellings, one automatic clothes washer connection shall be required per 20 dwelling units.

g In assembly and mercantile occupancies, a unisex toilet room shall be provided where an aggregate of six or more male or an aggregate of six or more female water closets are required. In buildings of mixed occupancy, only those water closets are required for the assembly and mercantile occupancy shall be used to determine the unisex toilet requirement.

h In recreational facilities where separate-sex bathing rooms are provided, a unisex toilet room shall be provided. Where each separate-sex bathing room has only one shower or bathtub fixture, a unisex bathroom is not required.

water can be retained in the basin. Lavatories are designed in various styles, shapes, sizes and colours and are made of various materials like porcelain, stainless steel, glass, plastic etc.

Lavatory trim: Lavatory trims consist of a faucet, draining pipe and plugging stopper. Faucets may be a single pillar cock or a single spout mixture faucet and are installed on the basin to control the water flow. A rubber stopper, attached by a chain to the bowl, or a push-type stopper is fitted to plug the drainage outlet at the bottom of the bowl.

Types of lavatories: There are a number of different lavatories available, with different styles for specific purposes of use, as mentioned below.

1 Wall hung basin
2 Pedestal basin
3 Counter top basin and
4 Corner basin.

9.3.1 Wall hung basin

Most basins are supported on the wall, specified as wall-hung basins, and some are supported by a pedestal which hides the waste pipe, called a pedestal basin. Some types of lavatories are designed to be installed on a countertop and so are named counter basins. A few basins are designed to be placed in corners to make the best utilization of space and are called corner basins.

The wall-hung lavatory is mostly fixed to the back wall with screws or bolts. Previously, it was installed by placing it on a pair of brackets screwed to the wall. Figure 9.1 shows diverse views of a wall-hung lavatory.

Installation: To install this fixture, the steps below are to be followed.

1 Mark the wall at the correct height for a lavatory and secure a hanger to the wall
2 Position the lavatory on the hanger
3 Install the lavatory faucets using a basin wrench
4 Fit the plug-drain or the pop-up type drain on the discharge opening of the basin

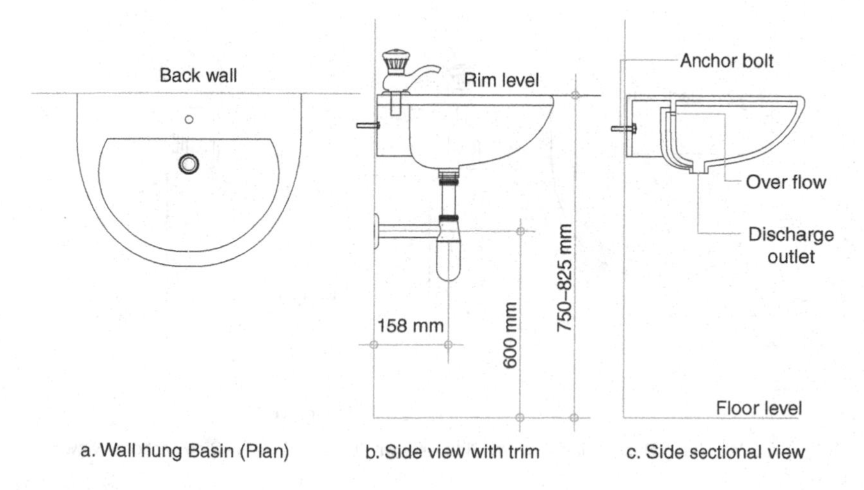

Figure 9.1 Wall hung basin.

5 Connect the water-supply lines to the faucet by a flexible connector pipe and
6 Connect the wastewater drain pipe under the discharge outlet of the basin.

9.3.2 Pedestal basin

For a pedestal basin, a pedestal is installed to hide the trim under the basin, as shown in Figure 9.2. Though the basin rests on the pedestal, it is fixed on the back wall by screws or bolts. In installing the pedestal basin, it is to be ensured that the basin is correctly positioned on the pedestal and that the back of the basin is in full contact with the wall.

9.3.3 Counter basins

Counter basins are installed on a counter slab, hence its name. There are three ways counter basins can be installed, as shown in Figure 9.3. According to the methods of installation, they are also named in different ways as below.

1 Countertop basins
2 Built in counter basin and
3 Undercounter basin.

Installation: All types of countertop basins are placed on the counter slab in such a way that its rim level is maintained the same height from the floor. For installing the basin in position, the discharge outlet or shape of the basin shall be perfectly marked on the counter slab and cut accordingly.

The rim of built-in counter basins is semi-recessed. The semi-recessed projected rim is supported on the counter top, as shown in Figure 9.3. A waterproof flexible mastic compound (silicone sealant) is to be applied to the bottom sealing surface of the rim of the basin, all around, to firmly fit the basin onto the counter slab-top.

The undercounter basin is installed just under the counter with the help of multiple metal brackets around the rim to hold the basin, and then screwed the bracket to the counter slab to support the basin, as shown in Figure 9.3. In this case, a waterproof flexible mastic compound

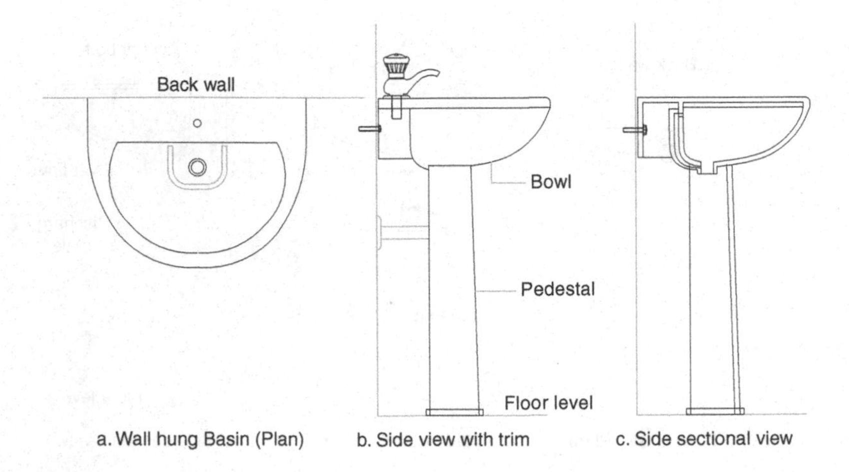

Figure 9.2 Pedestal basin.

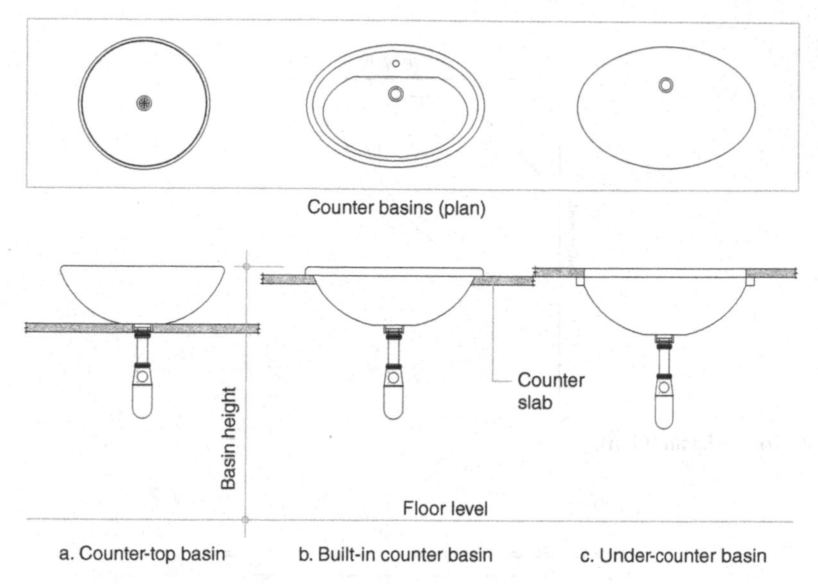

Figure 9.3 Counter basins.

(silicone sealant) is to be applied to all around the top sealing surface of the rim of the basin to fit the basin under the counter slab-bottom.

9.3.4 Corner basin

Corner wash-basins are preferred for restricted space applications. Corner basins are designed to be placed at the corner and mostly mounted on the walls, as shown in Figure 9.4. Pedestals can also be provided under these basins. The basins are mounted on the wall by metallic mount brackets and fixed with collapsible wing nut bolts. Generally, two fixing slots are provided on the back of the basin where the brackets fit in. The mounting brackets shall be accurately positioned against the underside rear of the bowl and against the wall.

9.4 Bathtubs

A bathtub is a receptacle, a long and rectangular type of trough, as shown in Figure 9.5, used for holding water and shaped to allow one or two persons to soak. A single user bathtub is mostly used, the water holding capacity of which depends on the size, which can vary in length from 1.5 to 2 m with a width of 750 mm. One end is termed a leg-end and the other a head-end. A standard bathtub holds about 150 to 180 litres of water to soak in. Bathtubs are manufactured from enameled cast iron or steel and fibreglass.

Bathtub trims: To get a water fixed type or telephonic shower head, or both, a hot and cold water mixture faucet – with a flow diverting knob to divert the flow of water to a faucet or shower head to supplying water – is installed on the wall at the leg or draining end side. A drainage outlet on the bathtub floor and an overflow outlet on wall on at leg-end side are provided for draining wastewater and overflowing water respectively. The overflow and drainage pipes are combined and then extended to the corresponding stack or floor drain pipe, whichever is shorter. The size of drainage pipe provided is at least 40 mm.

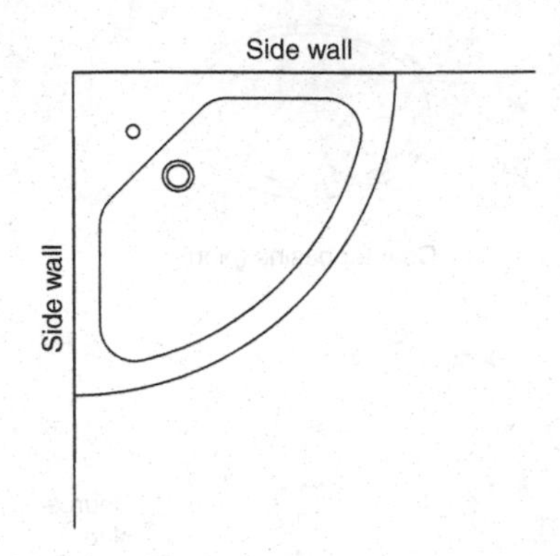

Figure 9.4 Corner basin (Plan).

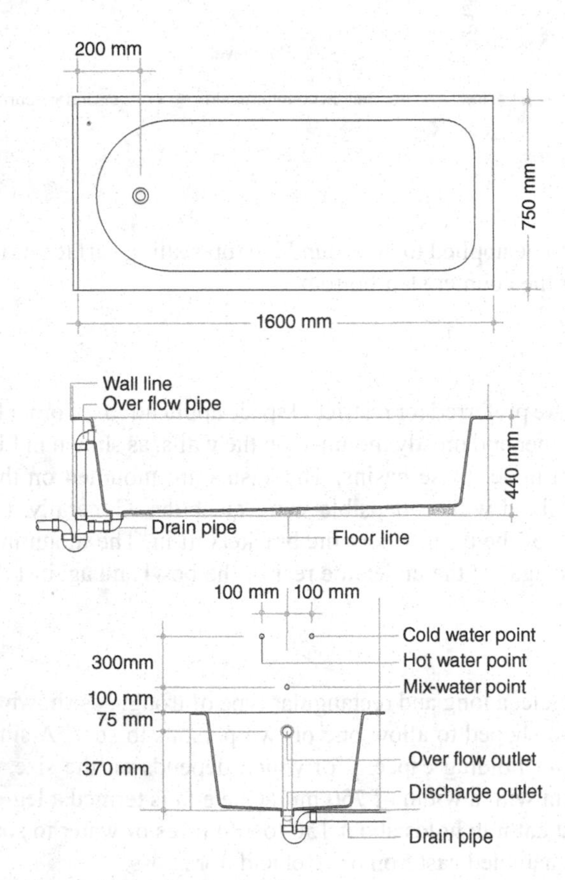

Figure 9.5 Bathtub and its trims installation details.

Installation: A bathtub is installed on the floor, generally in a corner of the bathroom. The position of the drain pipe is to be well marked and cut. After assembling wastewater overflow and the drain pipe with the tub, it is placed on the floor. On the floor, about 50 mm mortar is spread to embed the tub. The tub is secured to side walls. The outlet of the drain pipe, supported by a trap, is finally joined with the drainage piping system.

9.5 Bath trays

A bath tray is also a receptacle, as shown in Figure 9.6. It is used for containing bathing wastewater and draining through the discharge outlet into the drainage piping. Bath trays are designed for bathing a single person in a standing position. It is not intended for holding water. Bath trays are mostly designed as square in shape with a minimum area of 1024 sqin (0.66 sqm); if made rectangular, the minimum dimension shall be 750 mm. Size varies from 750 mm to 1000 mm square. In case of rectangular trays, the average width and length is typically between 700 mm and 1700 mm. The height is a maximum of 125 mm. Shower trays are also made of enamelled steel or fibreglass sheets. The floor of a shower tray is designed to be slip resistant.

In bath tray, there is a similar arrangement for water supply: in a side wall, as for a bathtub. The drainage outlet in the bath tray is kept in one corner. The size of the drainage pipe provided for a bath tray is at least 40 mm.

Installation: Bath trays can be installed in three positions: 1. above floor, 2. on floor and 3. at tray-rim at floor level. Generally, bath trays are installed on a floor; in this case, a hole is made in the floor for the drainage pipe to pass through. Installing a bath tray above the floor, about 300 mm from floor level, allows the drainage pipe to run above the floor, so no floor punching is required. In these cases, users need to step up to get onto the tray. To avoid stepping over, trays may be installed with the rim at floor level. In this case, the floor needs to be cut according to the size of the tray to fit.

In case of installing a bath tray on the floor, a mortar layer of fine sand and cement with an anti-shrinking agent to a suitable quantity shall be laid on the floor to cover the full area under the tray. The mortar layer is to be of sufficient thickness to ensure leveling and full support of the tray and shall not be less than 25 mm in any case.

9.6 Kitchen sinks

A kitchen sink is a shallow-depth bowl with a sloped flat bottom plumbing fixture and is used for cleansing dishes and utensils. There is a corrugated table or drain board on one or

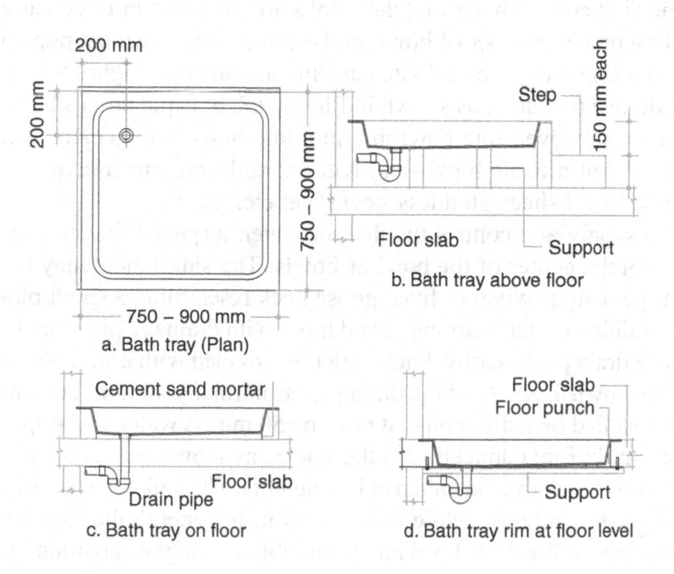

Figure 9.6 Bath tray installations.

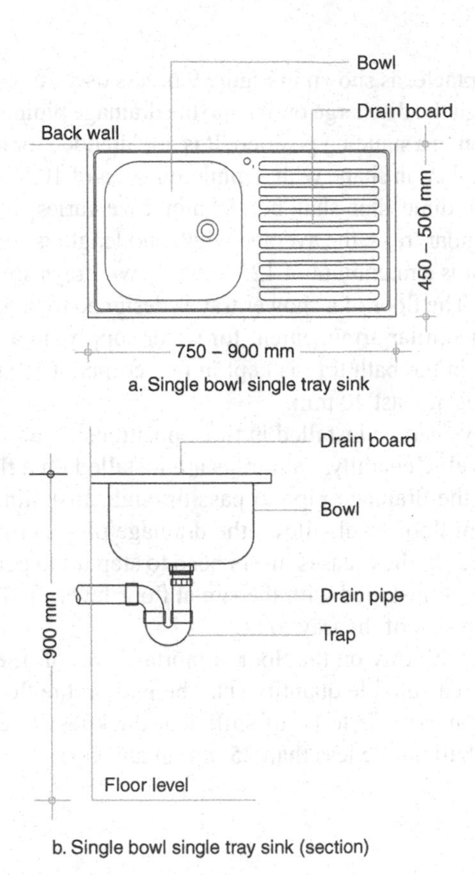

Figure 9.7 Single bowl single tray sink.

both sides of the sink bowl. Though kitchen sinks are available in large variety of sizes and shapes, depending on the number of bowls and drain boards, the most popular and used is a single bowl with a single drain board kitchen sink, as shown in Figure 9.7. Others resemble a double bowl, double tray sink, as shown in Figure 9.8, or triple bowl sinks etc. Bowl depth is usually 200 mm; however, one bowl in a multiple bowl arrangement may be deeper, as shown in Figure 9.9 of a triple bowl sink. Kitchen sinks are manufactured from enameled cast iron or pressed steel sheet, stainless steel sheet etc.

Sink trim: To supply and control the flow of water, a special faucet – called a sink faucet – is installed at the center of the bowl or bowls. The sink faucet may be a hot and cold water mixture type with a swivel or fixed goose neck resembling a spout pipe. At the centre of the bowls, a draining outlet is arranged and fitted with drainage pipe that leads to a branch drain pipe or floor drain pipe nearby. Each outlet is provided with a suitable rubber stopper to retain water in the bowl. The size of the drain pipe is normally 40 mm, but when a food waste grinder is to be installed on a drain pipe, it must be 50 mm. A water seal trap is to be installed on the drainage pipe before connecting to the stacks, as shown in Figure 9.7.

Installation: Sinks are never installed at low heights. The sink top shall be at such a height that the user need not bend over while working on it. In general, the sink top is fixed at not less than 900 mm above the floor level and preferably at a higher position. The sink may be installed on a back wall or corner walls with the help of brackets or on a counter slab.

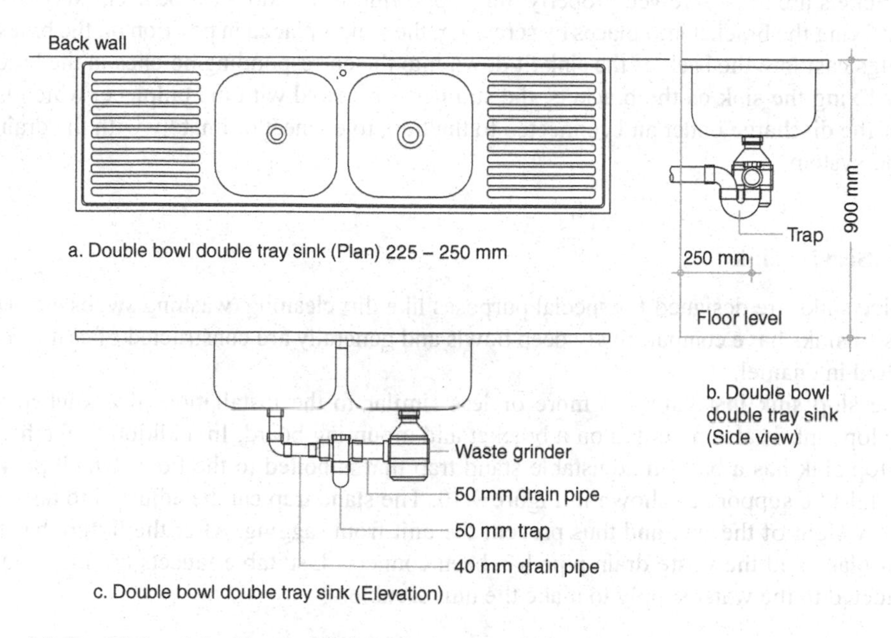

Figure 9.8 Double bowl double trays sink having waste grinder.

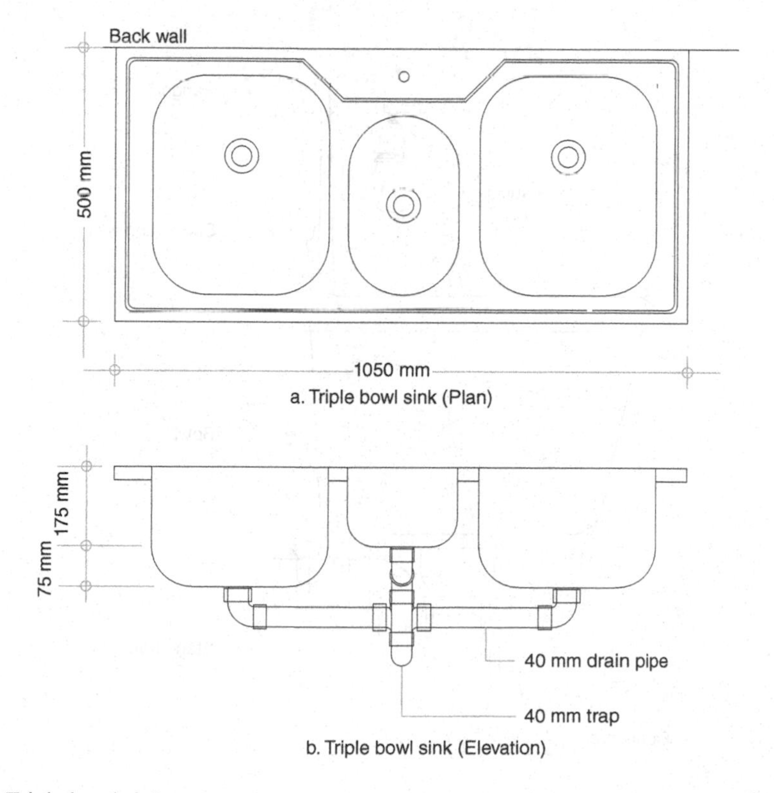

Figure 9.9 Triple bowl sink.

Brackets are to be screwed properly into supporting walls and must be accurately placed. After fixing the bracket into places by screwing, the sink is placed in position on the bracket – the lugs cast into the back of the sink fit down into the corresponding notches in the bracket. After fixing the sink on the brackets, the strainer is screwed with the tailpiece, which is fitted at the discharge outlet and connected to the trap, to connect ultimately with the drainage piping system.

9.6.1 Service sink

Service sinks are designed for special purposes like dirt cleaning, washing swabs etc. These types of sinks have comparatively deep bowls and generally are constructed of cast iron and finished in enamel.

The slop sink installation is more or less similar to the installation of a kitchen sink. The slop sink is also mounted on a bracket and mounting board. In addition to the hanger, the slop sink has a built-in adjustable stand trap that is bolted to the floor, which provides pedestal-like support, as shown in Figure 9.10. The stand trap can be adjusted to take most of the weight of the sink and thus prevent the unit from sagging. After the fixture has been set in place and the waste drain pipe has been connected, suitable faucets are installed and connected to the water supply to make the unit useable.

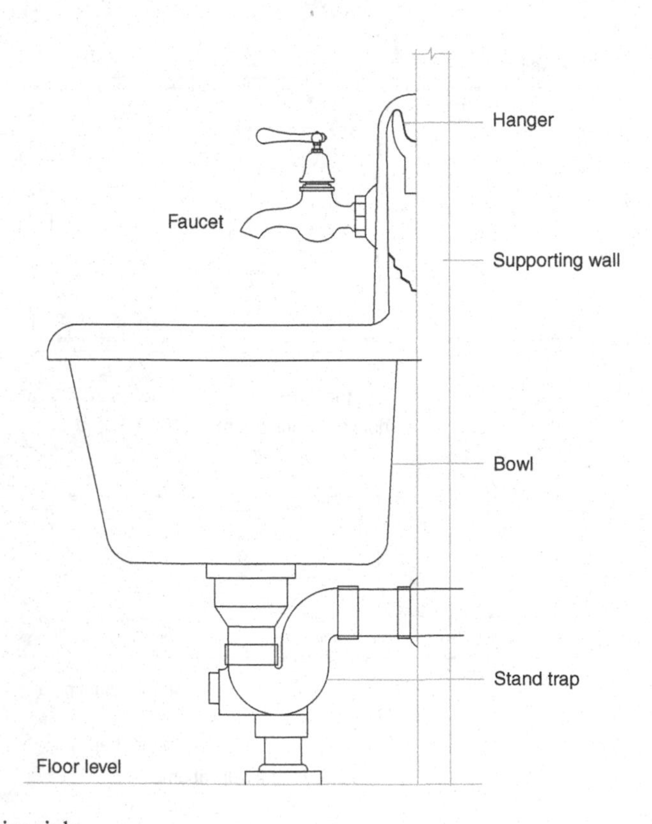

Figure 9.10 Service sink

9.7 Water closets

Water closets are oval-shaped bowls or elongated receptacle type plumbing fixtures that receive human excrement wastes and finally dispose of this in the drainage system by flushing water. The size of the drain pipe of a water closet is normally 100 mm, but there are some which are made with a 75 mm drain pipe. A 100 mm drain pipe can discharge about the twice the flow rate of a 75 mm drain pipe. Water closets are mostly made of porcelain with a vitreous china finish and may also be made of steel sheets for special purposes of use. Water closets are made in various classes, types and operating systems. Various types of water closets may be grouped according to the following factors.

1 The style of sitting for using water closets
2 The type of trap integrated
3 Operating principle of water closet bowl
4 Flushing mechanism
5 Making of water closet
6 Installation and
7 Shape.

9.7.1 Water closets of various styles of sitting

According to the style of sitting, water closets are typed as below.

1 Pedestal, also popularly known as the European type
2 Squatting, also known as the Indian type; historically referred to as Oriental type and
3 Universal, also called as Anglo-Indian type.

European water closet: European water closets are mostly an oval-shaped bowl type with a flat wash rim on top, as shown in Figure 9.11. A seat coupled with a cover is fitted to sit on the rim. The discharge outlet has an integral trap. A water closet bowl is supported by a flushing system.

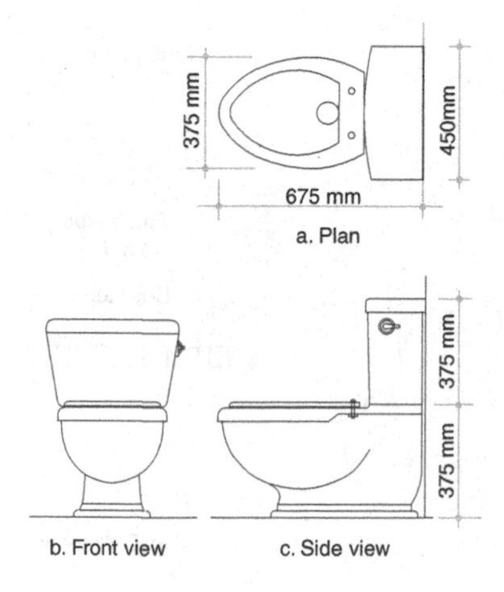

Figure 9.11 European water closet.

Squatting type water closet: This type of water closet has a sunken and elongated receptacle with a discharge outlet at the back end side, as shown in Figure 9.12. On the top edge of the receptacle, there is a rim all around. These types of water closet usually have a flat top surface, to which the foot rest is made integral. Otherwise, an extra foot rest is to be placed by the sides of the pan. With these water closets, there is no water seal trap made integral, so an extra trap is to be installed on the outlet of the pan and then connected to the drainage pipe.

Universal water closet: It is a water closet that is a combination of European and squatting type water closets, as shown in Figure 9.13. The top of the water closet is made like a squatting type, flat and wider, and the bowl part is similar to the European type. The top may be so designed that user can use this water closet in either a sitting position, as done with a

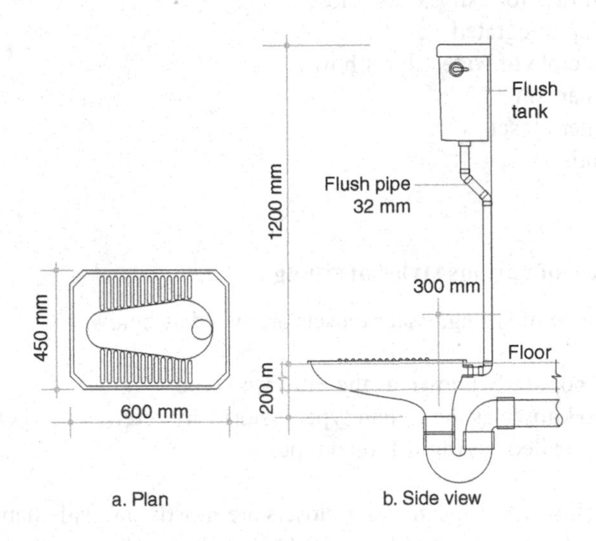

Figure 9.12 Squatting water closet.

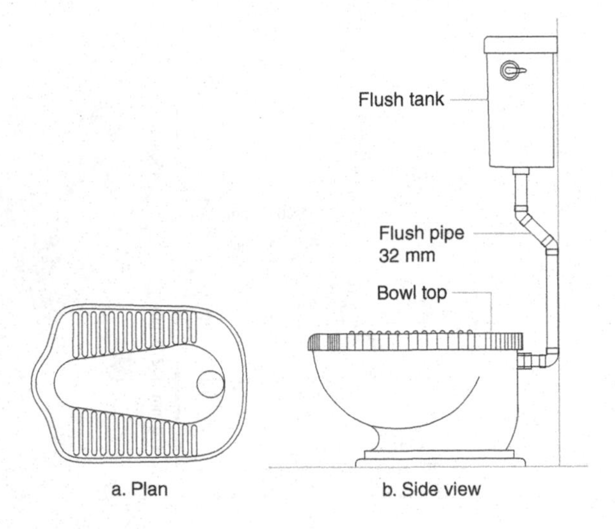

Figure 9.13 Universal water closet.

European water closet, or a squatting position. This type of water closet has the inbuilt water seal trap found in the European type.

9.7.2 Types of integrated trap

European water closets include a water seal trap as an integral part. According to the shape of the trap, which resembles some English letters, these water closets are specified as follows.

1 P-type
2 Q-type and
3 S-type.

P-type water closets have an English letter P-like shape of the outlet running above the floor, towards the back wall. So, installing of these water closets does not require the drainage pipe running under the respective floor. As a result, there is no need of punching the floor below.

Q-type water closets are supposed to have English letter "Q"-like shape with outlet above the floor is inclined at a 45° angle.

S-type water closets have an English letter "S"-like shape with the outlet bent towards the floor, requiring the drainage pipe to run under the floor and thus requiring the floor to be punched carefully at the correct position, centering the trap outlet.

All European water closets with various types of traps are shown in Figure 9.14.

9.7.3 Operating principles of water closet bowls

The performance of a water closet can be judged by the flushing efficiency to dispose of the wastes from the bowl. Flushing efficiency depends upon the manner of flushing, the water volume required and the sound generated during flushing. There are various ways of operation for delivering water into the bowl and for flushing off. Various techniques are adopted

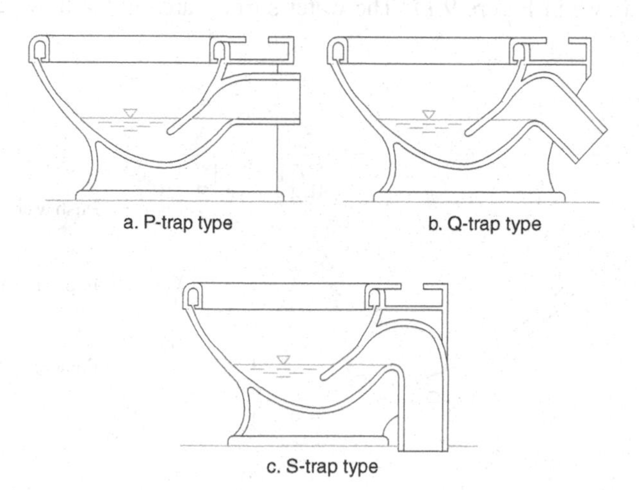

Figure 9.14 European water closets of various traps.

in creating the force for driving out the wastes. Depending upon these operations of water delivering and creating force for flushing, the European water closets are classified as below.

1 Siphon jet
2 Siphon vortex
3 Reverse trap
4 Wash down and
5 Blow out.

In all water-closet bowls, the presence of one item is very common: a wash rim. The rim is a hollow channel made to flow flush-water through the top of the bowl, with holes along the bottom to flow down. When the flushing mechanisms are activated, the rim fills with water, which drains into the bowl through the holes in the bottom of the rim. This effectively washes the inner sides of the bowl and creates pressure on wastes to get out with each flush.

Siphon jet: The flushing action in the siphon jet bowl, as shown in Figure 9.15, is accomplished by pushing the remaining water in the upper leg of the trap-way by directing jet through a jet hole at the bottom of bowl. As the water flows through the down leg of the trap-way, it drives away the air present in the passageway, thus creating a partial vacuum. Atmospheric pressure working on water in the bowl pushes it along with its waste content through the trap-way. The process is continued by siphonic action until all water along its content is pushed out. The flushing operation in this way is more efficient, quick and relatively quiet.

Siphon vortex: In this type of water closet bowl, shown in Figure 9.16, when the flush tank or valve is operated, water enters diagonally through diagonal punching under the rim, around the bowl, and directs a water-jet through an angled jet hole at the bottom of bowl. This creates the swirling action, resulting in a vortex of the contained water and thereby pushing the wastes through the trap passageway, producing siphonic action. This type of water closet is comparatively costly but very efficient, quiet and extremely sanitary. The bowl contains a large volume of standing water, and so the large water surface remains at a comparatively elevated level with respect to rim level.

Reverse trap: The flushing action of the reverse trap water closet bowl is similar to that of the siphon jet water closet. In this type, the bowl discharges into the trap-way at the rear of the bowl, as shown in Figure 9.17. The water surface area in the bowl, the passage way

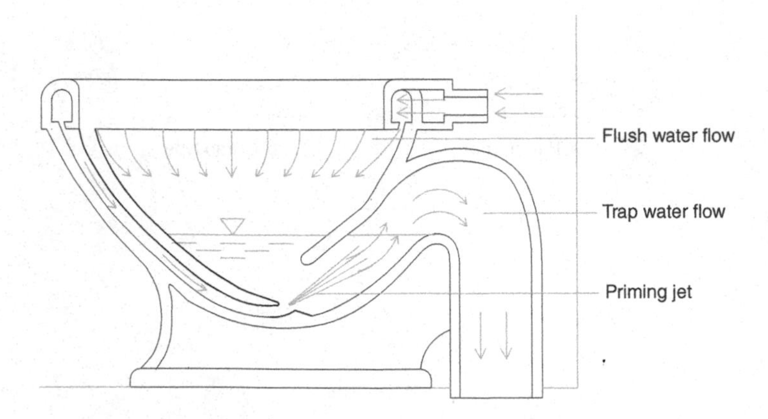

Figure 9.15 Siphon jet water closet.

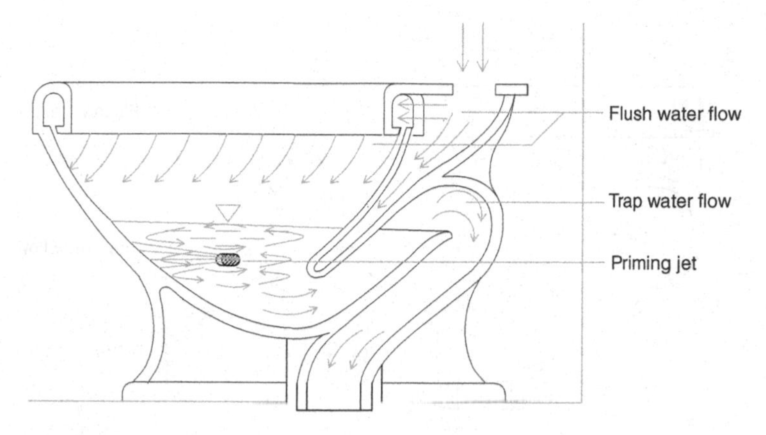

Figure 9.16 Siphon vortex water closet.

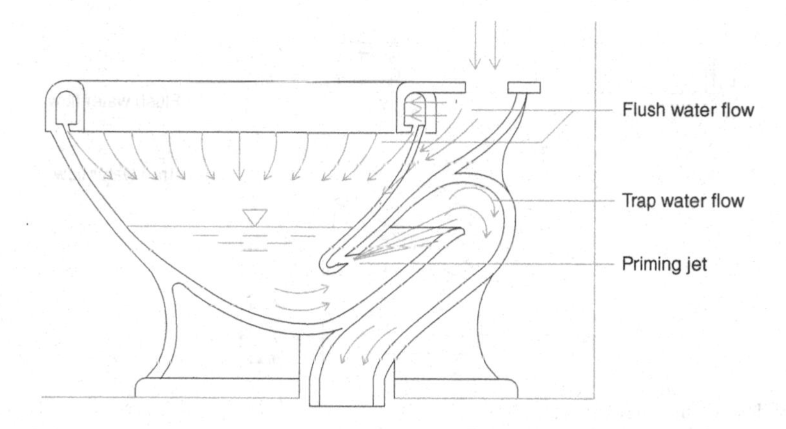

Figure 9.17 Reverse trap water closet.

and the seal is comparatively smaller than that of a siphon jet water closet. So, comparatively less water is required for the flushing operation. In this way, flushing is moderately noisy.

Wash down: In a wash down type water closet, the water from the flush tank pours into the bowl rim and washes the waste material down the discharge outlet. In this case, a thrust is produced only to cause the free flowing of water to take waste along with it through the trap-way, as shown in Figure 9.18. These water closets have small water surface areas in the bowls that are deep down in the bowl.

Blow out: In blow out water closets, the flushing action is accomplished by driving a large volume of water to spread from the rim surrounding the bowl and the accumulated waste and water into the trap passageway, as shown in Figure 9.19, instead of driving by siphonic action. This type of flushing operation is economical in terms of water use and also efficient, but the action is very noisy. This type of water closet is preferable for use in public places, bus and train stations, marketplaces etc. These types of water closets should be flushed using flush valves.

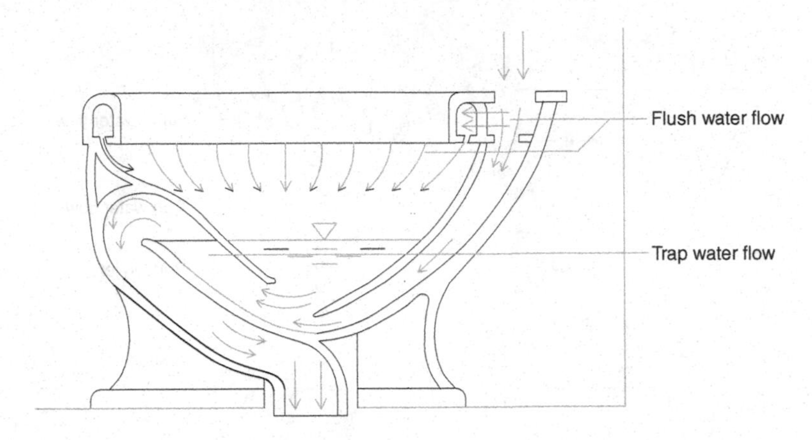

Figure 9.18 Wash down water closet.

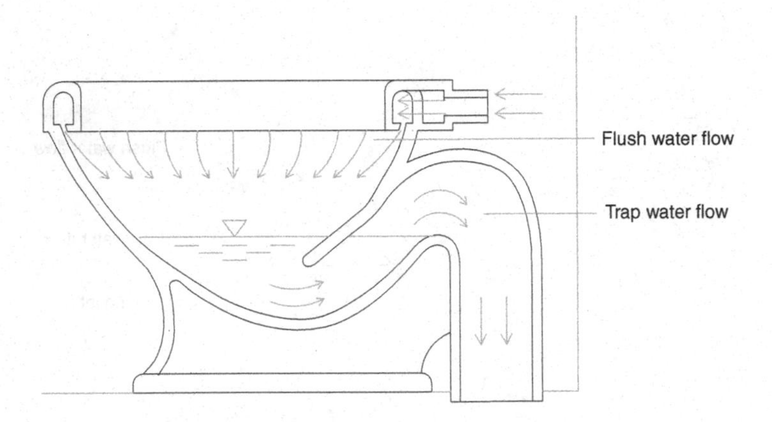

Figure 9.19 Blow out water closet.

9.7.4 Making of water closets

European water closets are also made in different ways as follows.

1 One piece
2 Closed coupled and
3 Uncoupled.

One piece: In a one-piece water closet, the flush tank and the bowl are molded as a single unit, as shown in Figure 9.20, and thus the height of the water closet is reduced.

Close coupled: In this type of water closet, the flush tank and the bowls are separate units, but the flush tank is placed on the bowl attached to the rim at the back. Figure 9.11 represents a close coupled water closet.

Uncoupled: This type of water closet consists of only the bowl, with provision for the connection of flushing pipe to flush tank or flush valve. The flushing tank is placed above the bowl and is connected to it by a flushing pipe or a flush valve is installed on the bowl rim at the back.

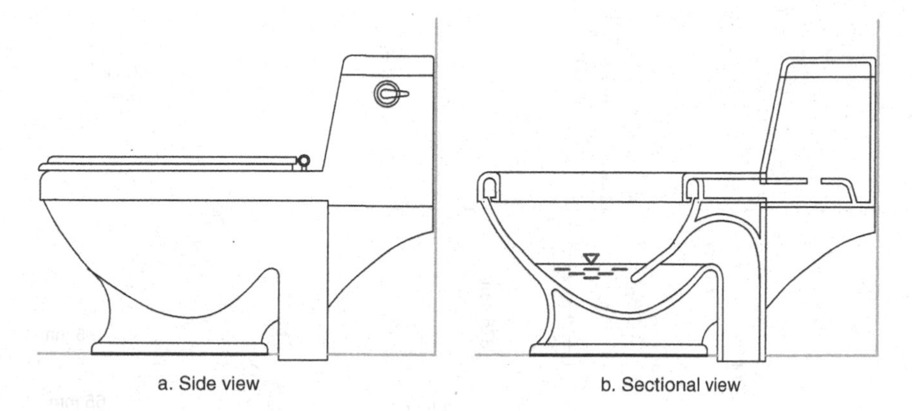

Figure 9.20 One-piece water closet.

9.7.5 Water closet type based on installation

On the basis of the support condition, European type water closets are grouped in two categories, as follows.

1 Floor mounted and
2 Wall mounted.

Floor mounted: Floor mounted European water closets are mostly used for strong stability and support on the floor. Installing these types of water closets creates some dirty space, behind and in the corner when water closets are placed in the corner. The position of the water closet and its pedestal obstruct cleaning operations in a narrow space, which makes the surrounding area dirty.

Wall mounted: Wall-mounted European water closets are chosen for facilitating easy cleaning of the total toilet floor surface as the total units remain hanging, supported on the back-wall, and thereby keep free space underneath to clean easily. The other advantage is that the height of the seat can be adjusted for the user's comfort. Wall-hung water closets also occupy less space. The supporting back wall must be strong enough to carry the weight of the water closet unit plus the body weight of the user, and the anchor bolts must be sized enough to withstand all the stresses that might be developed. Wall-hung water closet and their position of installation is shown in Figure 9.21.

9.7.6 Shape of water closet

European water closets are a bowl type, with the top mostly oval-shaped. The front of the oval-shaped top is again elongated, rounded or other curve shaped. In public toilets, an elongated bowl type is preferred to avoid unsanitary and objectionable condition of the bowl after use.

9.7.7 Water closet seats

An extra seat coupled with a cover is placed on the rim of the water closet bowl that supports and provides comfort to the user while sitting on it and keeps the bowl covered after use, for

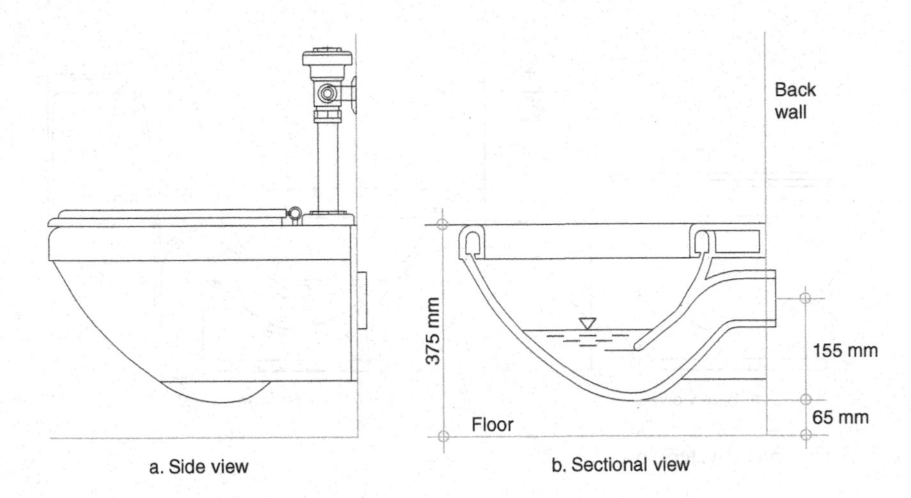

Figure 9.21 Wall-hung water closet.

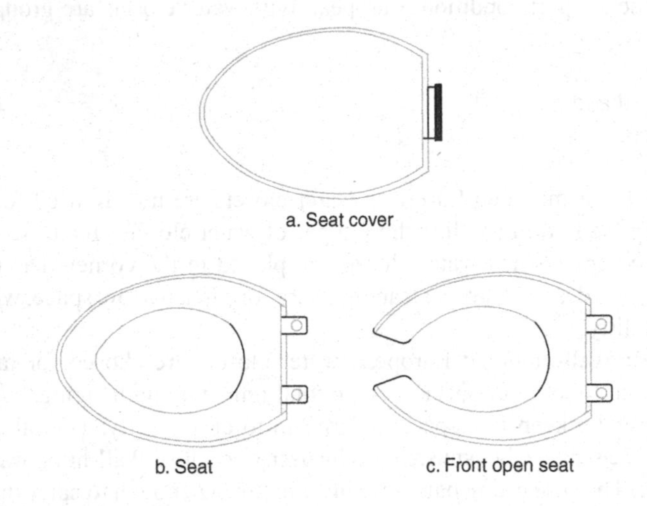

Figure 9.22 European water closet seat and cover.

safety. The seat is securely fastened with the help of screws at the rear side of the bowl top. The seat rests firmly on the rim and the cover is hinged with the seat at the rear side, resting on the seat to cover the bowl. The seats are mostly made of hard plastic, which should have a special surface finish offering protection from the growth of bacteria, molds, fungi etc. and resistance to chips and stains.

Two types of seats are available, as shown in Figure 9.22. Seats, with cover, are generally designed oval in shape with the front elongated or rounded, commonly termed as a closed front seat. These types of seats are mostly used in private toilets. The other types are front open seats with no cover. These types of seats are generally used in public toilets.

9.7.8 Water closet trims

Water closets are generally supported by a supply of water for flushing the closet and cleansing purposes. For cleansing purposes, water needs to be sprayed by a hand-held sprayer, for which a water supply pipe is installed with an outlet fitted with angle stop cock, generally on the right side of the closet and about 300 mm above the floor. The outlet of a stop cock is fitted with a hand-held sprayer. For flushing the closet using a flush tank, a water supply pipe is installed with an outlet fitted with an angle stop cock. The tank inlet is connected to the stop cock by a flexible connection pipe to supply water into the flush tank. The flush tank is connected with the water closet by a 40 mm pipe to flush water into the water closet. In the case of a flush valve, a 25 mm supply pipe is installed, which is connected to the flush valve. The flush valve is installed on the water closet to flush the toilet.

The discharge outlet end of the water closet is fitted on the drainage pipe, which is mostly a 100 mm diameter pipe. In case of a squatting type water closet, a water seal trap is fitted at the outlet of the pan and the drain pipe is fitted at the outlet end of the trap. The drain pipe is ventilated by installing a vent pipe on the drain pipe.

9.7.9 Installation of water closets

To install a water closet, the following procedures are to be carried out as a general guide.

1. Slip the water closet flange over the floor hole and slide it down until it is level with the finish floor
2. Slip the closet bolts on either side and move the lock washers down to fasten against the flange firmly and
3. Put a sealing (wax) ring directly on the flange or around the discharge outlet (horn) under the closet bowl, as shown in Figure 9.23
4. Set the water closet bowl over the flange, lining up the holes of the closet with the bolts and projecting the horn down into the flange
5. Put the washers over the bolts and hand-tighten the nuts alternately on both sides and
6. Check the closet-bowl for level front-to-back and side-to-side. After leveling, tighten the nuts on the closet bolts. Do not over tighten, as this may crack the base of the water closet.

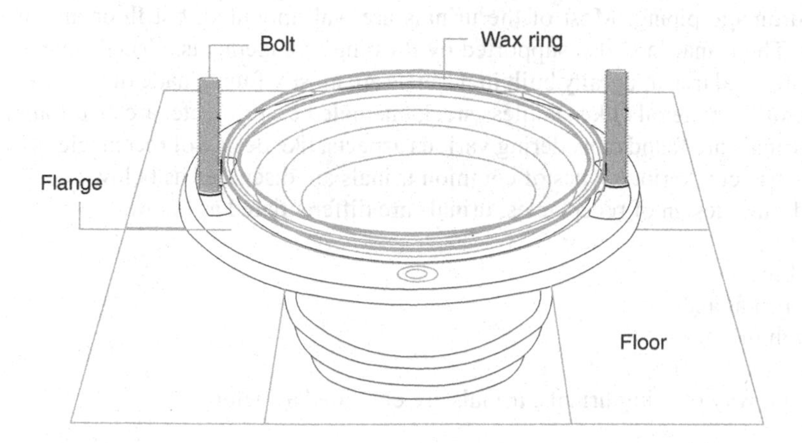

Figure 9.23 Flange and accessories for installing pedestal type European water closet.

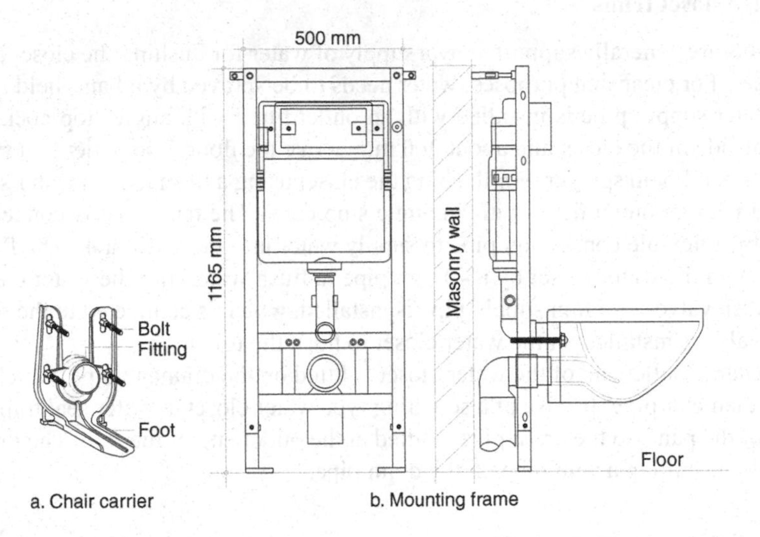

Figure 9.24 Wall hung water closets carriers.

A wall-mounted water closet is installed on a wall by a chair carrier, as shown in Figure 9.24a. The chair carrier is placed in position and bolted to the floor. The foot of the chair carries the weight of the entire closet independent of the walls and drainage connections. A standard fitting is used to connect the drain pipe and the closet-bowl outlet, after the chair carrier is bolted properly. To facilitate installation of the closet bowl, the bolt holes in the chair-carrier are slotted.

Wall-mounted water closets are installed on a wall by the support of a mounting frame, shown in Figure 9.24b. The frame provides the support for the water closet bowl and the cistern as well. The drain pipe fitting is also incorporated into the mounting frame, to connect to the outlet from the bowl.

9.8 Urinals

A urinal is a plumbing fixture designed to receive urine while urinating and finally discharge into the drainage piping. Most of the urinals are wall mounted, but floor mounted is also designed. The urinals are also supported by flushing arrangements. Urinals may or may not have a water seal trap integrally built in. Urinals are mostly found made of vitreous china, but urinals of other materials like stainless steel, enameled cast iron etc. are also found. Various types of urinals are found considering various aspects, like design of receptacle, way of using and water use etc. Various types of common urinals are discussed as below.

Considering design of receptacles, urinals are differentiated as below.

1 Bowl urinal
2 Stall urinal and
3 Trough urinal.

Considering way of using urinals, urinals are classified as below.

1 Pedestal urinal and
2 Squatting urinal.

Considering use of water for flushing or disposing of urine from the bowl, urinals are classified as below.

1 Ultra-low flush urinal
2 High-efficiency urinal and
3 Waterless urinal.

Bowl urinal: These types of urinal are bowl types, with a flat back for fixing on the wall at a convenient height. Two general shapes of these bowls are available; one is a round type and the other is a lipped type, as shown in Figure 9.25. The latter type is preferable for receiving urine to the last of dripping. Bowl urinals are provided with a flushing rim for thorough cleansing of the entire interior surface. Two methods of flushing are used in these types of urinals: wash down type and siphon-jet type. In the siphon-jet type, some standing water remains on the trap inlet, at the bottom of the bowl, after each flush.

Stall urinal: A stall urinal, shown in Figure 9.26, is supported on the floor attached to a wall of a toilet-room. It has a drip receiver at the bottom, usually larger with the sides

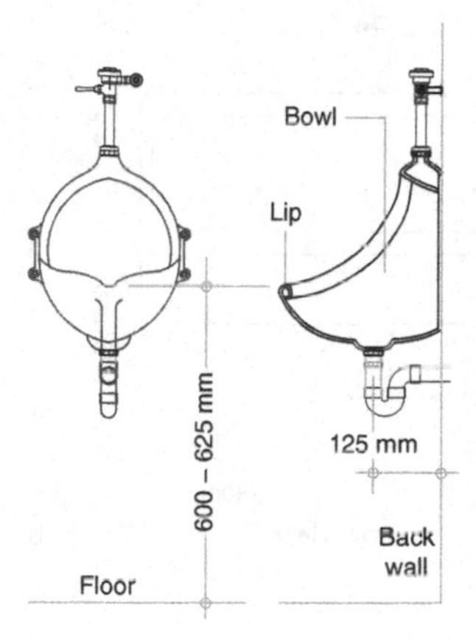

Figure 9.25 Bowl urinal

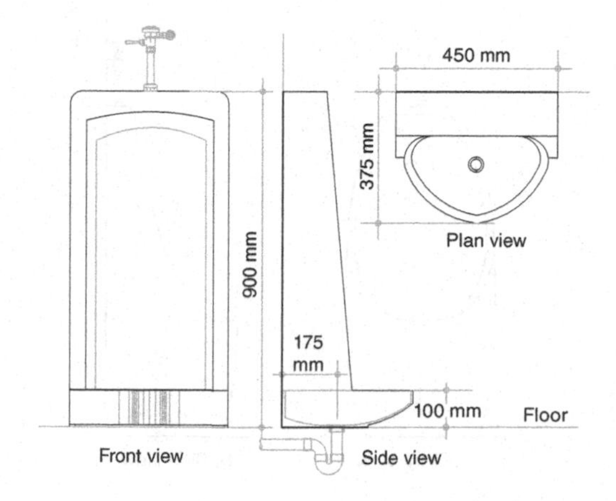

Figure 9.26 Stall urinal

straight, the top of which is flush to the floor. The waste outlet is in the middle of the drip receiver, which is attached with a 50 mm water seal P-trap underneath that is connected to a soil drain pipe.

Trough urinal: Trough urinals have a trough-like receptacle, as shown in Figure 9.27, designed to be used simultaneously by more than one person at a time. The urinal trough is fixed on a wall. These urinals are not supposed to be conducive for maintaining sanitary conditions. So, its use is limited to temporary installations. Trough urinals are generally made of enameled cast iron material for economy.

Pedestal urinal: Pedestal urinals are floor mounted, similar to the European type water closet bowl, as shown in Figure 9.28. These urinals are flushed, incorporating siphon-jet action, as described earlier. So, this type of urinal is considered to be the most sanitary in comparison to all other urinals.

Squatting urinal: These are a flat type urinal, as shown in Figure 9.29, on which persons urinate in a squatting position. These urinals are popularly used by Muslims as they are not supposed

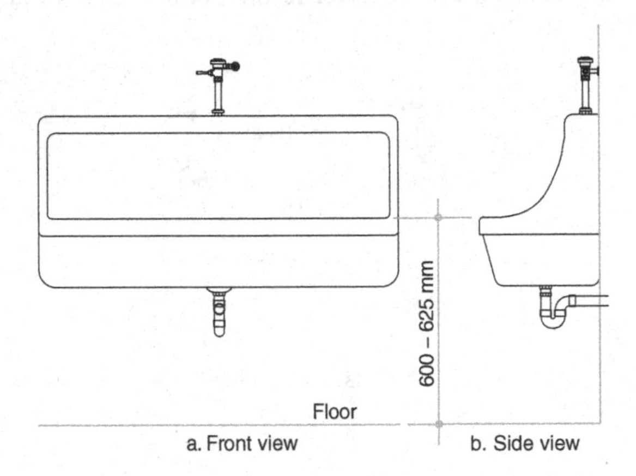

Figure 9.27 Trough urinal.

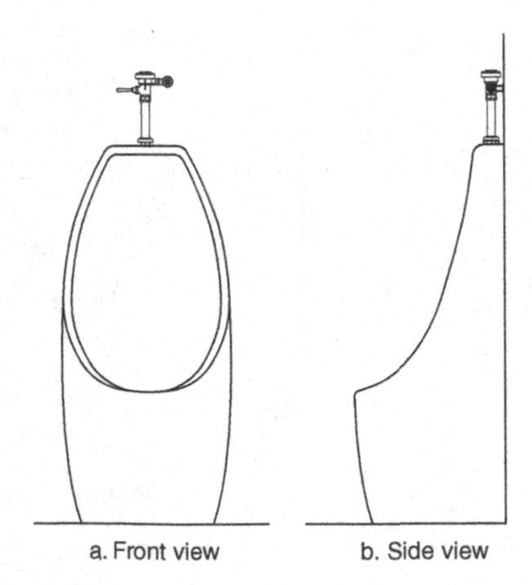

Figure 9.28 Pedestal urinal.

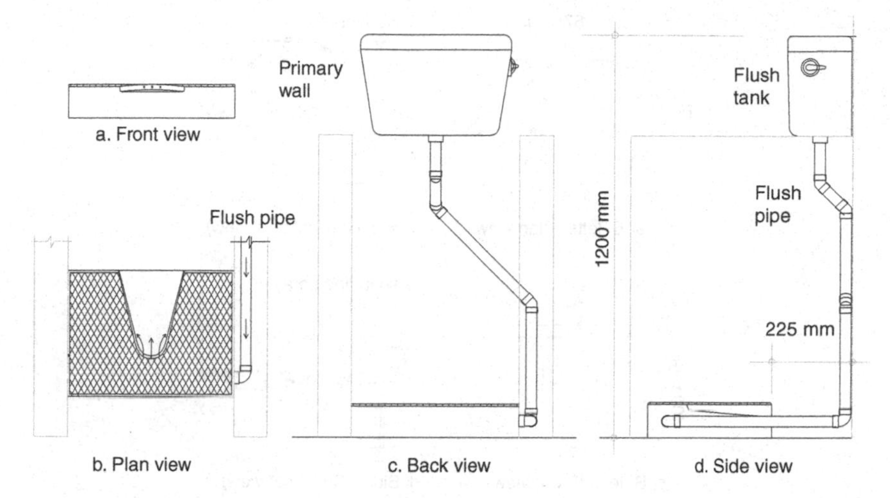

Figure 9.29 Squatting urinal.

to urinate in a standing position unless they become unable to due to any physical or spatial reason. In a battery of urinals in a row, partitions are provided with 600 mm intervals in between for privacy. In these types of urinals, there is an arrangement for flushing using a flushing tank.

Low flush urinals: Early age urinals used about 5.6 lpf, but these low flush urinals use about 3.8 lpf or less. These are preferred where water saving is one of the prime objectives in building development.

High-efficiency urinals: These urinals are supposed to be highly efficient because they use a flush valve with a discharge capacity 1.9 lpm or less [2]. The mechanism for flushing, the flushometer valve, is more or less the same as for conventional flushometers, but it requires comparatively higher pressure and higher velocity of the supplied water flow and a smaller orifice in the diaphragm of the flush valve.

Waterless urinal: Urine itself is liquid, which is supposed to be drained out due to its inherent property of flowing down. So, these urinals, similar in appearance to bowl urinals, require neither water nor a valve to flush. Instead, these fixtures use a cartridge which contains a sealant liquid which is less dense than water. Urine sinks down below the sealant liquid and the weight of the sealant liquid above pushes the urine below into the drainage pipe. When the sealant gets reduced due to depletion, the cartridge does not perform fully, resulting in unpleasant odour. During cleaning, excess amounts of water should not be used that will cause loss of functionality of the cartridge [2]. The drain pipe of waterless urinals is more prone to significant waste deposition and blockages.

9.9 Bidets

A bidet is a fixture used to help clean off the private body parts after toilet use. The bidet projects a gentle stream of water to cleanse the external genitals and posterior parts of the body exposed to the toilet, for personal hygiene. There are two types of bidets: 1. standalone bidet and 2. add-on toilet side rim bidets. Standalone bidets look something like a water closet bowl with a faucet, as shown in Figure 9.30. Add-on toilet side rim bidets are designed as built-in bidets that fit over the toilet side rim.

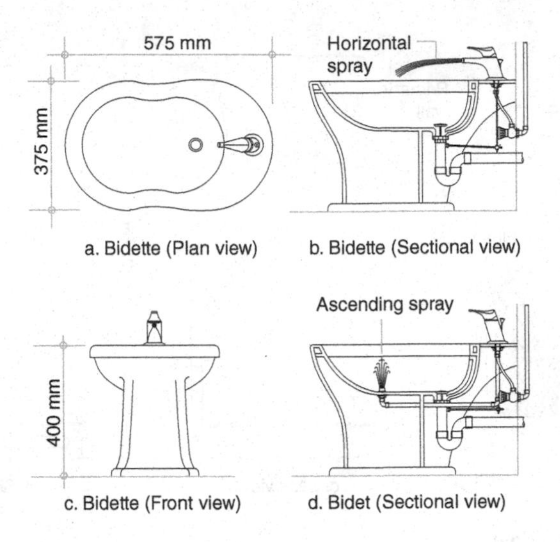

Figure 9.30 Bidet and Bidette.

When the bidet-faucet is activated, an arc of clean water from a nozzle is sprayed, either beneath the rear of the rim or deep within the fixture cavity. Users sit on the seat over the rim, facing forward or backward, based on the water jet configuration and the body part that needs cleaning. Instead of fixed faucets, a hand-held bidet sprayer hose can be set with a T-Adapter valve and bracket holder on the side or rear wall to get the water spray for the cleansing purpose.

A bidet uses comparatively little water and offers a hygienic alternative that eliminates the use of toilet paper for the purpose. Water jet flow can be controlled by the faucets. Hot and cold water can be sprayed through using the bidet mixture-faucet.

Installation: Bidets are mounted on the floor. The water connection to the over-rim faucet for horizontal spray can be connected to the water supply piping system. But for an ascending spray, water must be supplied from an independent water storage cistern with no other draw-off points. The wastewater generated is disposed of through soil drain pipes.

9.10 Flushing devices

Every water closet and urinal must be flushed by water after use. So, these fixtures need the support of additional devices to supply sufficient water at an adequate rate and velocity, for satisfactorily driving out of the wastes from the fixture bowl. About 6 litres of water is generally used per flush. There are also low-volume flushing devices that consume as little as 3.75 litres of water per flush for flushing urine [3].

For the purpose of flushing, two appliances are mostly used: 1. flush tanks or cisterns and 2. flush valves, discussed below.

9.10.1 Flush tanks

A flush tank or cistern is a container that holds water for flushing and equipment to cause flow through the tank outlet pipe, connected to the fixture inlet for flushing, at a desired rate and velocity. The tank size depends upon the volume of water needed per flush and the volume occupied by the equipment parts. About 6–9 litres of water are stored for flushing water closets once within 10 seconds. For urinals, a 4.5-litre capacity flush tank is sufficient. There is also a low flush or low-flow flush, which is a specially designed flush tank adapted in order to use significantly less water than a conventional full-flush tank [4]. Tanks are installed on the back wall, by screwing.

Basically, there are two techniques for the flushing mechanism incorporated inside a flush tank: one uses siphonic action and the other valve.

In a flush tank, flushing-flow is continued and stopped by applying the theory of siphonic action. Three types of siphonic flush tanks have been designed: the plunger type, the bell type and the flushing trough type. Nowadays, the bell type and the flushing trough type are seldom used.

Plunger type: In a plunger type flush tank, as shown in Figure 9.31, the float valve assembly controls the inlet flow of water in the tank, as described earlier. When the tank is full of water, to flush the water closet the handle is turned down or a button is pressed to lift the plunger, connected by wire or chain to the lever, out of the seat of the discharge outlet. The sudden rising up of the plunger causes water to fall into the overflow tube and flow fully down through the overflow tube with such a velocity that it creates siphon (partial vacuum). The greater atmospheric pressure on the water surface forces water up past the diaphragm washer to flow through the overflow tube connected to the flush pipe; the flow thus continues until the water level is low enough in the tank to allow air to enter into the siphon to break the siphonic action. With the breakage of siphonic action, the flushing action stops and brings the plunger back to the seat.

Bell type: In this type of cistern, there is a bell-shaped dome-like element which is raised up by a lever by pulling down a chain attached to the lever, as this type of cistern is placed at a height of 2 meters above the floor level. When the chain is released, the bell falls down and displaces an equal volume of water, which falls down the flush pipe, taking along some air from inside the bell and thus creating a partial vacuum. This causes siphonic action, resulting in continuous down-flow through the flush pipe connected to the water closet.

Trough type: Trough flushing cisterns have been designed to serve more than one fixture for frequent flushing.

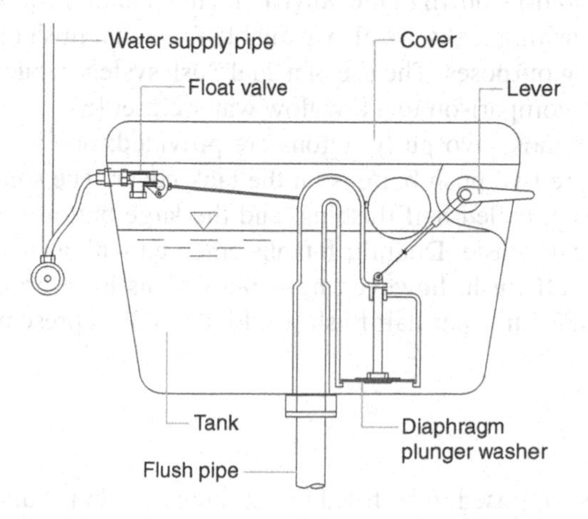

Figure 9.31 Flush tank: Plunger type

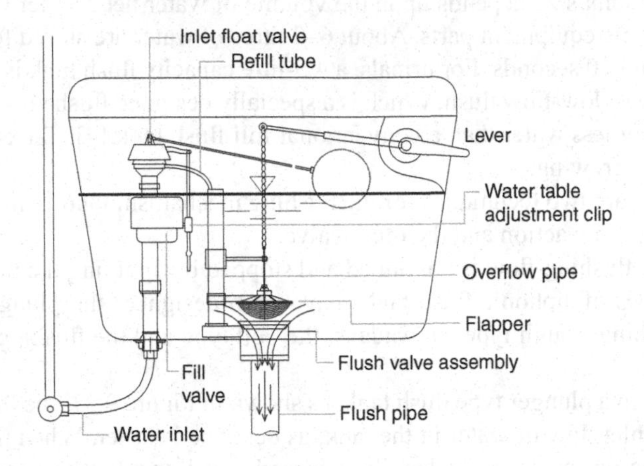

Figure 9.32 Flush tank: Valve type.

Valve type: In this type of flush tank, the following major components are incorporated for efficient flushing.

1 Float valve or ball-cock assembly connected to the water inlet pipe and
2 Flush valve assembly, including an overflow tube, valve seat and a rubber flapper or ball attached to the lever by chain.

In normal condition, the tank contains water at its highest level and the flapper on the flush valve seat, causing no flow of water. To flush, the lever is turned down or the button is pressed, which lifts the flapper out of the valve seat. The water in the tank flows down with such a velocity that siphonic flushing action occurs. As water flows down, the flapper or ball comes down onto the seat and flushing is stopped. There is an overflow pipe attached to the flush valve, as shown in Figure 9.32; supplies refill water in the closet bowl and prevent overflow of the tank. A refill tube is provided with the float valve assembly to supply a measured amount of water into the overflow pipe to flow down to the bowl and remain until the next flushing operation.

Dual flush tank: Flushing tanks are also available for a dual flush operation to save water consumption in flushing purposes. The use of a dual flush system reduces water consumption by about 68 percent in comparison to a low flow water closet [5].

In this type of flush tank, two push buttons are provided; one is comparatively smaller than the other. There are two push buttons on the tank cover. The small button is for flushing liquid (urinal) waste, called half flushing, and the large button is used for full flushing to drive out solid (faecal) waste. Dual flush tanks emerged with a capacity of 6 litres in full flush and 3 litres for half flush; however, new innovations have brought that down to 4.5 litres per full flush and 3 litres per half flush [6]. Figure 9.33 represents various major parts of a dual flush tank.

9.10.2 Flush valves

A normal flush tank is supposed to be filled in 2 minutes [7], but it may take longer if water pressure is not sufficient. The interval between successive flushing may be short in some

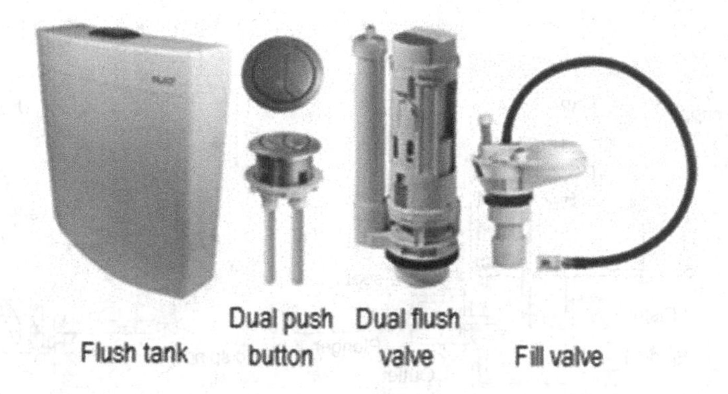

Figure 9.33 Flush tank, dual flush button and flushing equipment [8].

cases, particularly in assembly buildings, where use of a flush tank is unwise. In this case, a flush valve is recommended.

A flush valve is a device installed to supply a predetermined quantity of water directly from the water supply pipe, for flushing some sanitary fixtures. Because flush valves are directly connected to the water supply piping, no stored water is required for flushing purposes. For effective operation, the valve needs a minimum and maximum head of water of 3 m and 30 m respectively [9]. Flush valves of sizes 20 mm, 25 mm, 32 mm and 40 mm are available, with a flushing range between 3.8 lpf and 13.25 lpf. Generally, for saving water, 20 mm flush valves with a capacity of 3.8 lpf are mostly used for urinals and 25 mm flush valves with a capacity of 6.0 lpf are used in water closet flushing.

Flush valves are designed to incorporate two different technologies, but the operation principles of each type of valve are theoretically the same. The two types of flush valves are named according to their respective design methodology, as below.

1 Piston type and
2 Diaphragm type.

Each type of valve has an upper control and a lower supply chamber interconnected through a bypass. The bypass is a small hole or orifice measuring between 0.5 mm and 0.75 mm in diameter [10].

Piston type flush valve: A piston type of flush valve, as shown in Figure 9.34, has an upper control and a lower supply chamber interconnected through a bypass and separated by double-moulded cup acting as a piston. When the valve is in the closed position, the upper chambers remain full of water.

When the lever is turned down for operation, the plunger pushes the spindle, causing tilting of the relief valve and allowing water to escape from the upper chamber, thereby reducing pressure in the upper chamber. The reduced pressure in the upper chamber forces the piston assembly upward, allowing the water from the supply line to flow through the flush pipe connected to the fixture.

At the same time, water enters the upper chamber of the valve through the bypass, so pressure increases in the upper chamber. When pressure is in equilibrium in both chambers, the piston gets down to its seat and stops the water flow into the fixture to flush. The upper

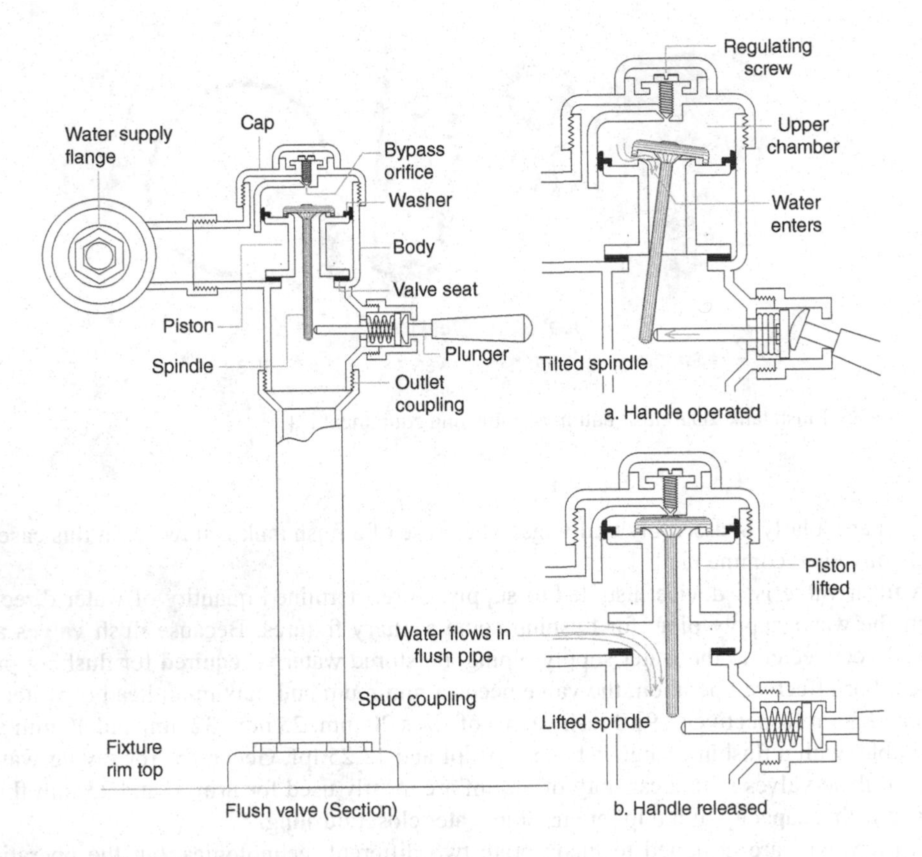

Figure 9.34 Piston type flush valve.

chamber empties in about 10 seconds. The closing time of the valve can be regulated by a screw that controls the span of time the valve remains open.

Diaphragm type flush valve: In this flush valve, a flexible diaphragm disc of rubber divides the valve into two chambers; the upper and the lower chamber, as shown in Figure 9.35. A bypass hole in the diaphragm makes the upper and lower chamber interconnected. Small amounts of the supply water enter into the upper control chamber through the bypass hole in the diaphragm. In normal conditions, both chambers are under equal water pressure on both sides of the diaphragm. When the handle is operated by turning up or down, it pushes the plunger, causing tilting of the relief valve and allowing water to escape from the upper chamber through the valve barrel. The pressure in the upper chamber, now lower, raises the relief valve disc and diaphragm as a unit, allowing water to flow from the supply line down through the flush pipe connected to the fixture.

During flushing, a small amount of water flows through the bypass porthole and into the upper chamber, gradually equalizing the pressure in both the chambers; the diaphragm returns to its normal position and the relief valve to its seat to close the flow of flushing.

Any sort of impurities, dirt, suspended matters etc. in the water may clog the bypass port. Plugging of the bypass port in this way will result in continuous flushing until the port is cleared.

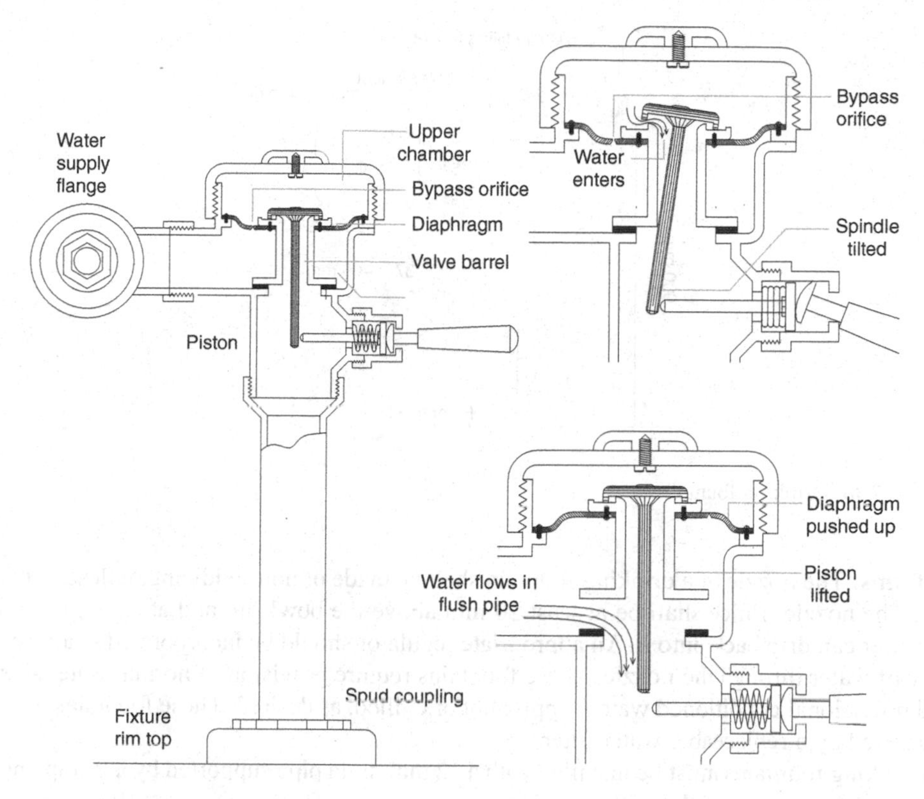

Figure 9.35 Diaphragm type flush valve

9.11 Drinking fountain

A drinking fountain, also called a bubbler, is an appliance designed to provide a small jet of water at a comparatively low discharge flow rate through a nozzle, to put water into the mouth for drinking directly. For drinking, the user needs to bend down to the stream of water. Figure 9.36 represents a drinking fountain.

These appliances are basically a shallow-depth basin with a specially designed tap called a bubbler faucet and may be a pedestal type, countertop type or wall-hung type. These is also another type of faucet, a bottle refilling faucet, which helps with collecting water in a bottle. Bubbler faucets are generally found with a push button and self-closing system. Drinking fountains are mostly for single users but some include an additional unit of a lower height for children and short adults.

Drinking fountains may be supported by water conditioners. To remove impurities from the water, a replaceable filtering system is incorporated, and most fountains have a built-in chiller to get water at lower temperatures. A hot water supply system may be installed. Cooling capacity of chillers is about 20 lph. Water temperature is regulated by thermostat.

Drinking fountains are manufactured for indoor and outdoor installation. These are mostly found made of stainless steel sheets with a thickness of 1.0 mm to 1.6 mm. Outdoor drinking fountains generally have a weather-resistant powder-coated finishing.

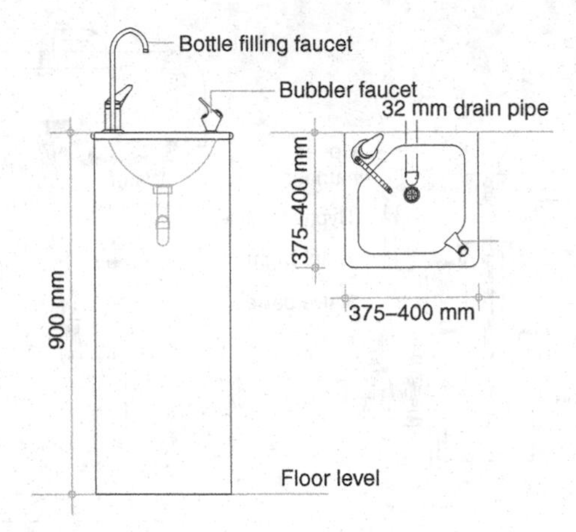

Figure 9.36 Drinking fountain.

Trims: The nozzle of a drinking fountain shall be made of non-oxidizing, unleaded material. The nozzle orifice shall be at least 20 mm above the bowl rim and at an angle so that no water can drip back onto it. An appropriate regulator should be incorporated to adjust the flow of water through the nozzle. These fountains require provision of normal water supply and provisional conditioned water supply: hot or chilled, as desired. These fountains may be supported by a replaceable water filter.

Drinking fountains must be installed with a 32 mm drain pipe supported by a P-trap underneath the wastewater outlet with a strainer.

Installation: The mounting of the fixture on the wall should be sturdy and strong enough to withstand more weight than that of the fixture itself. All types of drinking fountains should be installed with the nozzle orifice located within a range of 750 to 1000 mm above the floor, depending upon the average height of users. For electrically conditioned drinking fountains, an electrical outlet nearby for power supply will be required. Because of the many variations in style and optional services of drinking fountains, it would be wise to follow the manufacturer's installation guides and specifications in each case.

9.12 Washing machines

Washing machines are used to wash or clean dirty items quickly and automatically by using hot and cold water, and the resulting wastewaters are discharged into the drainage piping. Two types of washing machines are used for two categories of washing; those are incorporated in plumbing system. The types of washing machines are as follows.

1 Dishwashing machine and
2 Clothes washing machine.

9.12.1 Dishwashing machine

A dishwashing machine is an electro-mechanical appliance used for cleansing and rinsing dirty dishware and cutlery conveniently and automatically, saving manual labour and time. In comparison to manual dishwashing, the dishwasher provides smarter, quicker and more efficient way of cleaning and hygienic washing of utensils. A dishwasher cleans by spraying hot water, typically between 60 and 75°C, at the dishes.

Normal temperature water is fed from the supply pipe, which is heated by the electric heating element to produce hot water. From the dispenser, detergent is mixed with water, which is then pumped up to the rotating spray arms or paddles. The pumped water squirts up through the holes in the paddle. The rotating paddle makes the squirted water spin around, which blasts the dishes and removes dirt. Sprayed water bounced off the dishes falls back to a receiving plate at bottom of the machine, where it is heated and recirculated through pumping.

Once the wash is finished, the wastewater is drained out and more hot water is pumped in to start the rinse cycle. After the rinse cycle is finished, the wastewater is drained and the dishes are dried.

Trim: In a dishwashing machine, normal water is fed through a 12 mm diameter pipe. If there is no inbuilt water heating system in the machine, then an additional hot water supply of a temperature ranging from 60 to 75°C is needed for cleansing and 82 to 88°C for sanitizing [11]. The supply connection to washing machine must have a backflow preventer. A minimum flow pressure of 103.5 kPa is needed in water supply pipe feeding the machine. In case of excess pressure it must be maintained below 207 kPa by installing pressure reducing valve on the inlet pipe.

A flexible waste pipe of diameter of 20 mm is drawn from the back of the machine which may be connected to the drainage pipe via an up-stand pipe with a trap underneath. Generally, the dishwasher wastewater discharge pipe is connected to the nearby sink drainage pipe with the help of a special washing machine branch connector fitting placed above the trap.

Installation: The drain pipe of a machine, equipped with a drainage pump, may be connected to an adjacent sink waste pipe before its trap, subject to having a sufficient air gap between the hose outlet and connection pipe to the sink drain pipe. For this purpose, a high loop connection is made to prevent potential backflow of water into the dishwasher, as shown in Figure 9.37. Another method of connection is to use a dishwasher-air gap apparatus; this device is usually a small cylinder installed on a cabinet or counter top near the sink and

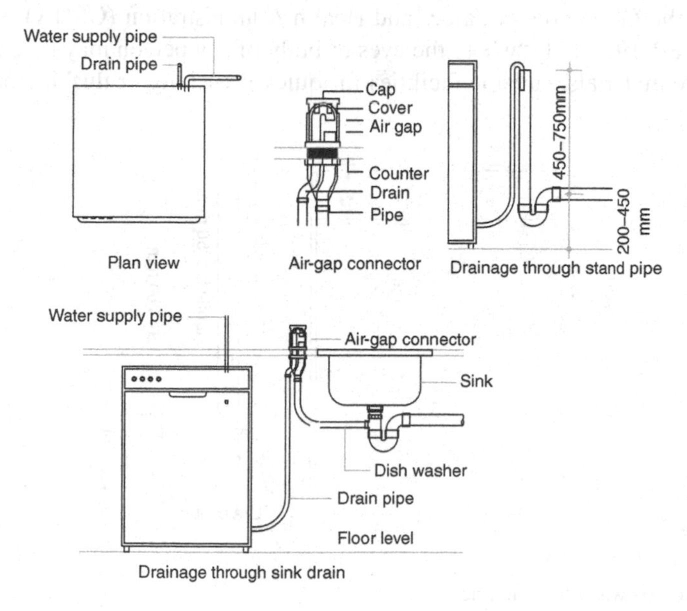

Figure 9.37 Dishwashing machine plumbing.

faucet. As the dishwasher cleans dishes, the wastewater is pumped out of the dishwasher and carried by a hose to the air gap.

9.12.2 Clothes washing machine

A clothes washing machine is an electro-mechanical appliance used for wet washing of laundry such as clothing and sheets. The washing machine has a large drum. The inner part of the drum is called the basket, into which clothes are put to wash; this is perforated for water to drain out. The outer part of the drum is the tub, a solid compartment surrounding the basket that keeps water in or allows it to drain out.

When the basket is filled with clothes for washing, the machine fills the tub with water and then stirs the clothes using an agitator. After agitating for some time, the washer drains the water and then spins the clothes to remove most of the water. Then, it refills and agitates the clothes some more to rinse out the soap. Afterward, it drains and spins again. To agitate the clothes, the wash tub is moved back and forth and the basket is spun to force the water out.

Trims: A clothes washing machine has two water line connections of 12 mm at the back, one for hot water supply and the other for normal water. At 40°C, about 72 litres of water are consumed for 2.75 hours of washing of 7 kg of cotton clothes. The water pressure requirement is a minimum of 1 bar and maximum of 10 bar. Water flow requirement is about 8 lpm.

Washing machines have a drain hose pipe to drain wash-water. The hose pipe shall not be drained directly connecting to any drainage pipe. It must be drained through a stand pipe, as shown in Figure 9.38. A stand pipe is 50 mm diameter and at minimum a 750 mm-high pipe fitted on a trap which is installed 300 mm above the floor level.

Installation: The legs of a clothes-washing machine are to be placed on neoprene rubber pads to minimize transmission of solid-borne noise to the building structure. Leveling screws are to be adjusted accurately to prevent the machine from wobbling [11].

9.13 Emergency fixtures

According to the Occupational Safety and Health Administration (OSHA) regulation contained in 29 CFR 1910.151, "where the eyes or body of any person may be exposed to injurious corrosive materials, suitable facilities for quick drenching or flushing of the eyes and

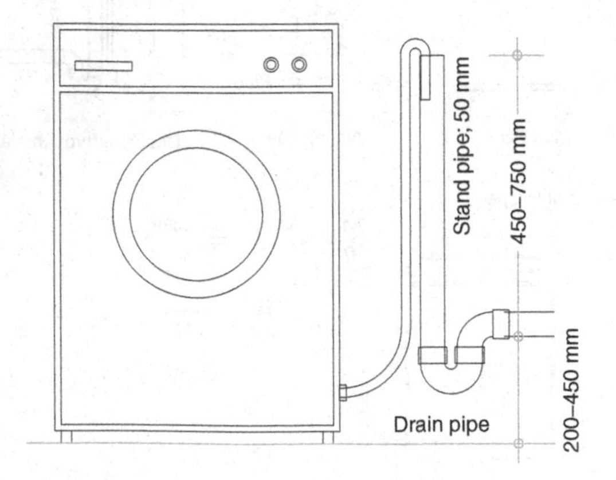

Figure 9.38 Clothes washing machine.

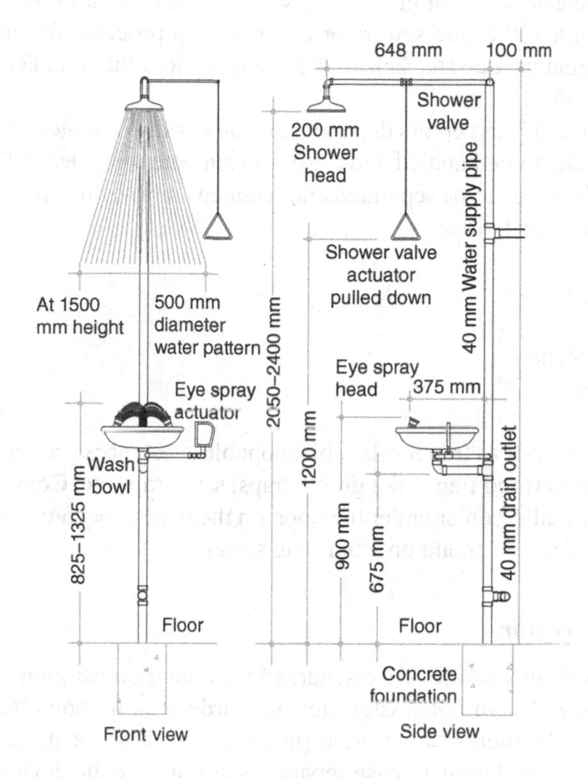

Figure 9.39 Emergency fixture including shower and eyewash spray.

body, shall be provided within the work area for immediate emergency use". With a view to providing such facility, emergency showers and eyewash stations are to be installed, as illustrated in Figure 9.39, near possible locations for accident occurrences. This station provides supply of sufficient water to flush the whole body and eyes, to remove hazardous substances that can cause injury to the user. Emergency showers are designed to flush the user's head and body and the eyewash sprayer is for cleansing eyes. About 75.7 lpm water flow is to be maintained at least for 15 minutes. The shower is designed in such a way that it is activated in less than 1 second and remains operational without the intervention of the user.

The eyewash faucet is designed to deliver water to both eyes simultaneously at a flow rate of not less than 1.5 lpm to run at least 15 minutes. The emergency stations with a combination eye and face wash require a flow rate of 77 lpm.

Trims: To supply sufficient water to the emergency fixture with eye washing provision, there shall be a water supply pipe of 40 mm diameter, and for draining wastewater, a drainage pipe of 50 mm diameter shall be connected to the wastewater drainage system specially managed for this purpose.

9.14 Interceptors

An interceptor is a device that separates and retains objectionable suspended materials; these might be present in wastewater and not permitted for discharging, while allowing discharge of acceptable quality of wastewater into the sewer system. The majority of plumbing codes do not recommend the presence of grease, oil, sand and other various materials in wastewaters that are harmful to either the sewers or the treatment process. To remove these objectionable materials from wastewater before they entering into the drainage or sewer system, interceptors are provided.

The design and size of interceptors depend upon the density of materials to be intercepted, flow rate of wastewater to be handled and the corrosion potential. Depending upon the characteristics of the substances to be separated and retained, various interceptors generally used in plumbing system are as below.

1. Grease interceptor
2. Oil interceptor
3. Sand interceptor and
4. Hair interceptor.

Interceptors basically work as trap for the objectionable substances present in wastewater, so these devices are also termed traps, like grease traps, sand traps etc. Commercially available interceptors can be installed on or under the floor, on the drain pipe, but large interceptors are generally constructed underground on a building sewer.

9.14.1 Grease interceptor

Grease is found mostly in wastewaters discharged from food preparation areas and kitchens. Grease remains suspended in hot wastewater but hardens as it cools. Relatively hardened grease deposits on and adheres to the drain pipe bottom, subsequently restricting the flow with an increase in accumulation. Grease separators or traps are the devices used to separate and retain grease present in wastewater while flowing through. A typical grease separator is illustrated in Figure 9.40. An internal baffle restricts the flow of floating grease for periodic removal while free wastewater flows through the outlet and discharges into the drainage

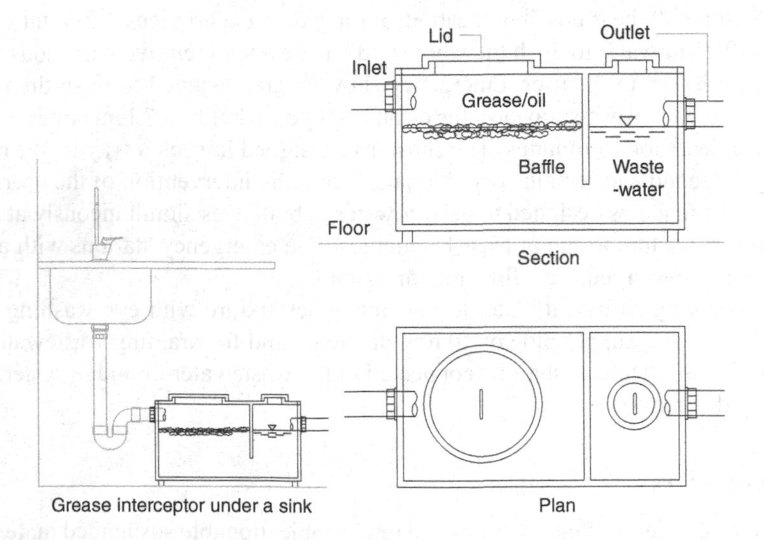

Figure 9.40 Grease interceptor under a sink.

Table 9.2 Sizing of grease interceptors, Converted in SI unit from source [12].

Sl. No	*Drainage fixture unit (DFU)*	*Size in cum*
1	8	1.9
2	21	2.84
3	35	3.8
4	90	4.73
5	172	5.68
6	216	7.57
7	307	9.46
8	342	11.36
9	428	15.14
10	576	18.93
11	720	28.39
12	2112	37.85
13	2640	56.78

piping. Grease removal may be done manually or automatically. The size of the grease interceptor to be used, which can be found in Table 9.2, is based on the drainage fixture unit to be served considering a 30-minute retention time. Drainage fixture units is discussed in later chapters.

9.14.2 Oil interceptor

Oil interceptors are devices designed to separate and retain floating oil and flammable or volatile liquid that may be present in wastewater, having been discharged from garages, car washing facilities etc. Oil and other objectionable liquids, being lighter than wastewater, float and are retained while flowing through a series of baffles within the separator. For being liquid in nature, the floating substances are drawn off continuously by separate overflowing arrangements leading to remotely placed oil storage tanks. In case of volatile or flammable liquids, the interceptor must be provided with a vent pipe, extending above the structure, to expel vapours into the open air.

9.14.3 Sand interceptor

Sand interceptors are constructed to separate and hold solids or semi-solids that may present in wastewater. These interceptors allow accumulation of solids at the bottom of interceptors for periodical removal mechanically or manually. The outlet pipe should be located sufficiently above the base to prevent discharge of solids along with wastewater and allow more accumulation of solids underneath. The inspection pits of storm or rainwater also act as sand interceptors.

9.14.4 Hair interceptor

Hair interceptors are mostly installed inline, instead of as traps, to prevent hair from flowing with wastewater and allow periodical removing from the interceptors, to ensure continuous flow through. These interceptors use a perforated bucket strainer placed inside a housing, as

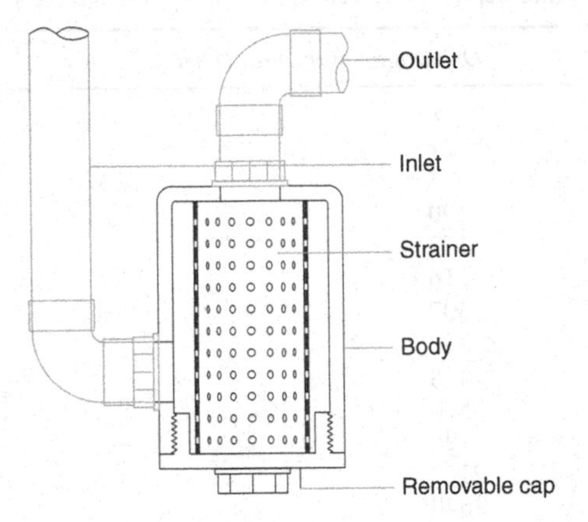

Figure 9.41 Hair interceptor.

shown in Figure 9.41, to trap the hairs. The strainer can be removed by opening the housing and replaced after cleaning the accumulated hair. The strainer is usually made of stainless steel and the interceptor may be made of steel, brass or plastic. The inlet and outlet pipe size re generally made 40 mm diameter.

9.15 Floor drains

Floor drains are considered a plumbing fixture which provides an inlet in the floor to receive wastewater generated on the floor, and finally discharges into the drainage piping system. In any floors where there is a possibility of wastewater accumulation, a sufficient number of floor drains of adequate size should be provided at ideal locations. The size of a floor drain is considered based on the size of its outlet diameter. The minimum recommended size of floor drain is 75 mm [12]. The inlet of a floor drain is generally provided with removable grating. The size of openings in gratings should be such that in all likelihood valuable things cannot get into it. For indoor gratings, the summation of openings shall be at least 1.5 times the cross-sectional area of the outlet drain pipe.

Floor drains are categorized into two groups: floor drains with integral water seal trap and without trap. Floor drains, ultimately draining into a sanitary or combined sewer, must be provided with a water seal trap. Floor drains to be installed in floors below ground level, where there is chance of backflow, must be equipped with a backflow preventer. There are other varieties of floor drains which include sediment buckets, as shown in Figure 9.42. In sediment buckets, slots are provided along the side to permit complete draining of wastewater while retaining coarse sediments and solids carried with wastewater on the bucket base.

Installation: The floor drain inlet shall be located at a place which is readily accessible and where maintenance work can be done easily. Furthermore, its location should be in such a position where water from the very remotest point of the floor can reach the drain inlet in the fastest possible time to make the floor dry very quickly.

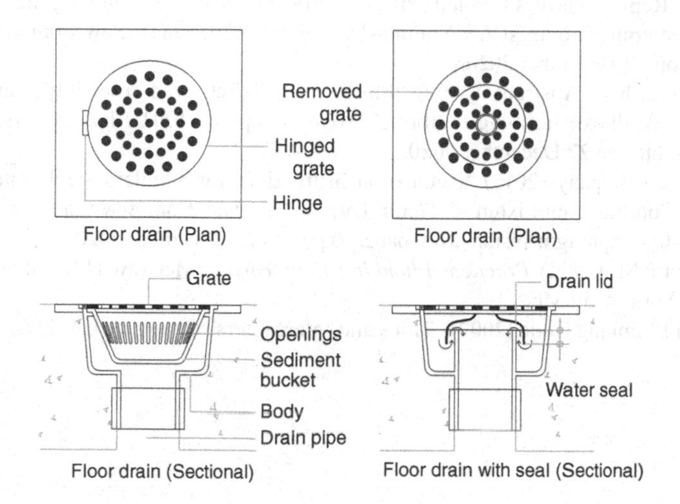

Figure 9.42 Floor drain with bucket and with water seal.

The floor shall have mild slope evenly maintained towards the floor grating so that there is no puddle of water on the floor or around the inlet. The slope of the floor shall be made in the ratio of 100:1 if it is roughly finished and 200:1 if it is smooth finished.

References

[1] International Plumbing Code (IPC).

[2] Martin, J., and Heaney, J. (2008) "Conserve Florida Water Clearinghouse" *Dept. of Environmental Engineering Sciences*, University of Florida, Retrieved from https://sswm.info/sites/default/files/reference_attachments/MARTIN%20and%20HEANEY%202008%20Water%20Use%20by%20Urinals.pdf on 27 December 2020.

[3] Koeller, J. (2007) "High Efficiency Urinals – US and Canada" California Urban Water Conservation Council, Retrieved from www.cuwcc.org/urinal_fixtures/HEU-07-06-14.pdf (accessed 5 September 2007); Found in "Water Use by Urinals by Jacqueline Martin and James Heaney, University of Florida" (2008) Retrieved from https://sswm.info/sites/default/files/reference_attachments/MARTIN%20and%20HEANEY%202008%20Water%20Use%20by%20Urinals.pdf on 22 May, 2021.

[4] Stauffer, Beat, "Low-flush Toilets" Retrieved from https://sswm.info/water-nutrient-cycle/water-use/hardwares/toilet-systems/low-flush-toilets on 22 May, 2021.

[5] Green Building Supply (UD) "Dual Flush Water Saving Toilets Never Clog" Retrieved from www.greenbuildingsupply.com/Public/Energy-WaterConservation/WatersavingToilets/CaromaDualFlushToilet/index.cfm on 10 May 2008, Retrieved from http://home.howstuffworks.com/dual-flush-toilet2.htm on 27 December 2020.

[6] Wikipedia.org, "Dual Flush Toilet" Retrieved from https://en.wikipedia.org/wiki/Dual_flush_toilet on 27 December 2020.

[7] Blower, G.J. (1996) *Plumbing: Mechanical Services*, Book 2, Addison Wesley Longman Ltd, England, p. 222.

[8] Publishers Representatives Limited (PRL) (2020) "Xiamen NAT Plumbing Inc" Retrieved from www.globalsources.com/si/AS/Xiamen-NAT/6008842023981/Showroom/3000000149681/ALL.htm on 27 December 2020.

[9] Commonwealth of Australia (2009) "Flush Valves" Retrieved from https://emedia.rmit.edu.au/dlsweb/Toolbox/plumbing/toolbox12_01/units/cpcpwt4001a_hot_cold/00_groundwork/page_006e.htm on 27 December 2020.

[10] Sloan Valve Company (2013) "Installation Instructions for Stainless Steel Water Closets, Urinals and Combination Fixtures" *Sloan Valve*, Retrieved from www.sloan.com/sites/default/files/2016-01/diaphragm-piston-whitepaper_0.pdf on 27 December 2020.

[11] Harris, Cyril M. (1991) *Practical Plumbing Engineering*, McGraw-Hill Publishing Company Ltd, New York, p. 5.8Mc.

[12] California Plumbing Code (2007) "Traps and Interceptors" Chapter 10, p. 155.

10 Sanitary drainage systems

10.1 Introduction

Drainage is the process of collecting fluid wastes and subsequently conveying them to an intended location. Drainage in buildings is the process of conveying foul and wastewater generated from a building to the legal point of disposal. But sanitary drainage in a building is the hygienic and adequate disposal of all sorts of foul wastes generated in the building, conveyed mostly through pipes, using water as a medium, to a legal point of disposal, safely and sanitarily.

The primary objective of installing a sanitary drainage system in a building is to provide efficient and sanitary disposal of human excreta, ablutionary wastewater, laundry and kitchen wastewater etc. from the fixtures to the public sewer. Sanitary drainage from a building can be done in various ways. In this chapter, the ways of disposing of all sorts of wastewater from buildings and associated works are discussed.

10.2 Basic principles of drainage systems

The sanitary drainage system in a building shall be planned, designed, developed, operated and maintained on the basis of the following principles.

1 The drainage system shall be planned following the local code of practice
2 Piping shall be designed and installed on the basis of sound hydraulic and engineering principles
3 The system shall be efficient, economic, durable and easy to maintain and
4 The system shall ensure sanitation and healthy environment.

10.3 Basic requirements for drainage systems

With a view to achieving the objective of developing sanitary drainage systems, and following the principles of planning, designing, operating and maintaining the plumbing system, some basic requirements are to be fulfilled. Following are those basic requirements.

1 The type of drainage system shall be chosen according to the local by-laws
2 Drainage pipes shall be adequately sized and graded properly for effective disposal of waste
3 Pipe run shall be kept as short as possible
4 Bends and junctions should be smooth, rounded and long sweep

DOI: 10.1201/9781003172239-10

5 At bends, junctions and at some critical locations, access doors or clean-out plug shall be provided
6 Pipe material selected shall be strong enough to withstand internal and external pressure and able to withstand corrosive waste
7 All fixtures shall be provided with an appropriately sized water-seal trap and
8 The piping system shall be fully ventilated.

10.4 Categories of sanitary drainage piping

Sanitary drainage pipes are named and grouped in different ways depending upon the type of content conveyed, location etc.

10.4.1 Nomenclature of drainage piping

Sanitary drainage in buildings mostly covers the piping system conveying various kinds of waste and associated products sanitarily. All pipes engaged in the drainage system can be categorized in the following ways, depending on the type of waste conveyed, purpose of use and location.

1 Soil pipe
2 Wastewater pipe
3 Vent pipe
4 Building drain and
5 Building sewer.

10.4.2 Grouping of drainage piping

Drainage systems in a building can be divided in three groups according to the location of the drainage piping and the methodology of drainage adopted.

1 Drainage system above ground
2 Basement drainage system and
3 Drainage in ground.

10.5 Drainage system above ground

Drainage of wastes – soil waste and wastewater – from all floors, including the ground floor, of a building can be developed in one of the following methodologies.

1 Two-pipe system
2 One-pipe system and
3 Single-stack system.

The selection of drainage method from a building primarily depends upon the drainage facilities available around the building site. There are other criteria on which the selection depends. These are local by-laws regarding waste disposal from buildings in a particular area, building occupation type and configuration, position of toilets included and availability

Table 10.1 Suggested system selection for different types of buildings [1].

Sl. No	*Type of building*	*Drainage system (Y = acceptable, N = not recommended)*		
		Two-pipe system	*One-pipe system*	*Single-stack system*
1.	Housing			
	a 1- or 2-storey bungalow	Y	Y	Y
	b 3- or 4-storey group housing	N	Y	Y
	c Multi-storey	Y	Y	N
2	Hospital			
	a 2- or 3-storey			
	i For hospital fixtures	Y	Y	N
	ii For toilets	Y	Y	Y
	b Multi-storey	Y	Y	N
	c Housing, hostel and administration building for hospital complex	Y	Y	Y
3.	Hotels			
	a 1- or 2-storey bungalow	Y	Y	Y
	b 3–4 storey	N	Y	Y
	c Multi-storey	N	Y	N
4.	Office building			
	a 6- to 8-storey	N	Y	Y
	b Multi-storey	N	Y	N

of appropriate pipe fittings. The guideline for selecting fixture layout and the drainage methodology for a particular type of building is furnished in Table 10.1.

10.5.1 Two-pipe system

In a two-pipe drainage system, two separate branches of piping, stacks, building drains and building sewers are installed for conveyance of all wastes generated from all floors, on and above ground, of a building. Excreta waste, generating from water closets, urinals and bidets, and other very obnoxious wastes are conveyed through drain pipes known as soil pipes, leading to the vertical drain pipe termed soil stack. The soil stacks are connected to the sanitary building drain or sewer to dispose of waste to the septic tank, sanitary sewer or combined sewer, which is available and applicable.

On the other side, wastewater generated from basins, bathtub or trays, sinks and floor drains are disposed of through separate wastewater pipes and conveyed to a vertical drain pipe termed wastewater stack. The wastewater stacks are connected to the building's sanitary drain or sewer to be finally disposed of in the public sanitary or combined sewer.

Both the soil and wastewater pipes and stacks are supported by ventilation pipes and stack. Both soil and wastewater stacks are extended above the roof for stack venting and terminate in inspection pits at ground level, as shown in Figure 10.1, or into the building drain.

Advantages of installing two-pipe systems are as follows.

1 Avoids danger of backflow of sewage in wastewater generating fixtures
2 Enables use of wastewater for irrigation, gardening and recycling or reusing and
3 Reduce waste load for off-site sewage treatment system.

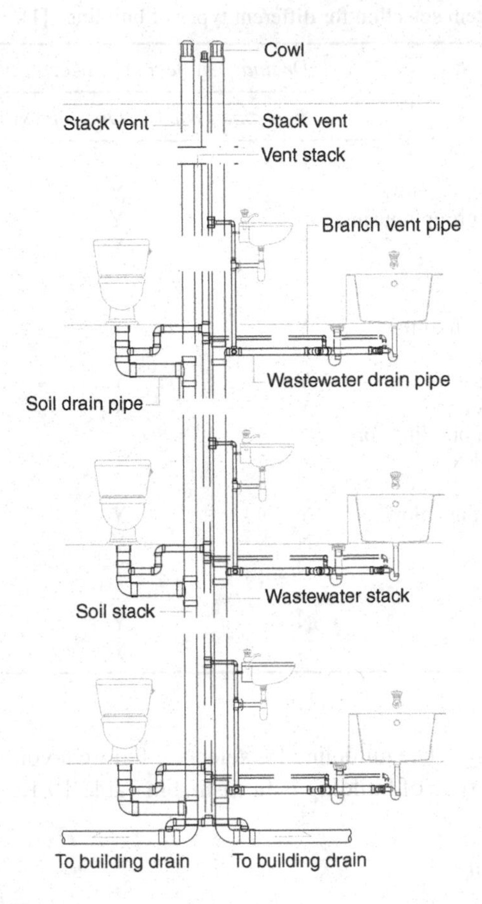

Figure 10.1 Two-pipe drainage system.

Disadvantages are as follows.

1 Requires comparatively more pipe and fittings, so costly
2 Requires more space for housing more pipes, or closely installed pipes offer difficulties in maintenance and repair works and
3 Use of more pipes invites more joints, increasing chances of more leakages.

Though there are some disadvantages to installing a two-pipe system of drainage, this system is developed where local code guides. Furthermore, where there is no public sewer or the sewage management authority advices for a septic tank in the building's waste disposal system, then a two-pipe system development is the only option.

10.5.2 One-pipe system

In a one-pipe system, both the excreta wastes and wastewater are disposed of through the same drain pipe leading to a stack, as shown in Figure 10.2. The stack ultimately disposes of the waste into the building drain or sewer, whence it finally goes into to the nearby public

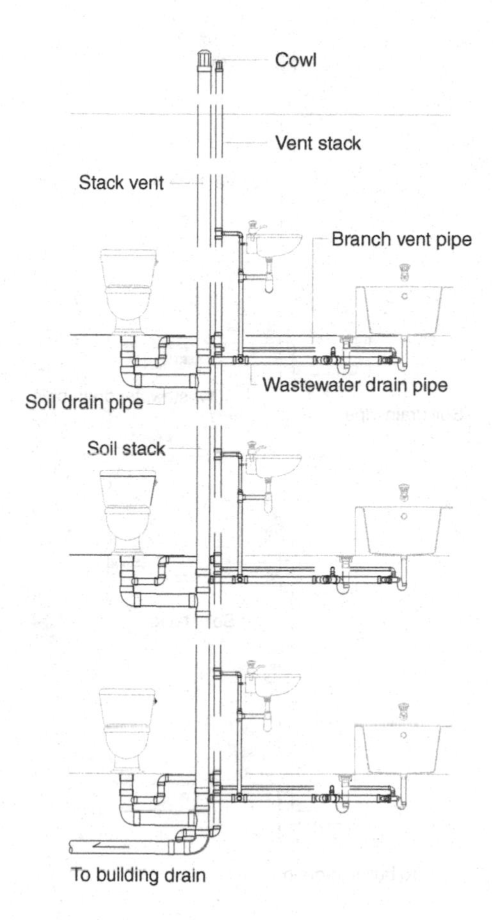

Figure 10.2 One-pipe drainage system.

sanitary sewer or combined sewer. Extra vent piping is provided for ventilation of soil and wastewater pipes along with trap ventilation. Obviously, this methodology of building drainage is advantageous in many respects, but where local by-laws in this regard do not permit a one-pipe system, and there is an obligation or necessity for wastewater recycling or reusing, in those cases a one-pipe system cannot be developed. This system should not be adopted where a septic tank system is to be built due to not having public sewer.

One-pipe systems can be successfully developed in high-rise buildings. In high-rise buildings, all fixture traps are more or less subject to the risk of losing water from traps due to developing siphonage. So, an efficient ventilation system is to be incorporated with one-pipe drainage piping.

10.5.3 Single-stack system

In this system, all the fixtures of a toilet battery discharge into a common single stack, as shown in Figure 10.3. Thereby, all soil and wastewater generating fixtures are connected directly to a stack receiving all sorts of waste discharged into the sanitary drain and sewer. The outstanding feature of this system is the complete absence of branch drainage and

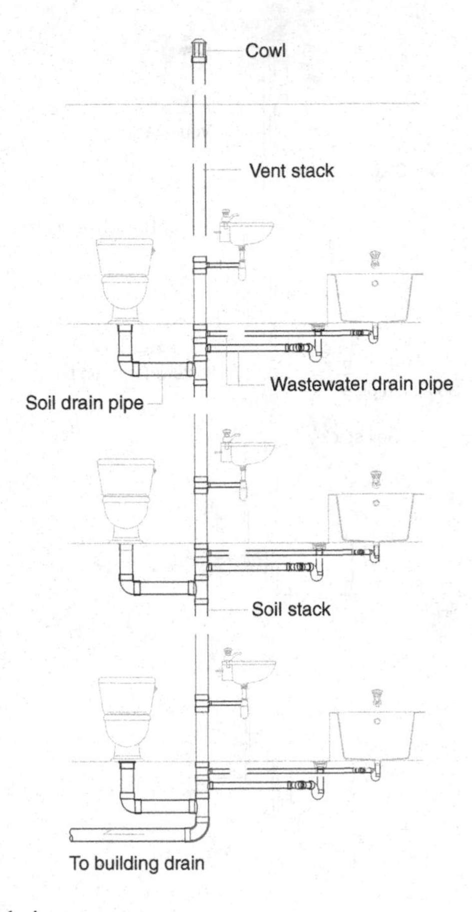

Figure 10.3 Single-stack drainage system.

ventilating pipes, except in very special circumstances. But stack venting is to be provided as usual. This drainage methodology can be adopted in some limited conditions. The limiting conditions are as follows.

1 Applicable for buildings not more than 5 storeys
2 Fixture drainage pipe shall be directly connected to the stack
3 Fixtures shall be within 1.5 m around the stack
4 Fixture layout should be repetitive vertically and
5 No offset in the stack can be allowed.

For a multi-storey building, a sovent system has been developed and successfully implemented in 20 to 30 storey buildings [2].

10.6 Sovent single-stack drainage system

The sovent single-stack drainage system is a patented improved and simplified single-stack system, which eliminates all of the limitations of conventional single-stack systems discussed earlier. This system is effectively applicable for multi-storey buildings more than five storeys.

The sovent single-stack drainage system incorporates two special fittings: the aerator and the de-aerator fittings. These fittings have self-venting features which eliminate use of an extra vent pipe for the purpose. An aerator is installed at every junction where a soil branch pipe is connected to the stack. On the other hand, a de-aerator is installed at the bottom part of the stack and at every offsetting of the stack, as shown in Figure 10.4.

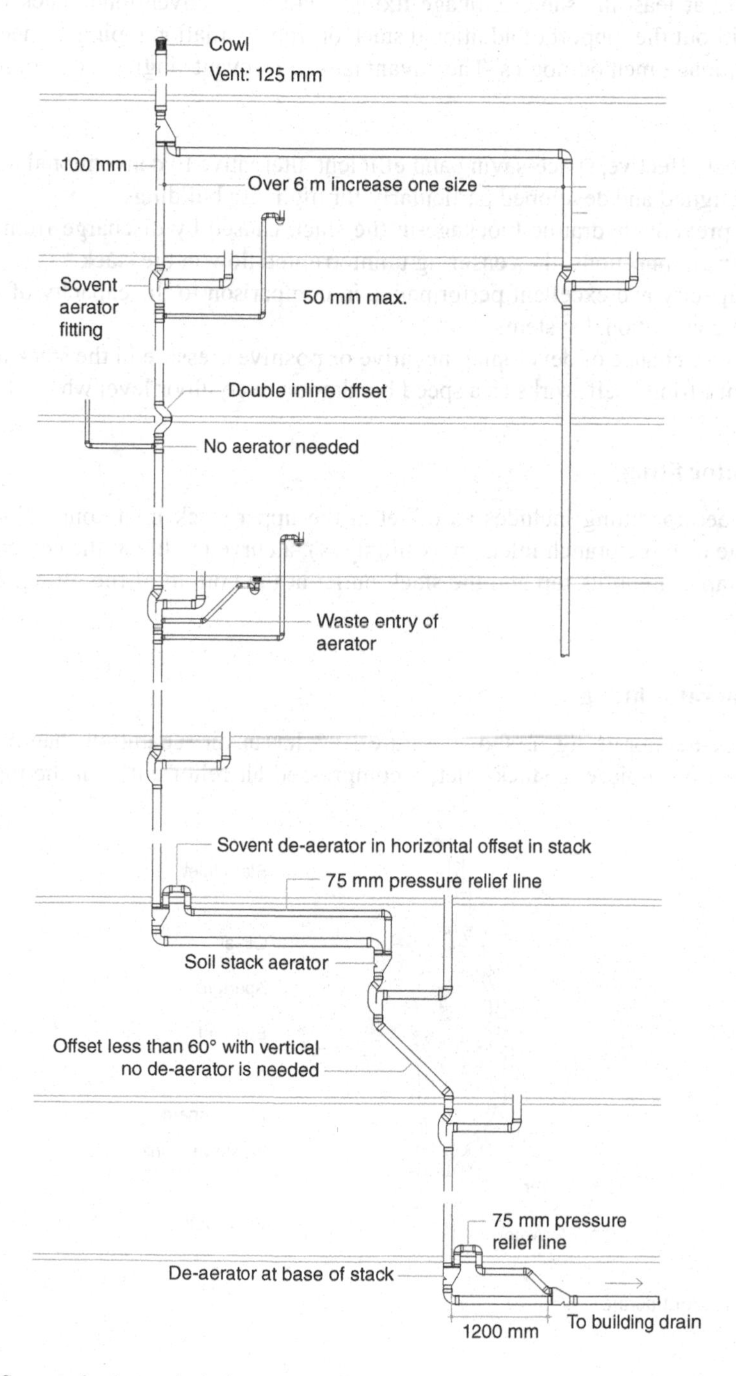

Figure 10.4 Sovent single-stack drainage system.

No aerator fitting is required where a 50 mm soil branch pipe is connected to the 100 mm stack. On any floor, if no soil branch pipe is connected, then only a double inline offsetting is installed.

The sizing of a sovent single stack follows the same procedure in sizing stack in one-pipe drainage systems, believing that the capacity of a sovent single-stack of particular diameter would handle at least the same drainage fixture load as a conventional stack of the same diameter, without the support of additional stack or trap ventilation piping, as needed in conventional drainage methodologies. The advantages of a sovent single-stack drainage system are as follows.

1. More cost effective, space-saving and efficient alternative to conventional drainage systems designed and developed particularly for high-rise buildings
2. Sovent prevents hydraulic blockage in the stack caused by discharge from the branch drain pipe connections, thus ensuring uninterrupted flow in the stack
3. High capacity and excellent performance in comparison to the capacity of a same size stack in conventional systems
4. There is no chance of developing negative or positive pressure in the stack and
5. A sovent fitting itself works as a speed breaker on every floor level where it is installed.

10.6.1 Aerator fitting

The sovent aerator fitting includes an offset at the upper stack inlet connection, a mixing chamber, one or more branch inlets (maximum six), a curved baffle at the center of a chamber with an aperture at its top and the stack outlet at the bottom of the fitting, as shown in Figure 10.5.

10.6.2 De-aerator fitting

The sovent de-aerator fitting, as shown Figure 10.6, has an air separation chamber supported by an internal nose-piece, a stack inlet, a compressed air relief outlet at the top end and a

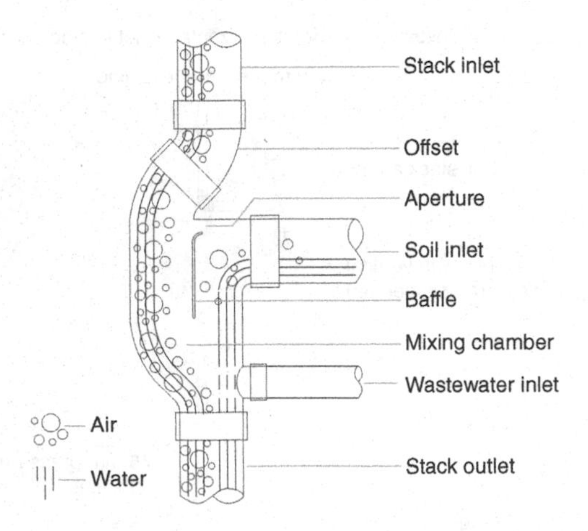

Figure 10.5 Sovent aerator.

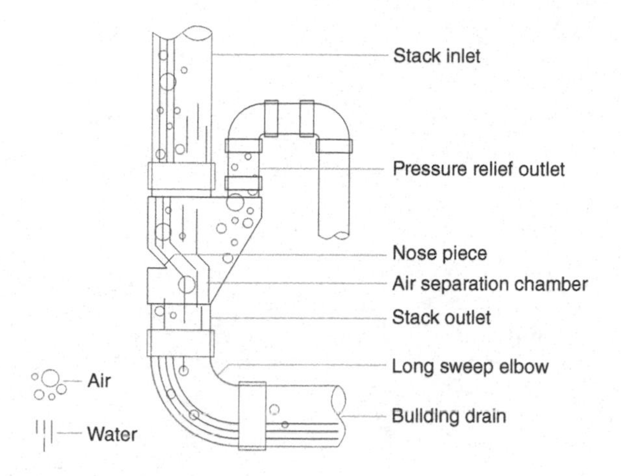

Figure 10.6 De-aerator fitting

stack outlet at the bottom end. A de-aerator fitting is to be installed in the following two conditions of stacks.

1 When the bottom of stack is to be bent 90° and the discharging end is more than 1.2 m away from a bend and
2 When a stack is offset.

For the first condition, a de-aerator fitting is to be installed just on the elbow, and the pressure relief line extension is to be connected at least 1.2 m downstream onto the top of the building drain. For the second condition, a de-aerator fitting is to be installed just on the top of the upper elbow and just below the lower elbow of a stack offset. The pressure relief line is to be extended from the upper de-aerator outlet to the lower de-aerator pressure relief outlet.

10.7 Building drain

The building drain is that part of a drainage systems that remains at the lowest level of a drainage piping to receive the discharge of all stacks inside; it extends to a maximum of 2.0 m beyond the exterior walls of the building and conveys the draining wastes to the building sewer. So, a building drain is also referred to as the collection pipe from the stacks inside the buildings. In a two-pipe system, two types of building drains are installed: the wastewater building drain receives wastewaters from all wastewater stacks, while the soil building drain collects soil wastes from all soil stacks. In a one-pipe system, one type of building drain, a sanitary building drain, receives waste from all stacks.

Building drains are generally run under a suitable floor, above the building or at public sewer level, under which a downward extension of stacks is not required or cannot be done for any valid reason. In most cases, these drain pipes are run under the ground floor or first floor slab, as planned, installed by keeping suspended from the slab above of respective floor, as shown in Figure 10.7.

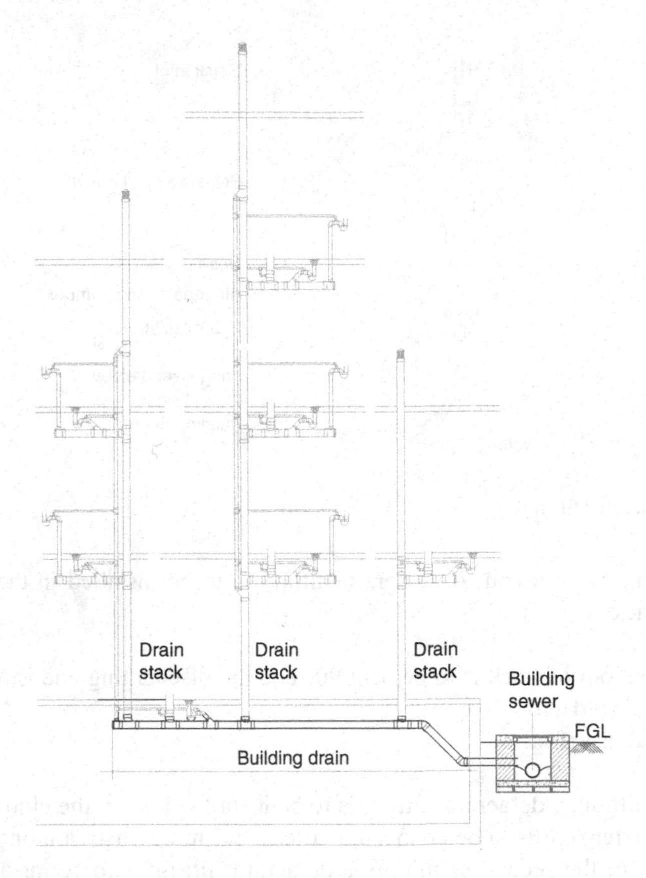

Figure 10.7 Building drain and sewer.

10.7.1 Types of building drain

Building drains are classified into three categories, depending upon the types of waste carried. The various drains are named as below.

1 Sanitary drain
2 Combined drain and
3 Storm drain.

Sanitary drain: The drain which receives and conveys all the discharge from the soil and wastewater stacks of a building is termed a sanitary building drain. So, sanitary drains convey all sorts of wastes from a building except storm water.

Combined drain: This drain receives all sorts of wastes, i.e. soil, wastewater and storm water from respective stacks or rainwater down pipes, and conveys them to the combined sewer.

Storm sewer: This type of drain receives only rainwater from the roof and other surfaces exposed to sky, clear water from any sources, except any sort of sanitary wastewater, and conveys them to the storm sewer.

All types of building drain are illustrated in Figure 10.8.

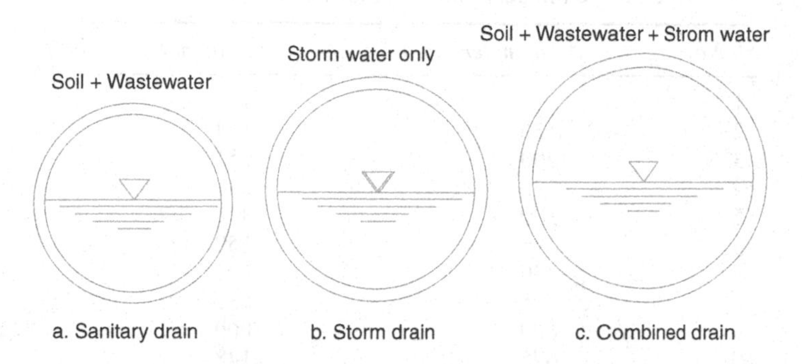

Figure 10.8 Various building drains.

10.8 Grading for building drains

Grade or pitch is the fall (slope) of a pipe in reference to a horizontal plane. The efficiency of horizontal drainage pipes depends upon the velocity of flow. Velocity of flow increases with the increase of grade of pipe and decreases with the increase of diameter, and vice versa. The maximum velocity in a building drain shall be 146 mm/min and the minimum velocity shall be maintained at 36 mm/min. In unavoidable circumstances, steeper grade is to be provided.

To maintain minimum velocity in any drain, the grading must be provided in accordance to Table 10.2.

The grade of a building drain of particular size that would generate maximum velocity of flow is furnished in Table 10.3.

10.9 Offsetting of stacks

A graded offset is any offset made at an angle of less than 45° to the horizontal. An offset or any horizontal drain pipe at an angle more than 45° to the horizontal is considered to be a vertical drain pipe. If any offset is to be made at an angle lower than 45° to the horizontal, the required diameter of the offset and the stack below shall be determined as for a building drain. The minimum grade of a graded offset shall be in accordance with Table 10.4.

In case of any graded offsetting, there is restriction in connecting branch drain pipes in certain portion of the stacks above the offsetting point. The restricted zone is shown in Figure 10.9. The conditions are as follows.

For graded offsets, no connection shall be made within following zones.

1 600 mm above the bend, when the stack height is not more than five floor levels above the offset
2 1 m of the bend when the stack height is more than five floor levels above the offset or
3 2.5 m above when there is chance of developing foaming.

10.10 Basement drainage system

Basement floors of a building, which remain below ground level, may receive waste from various sources, such as other floors above ground. Furthermore, wastewater from car washes, leaking water from various machinery, pipes running through basement floors,

Table 10.2 Approximate minimum gradients for drains and sewer [3].

Sl. No	*Diameter (mm)*	*Gradient (length/fall)*
1	100	105
2	150	174
3	200	251
4	225	288
5	300	417
6	375	550
7	450	692
8	525	832
9	600	1000
10	675	148
11	750	1318

Table 10.3 Approximate maximum gradients for drains and sewer [4].

Sl. No	*Diameter (mm)*	*Gradient (percent)*
1	100	21.7
2	150	12.6
3	250	6.4
4	300	5.0

Table 10.4 Minimum grade of offsets.

Sl. No	*Size of offset pipe (mm)*	*Min. gradient (percent)*
1	80 or less	2.50
2	100	1.65
3	125	1.25
4	150	1.25
5	225	0.60
6	300	0.40

accidental flooding due to ingress of water from outside etc. may accumulate on basement floors and need an appropriate system for draining out efficiently. Generally, it is found that the basement floor level remains below the bottom level of nearby public sewer or drainage systems. In such a case, drainage from a basement cannot be done by gravity. Furthermore, if the outside sewer is found below the lowest basement floor, drainage from the basement by gravity may cause a flood of sewage in the basement due to backflow from external sewers, for getting blockage in sewer or flooding outside causing the sewer system submerged under water.

So, generally for basement floors, drainage is accomplished through pumping. To facilitate pumping, a sump pit is constructed at suitable locations where the sump pump is installed for pumping out the accumulated wastewater on the basement floor. To direct accumulated wastewater on the floor to the sump pit, floor finish is made sloppy towards the sump pit. In a case of very large basement floor, to minimize tributary area and sloping surface from being too long and thereby too thick, a considerable number of sump pits or floor drains are provided at suitable locations in bottommost basement floors. In building codes, it is found

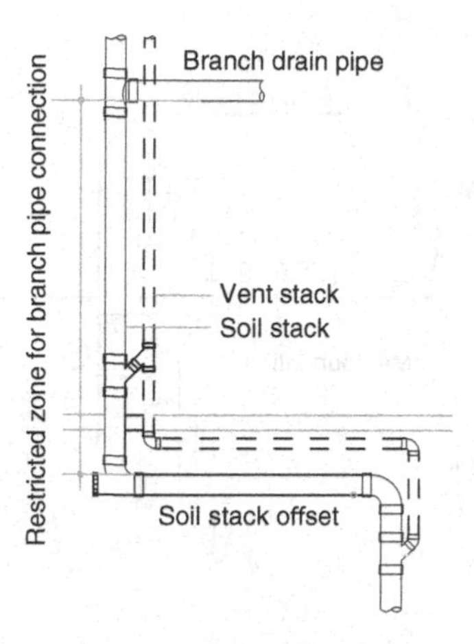

Figure 10.9 Restricted connection zone above the graded offset.

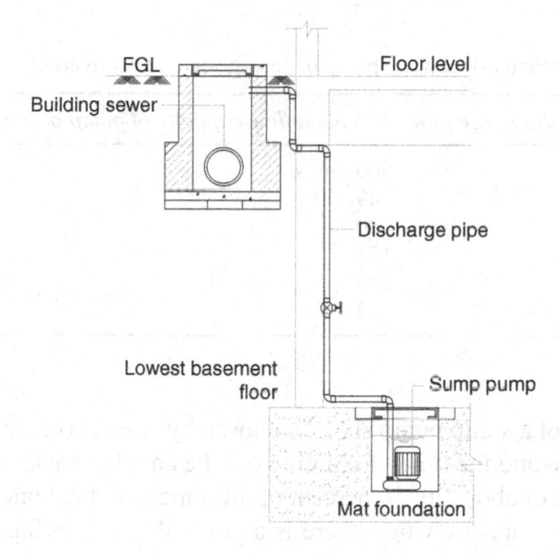

Figure 10.10 Pumping, wastewater from master sump pit in the lowest basement floor, to building sewer.

that for every 1,000 sqm, one sump pit should be provided [5]. From these floor drains or sump pits, wastewater is drained towards a master sump pit though interconnecting drainage pipes, laid in or below the floor. The master sump pit, from where pumping is done, is provided with an automatic sump pump which starts pumping when the pit gets filled up by accumulated wastewater to a predetermined level. The discharge pipe of a sump pump is connected to the building drain or building sewer system, as illustrated through Figure 10.10.

10.10.1 Sump pump

Generally, two types of sump pumps are used: submersible type or column or vertical type. In column or vertical type, the motor remains above the floor level and the pump on the sump pit bed, as shown in Figure 10.11.

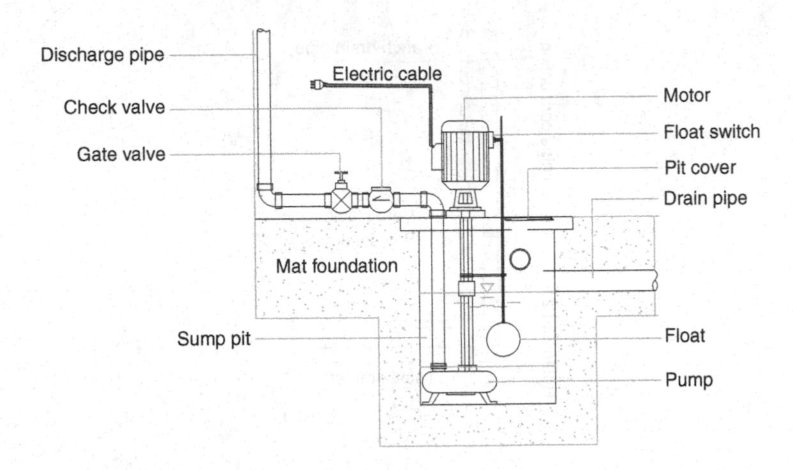

Figure 10.11 Vertical (column) sump pump on sump pit

Table 10.5 Minimum capacity of sewage pump or sewage ejector. Converted in SI unit from source [6].

Sl. No	*Diameter of the discharge pipe*	*Minimum flow capacity of pump or ejector in lpm at velocity*	
	mm	600 mm/s	1950 mm/s
1	40	49	160
2	50	80	258
3	75	174	568
4	100	303	977
5	150	682	2215

The flow capacity of a sump pump shall be chosen by considering the anticipated volume of water accumulation and the time of draining out the anticipated accumulated wastewater. An additional volume of about 20–50 percent of the anticipated volume might be considered in the pump discharge capacity where there is a possibility of gusting of wastewater in the pit while pumping. The corresponding discharging pipe size can be readily determined by consulting Table 10.5.

10.11 Ventilation

A ventilation system is an essential part of the drainage system, not only for its efficient functioning but also for protecting health and environment, as ensuring the safety of the occupants is one of the vital objectives of plumbing development. Ventilation in drainage piping is to be provided basically for maintaining balanced atmospheric pressure inside the drainage piping, by permitting the admission or emission of air with a view to achieving the following objectives.

1 Allowing uninterrupted, rapid and silent flow of wastewater in the drain pipe
2 Allowing foul gases to be emitted outdoors safely and
3 Preventing water from being lost from the traps by siphonage or blowing.

With a view to developing proper ventilation in drainage piping, every trap and trapped fixture shall be vented, so that the water seal in a trap is not affected in any way. The water seal of any fixture trap shall not be affected by a pneumatic pressure differential of more than 250 Pa of water column.

Various ventilation methodologies: Venting predominantly is governed by the manner in which the plumbing fixtures are located and grouped. A complete vent pipe system is a combination of several adopted methods.

Two broad groups of venting, based on their principal function, are as follows.

1 Stack ventilation and
2 Trap ventilation.

10.11.1 Stack ventilation

Ventilation of the soil and wastewater stacks is only to maintain a balanced atmospheric pressure in the drainage stacks. Methods of such venting are as follows.

1 Stack vent
2 Vent stack
3 Relief vent and
4 Yoke vent.

Stack vent: A stack vent is the extension of a soil or wastewater stack above the connection of the highest horizontal drainage pipe to the corresponding stacks, as shown in Figure 10.12. Stack vents serve as a passage for eliminating objectionable gases produced in the drainage piping, which also acts as the terminal of the vent stack. The terminal of stack vents is capped by a cowl.

Vent stack: A vent stack, also known as main vent, is the principal vertical vent pipe of the venting system to which all branch vent pipes are connected. A vent stack is installed connecting to every soil and wastewater stack. The base of the vent stack originates at the

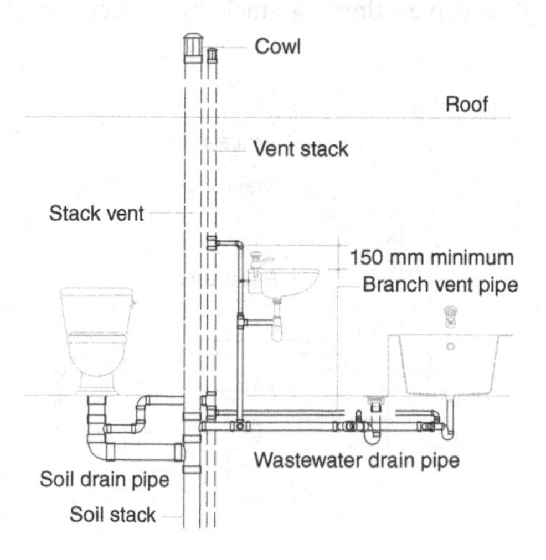

Figure 10.12 Vent-stack and stack-vent

bottom of the drainage stacks, joining by wye-fitting, below the bottommost branch drain pipe connection. The top of the vent stack is either connected to the soil stack or extended, undiminished in size, to the atmosphere. If connected to the stack, then the connection shall be 150 mm above the water level of the highest fixture of the topmost floor level.

Relief vent: In stacks, particularly in high-rise buildings, development of higher pressure than atmospheric or fluctuation of pressure is very common in some portions of the stacks and drain pipes. The vent stack connection at the base of the drainage stack and the branch vent connections to the branch drains may not eliminate these fluctuations. To limit and control such pressure development and fluctuation, relief vents are installed at particular locations prone to high pressure development or fluctuation. Relief vents are provided by connecting the stack or drain pipe to the vent piping system for good circulation of air, thereby releasing pressure. The diameter of a relief vent shall be same as the diameter of a soil stack or vent stack, whichever is smaller.

The locations where relief vents are to be provided, for preventing pressure development, are as follows.

1. Excessive pressure may occur in the lower part of stacks, particularly when large wastewater down-flow causes full flow in the building drain below the stack. At that time, any full flow of waste downward will compress the air and gas trapped inside. The compressed gas will then try to get released by pushing the water in the trap of fixtures in that pressurized zone. A relief vent at the lower part of stack, as shown in Figure 10.13, will prevent high pressure from developing in this way.
2. When a horizontal branch is located more than four branch intervals from the top branch of the stack, the lower branch pipe shall be provided with a relief vent connected to a vent stack or stack vent, or extended outdoors to the open air.
3. Long run drain pipes of more than 30 m shall be provided with additional relief vents located at intervals of not more than 30 m to equalize pressure in the system [7].

Yoke vent: A yoke vent is also a kind of relief vent that is provided by connecting a soil or wastewater stack to a higher location of a vent stack, angularly upward, in order to prevent pressure changes in the drain stacks, as shown in Figure 10.14. In high-rise buildings, excessive interference among air driven down a stack by wastewater flowing downward from

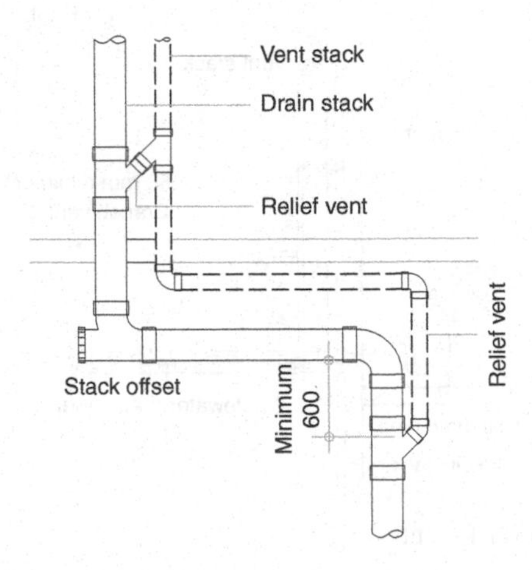

Figure 10.13 Relief venting of draining stack

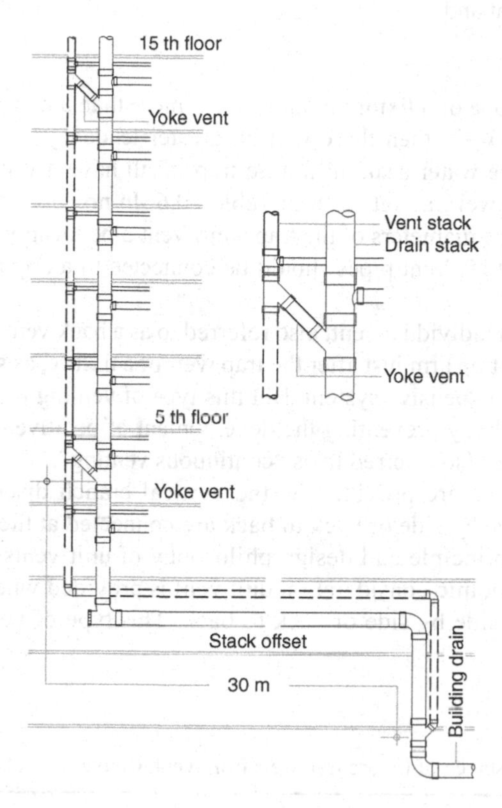

Figure 10.14 Yoke ventilation.

upper floors may occur when additional wastewater is simultaneously discharged into the same stack at a high rate of flow through lower drainage branches. Additionally, in offsets of draining stacks – made at an angle less than 45° with respect to horizontal and located more than 1.2 m below the highest drainage branch connection to the stack – are subject to extremely turbulent flow. To overcome these problems for smooth flow in stacks, it is recommended that at each tenth storey, counting down from the top, a yoke relief vent is to be provided.

Yoke relief vent or yoke vent is constructed by using a 45° wye-fitting on a soil stack or wastewater stack. The wye points upwards and connects using another second 45° wye-fitting on a vertical vent stack pipe, and the wye points downwards, at a point usually above any other branch vents that connect to the same vent stack. The yoke vent pipe diameter is typically the same as vent stacks to which it connects.

10.11.2 Trap ventilation

To protect from loss of water seal in the trap of fixture, ventilation is to be provided in addition to venting of stacks. Such ventilation protects the trap seal against siphonage and back pressure. These types of venting are as follows.

1 Individual or back vent
2 Unit or common vent

3 Circuit or loop vent and
4 Wet vent.

Generally, when the slope of a fixture drain is made more than a one-pipe diameter between vent opening and trap weir, then there will be greater tendency for self-siphonage of the trap seal. To protect the water seal, all fixture traps shall have a vent installed at a certain distance from the trap weir, as set forth in Table 10.6. In no case should the vent pipe be installed within two pipe diameters of the trap weir. Vent pipe location with respect to trap is illustrated in Figure 10.15. Vent pipes should be connected to a trap arm pipe or drain pipe by the tee-fitting.

Individual vent: An individual vent, also referred to as a back vent, is the single vent pipe which is provided on a trap arm just after the trap weir of a fixture, as shown in Figure 10.16. It is therefore the most expensive system. But this type of venting protects the water seal in the trap most effectively by preventing the development of positive or negative pressure in the trap. This venting is also referred to as "continuous venting".

Unit vent: Unit vents are provided on the vertical branch discharge pipes, in which two fixtures placed side by side or back to back are connected at the same level, as shown in Figure 10.17. The principle and design philosophy of unit vents is similar to the back vents but only for economic consideration unit vent is provided where similar fixtures are found closely located side by side or back to back. This type of venting is also termed a "common vent".

Table 10.6 Maximum distance of fixture trap weir from vent. Converted in SI unit from source [8].

Size of trap (mm)	*Distance from trap wier (m)*
32	0.75
40	1.45
50	1.5
75	2
100	3

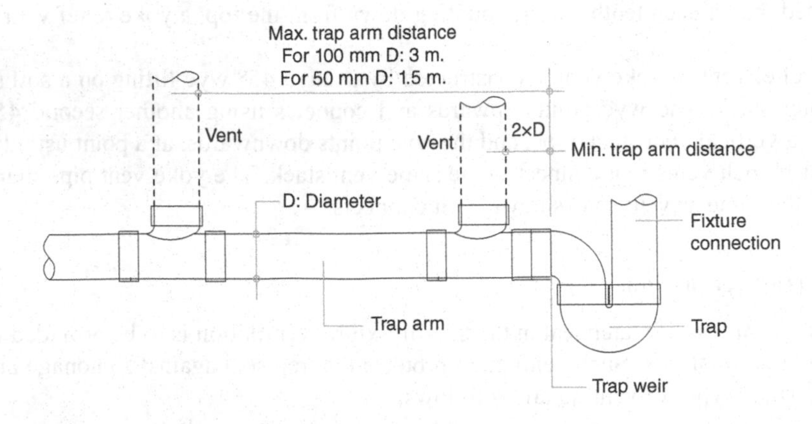

Figure 10.15 Venting locations with respect to trap position.

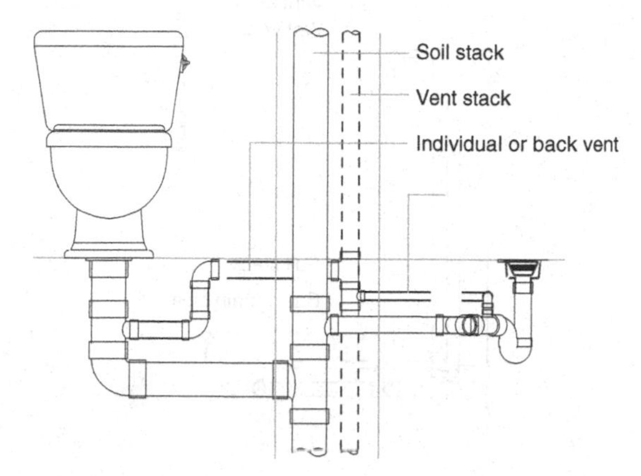

Figure 10.16 Individual or back ventilation

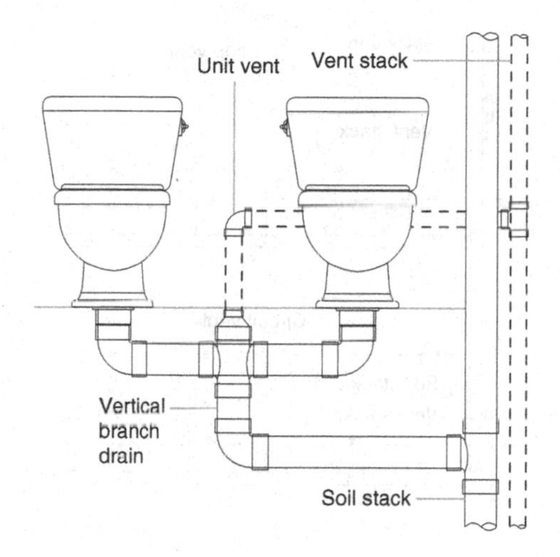

Figure 10.17 Unit venting

When the fixtures are connected at a different level, if the fixture drains are of the same size, then the vertical drain pipe shall have to be enlarged by one size. If the fixture drains are of different sizes, the smaller drain pipe shall have to be connected above the larger drain pipe, maintaining the size of the vertical branch pipe up to the top connection [9].

Wet Vent: The wet vent is the ventilation pipe which starts from the top of fixture trap arm connection to the fixture drain pipe and terminates to the vent stack or stack vent, as shown in Figure 10.18. In another way, a wet vent is the extension of the vertical offset drain pipe of a fixture, in which the lower portion serves as a drain pipe and the extension part as a vent pipe. This is the most economical method of venting generally provided for individual basin and sinks.

Circuit vent: A circuit vent is a vent that is provided to serve multiple fixtures on a horizontal branch drain pipe run on a floor interval. Generally, a circuit vent is employed on a horizontal soil or wastewater drain pipe, which serves two or more fixtures connected to a

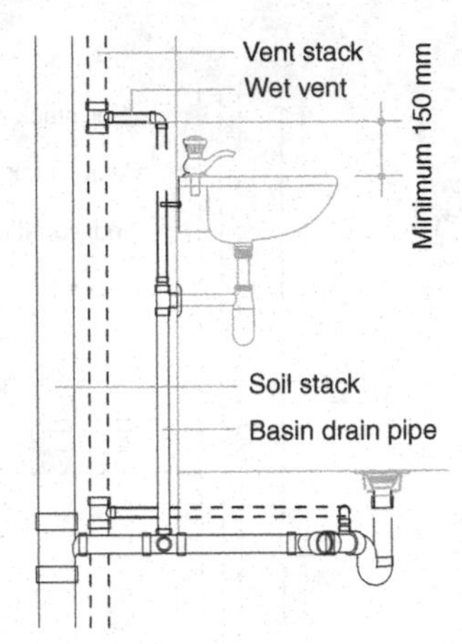

Figure 10.18 Wet vent.

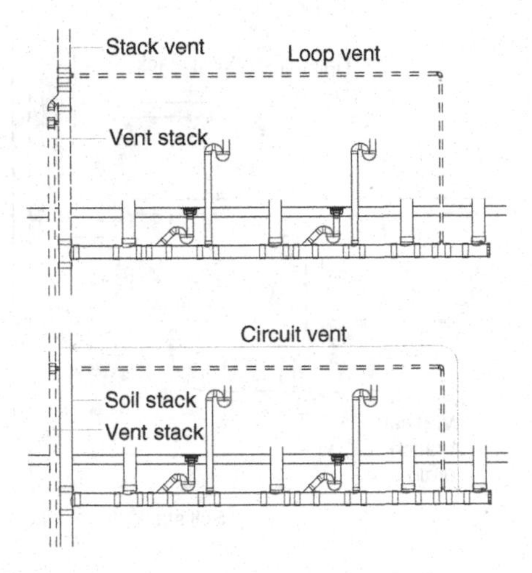

Figure 10.19 Circuit and loop venting

battery. The number of fixtures to be served by a circuit vent depends upon the number of fixtures served by the particular drain pipe size. For this system, the number of fixtures or fixture traps shall not be more than 8 in number on a 100 mm drain pipe and 24 in number on a 150 mm drain pipe. The fixtures to be connected to the drain pipes are water closets, shower stalls, urinals, basins and floor drains, except blowout water closets [9], in battery arrangement, which should be a floor-outlet type.

The circuit venting is provided on the drain pipe at a point between the two most upstream fixtures, as shown in Figure 10.19. Lavatories and similar fixtures may be connected to the circuit vent pipe provided the fixtures are individually vented. The limits of fixtures connected to a drain pipe of a particular size with circuit venting can be increased one and a half times by providing a relief vent on the drain pipe, connecting to the circuit vent pipe, before the first fixture connection most downstream and after the connection to the stack, as illustrated in Figure 10.20.

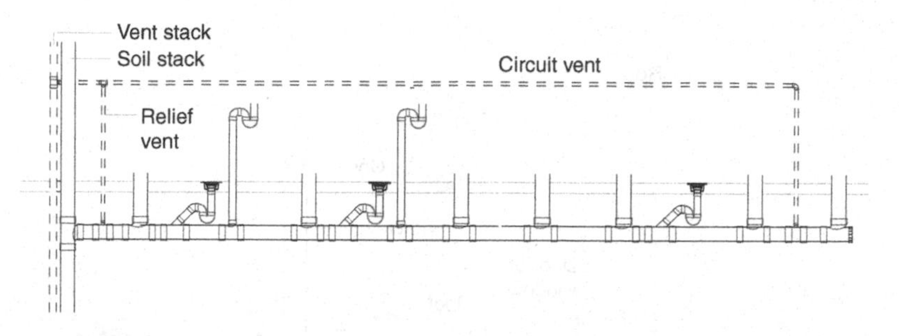

Figure 10.20 Circuit vent supported by relief vent.

A circuit vent is termed a "loop vent" when it is employed on the topmost floor serving fixtures allowed in circuit venting and is connected to the vent extension of the drainage stack instead of to the vent stack, as shown in Figure 10.19.

10.11.3 Location of vent stacks or stack vents terminal

A vent stack or stack vent terminals from a drainage system shall not be located directly beneath any door, openable window or other air intake openings of the building or of an adjacent building, and any such vent terminal shall not be within 3 m horizontally of such an opening and will be at least 600 mm above the top of such openings.

In roofs with restricted access, vent pipes should be extended vertically by not less than 150 mm above the surface to the outdoors and at least 300 mm from any vertical surface or wall. Where there is a snow-cover risk, the vent should extend 600 mm above the sloped roof surface.

For all open vent pipes and stack vents that extend through a roof which is used for any purpose other than weather protection, the vent stack extensions shall be run at least 2 m above the roof surface [5], as shown in Figure 10.21.

Vent terminals shall be 5 m away in any direction from intake of any air duct. The vent terminals shall be 600 mm above any eaves, coping or parapet wall that is installed within a horizontal distance of 600 mm from the vent [10].

10.11.4 Grade of horizontal vents

It is recommended that horizontal vents be installed at a minimum grade of 1.25 percent (1 in 80) so that any condensation or other liquids that form in or enter the vent can be drained out to the sanitary plumbing or sanitary drainage system easily [10].

10.12 Sanitary drainage in skyscrapers

The sanitary drainage systems of skyscrapers and taller buildings will obviously be different in many respects from the design of high-rise buildings. The most important factor to be addressed is the increased air pressure in the stacks. An interesting fact is that increased velocity is not a big concern, as it remains more or less the same up to a 100-storey building. So, there is no need for velocity breaking in buildings falling in this range of height.

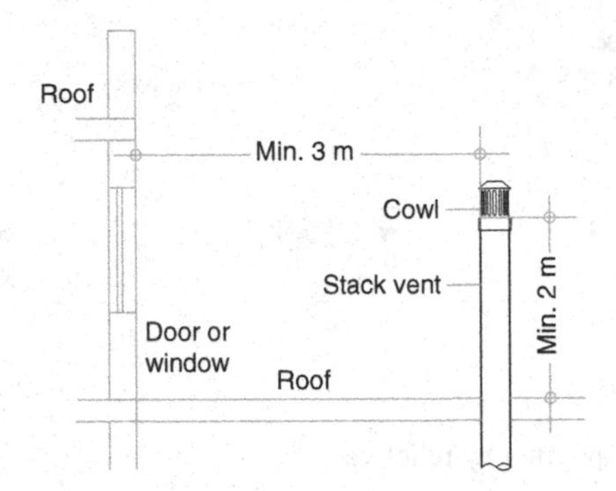

Figure 10.21 Stack vent terminal position.

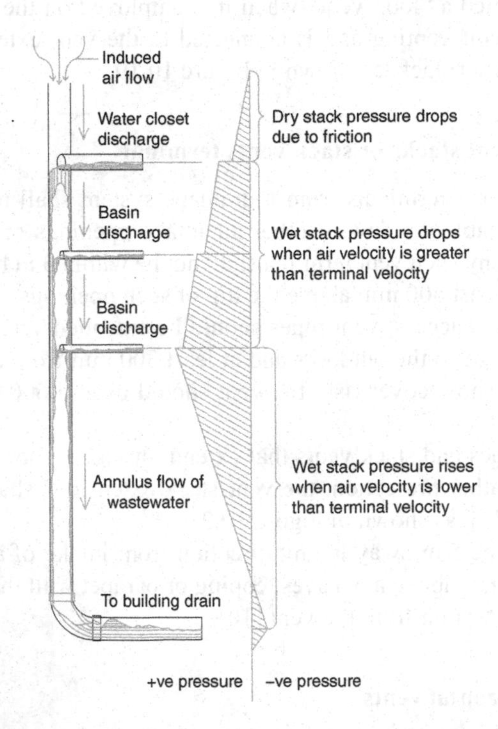

Figure 10.22 Schematic diagram of the water and entrained air-flows in a drainage stack, showing the possible pressure regime developed under steady flow conditions.

The pressure development pattern in the drainage stack of a multi-storey building is shown in Figure 10.22.

When wastewater is discharged into the stacks, air is entrained along with it, which is about 8 to 15 times the volume of wastewater entrained into the stack. The taller the building, the further the fresh air has to travel up, and the resistances generated due to friction at the wall of the stack result in increasing negative pressure. As the number of storeys increases, the length of the vent pipe increases, and as a result the negative pressure increases.

Table 10.7 Minimum distance from the base of the stack to the lowest branch connection [11].

Sl. No	*Application*	*Minimum height*
1	Single dwelling up to three storey	450 mm
2	Up to five storey	740 mm
3	More than five storey	One storey
4	More than 20 storey	Two storey

To eliminate positive pressure, the minimum distance from the base of the stack to the lowest branch connection varies depending on the height of the stack, as shown in Table 10.7. It is further guided that at the base of a stack there should be two 45° bends or a bend with a radius of 200 mm or greater.

In skyscrapers and very long or complex drainage systems, there occurs a continuous increase of flow loads into the stack; pressure thereby also increases, and the distances of the relief of the pressure regime become greater in time and distance.

Pressure starts falling and falls below atmospheric pressure (-ve pressure) immediately below the open-top of the stack. As wastewater flow down, negative pressure increases into the stack due to friction. Further, a pressure drop occurs where stack cross-section is restricted by additional branch flow into the stack. Below the lowest discharging branch, drain pipe pressure gradually increases. So, there is a continuous occurrence of pressure fluctuation in a particular drain pipe due to discharges from other drains or fixtures into the stack. When the system pressure exceeds +/− 40 mm WG (400Pa), the water trap seals can be lost due to development of induced or self-siphonage in the trap of a particular drain pipe.

Another problem occurs at the base of the stack or base of offsetting of the stack due to increase in pressure. Pressure increases above atmospheric pressure can cause a "blow out" of the water seal from the trap and thus cause loss of seal in a trap.

For high-rise buildings, the traditional method of controlling pressure fluctuation in a stack is to provide a secondary ventilation stack system in which the separate secondary ventilation stack and branch pipe are installed alongside the main stack, as discussed earlier. In skyscrapers, flow is likely to increase continuously at different locations, resulting in increases of volume of entrained air at those locations. The reality of drainage in stacks of skyscrapers is that the flows are inherently unsteady, and the flow rate, annular down-flow thickness, entrained airflow and suction pressure all vary with time. So, a traditional secondary venting system might not be found an effective and efficient system in skyscrapers. In some cases, larger diameter vent stacks would be needed considering the height of vent stacks in relation to building height, or rather in relation to the wastewater load to serve.

Instead of using a secondary ventilation stack system, now there is an active drainage ventilation system incorporated in the drainage of skyscrapers building, in which mechanical devices are used to control the pressures developed in drainage systems. The devices are as follows.

1 Positive air pressure attenuator (PAPA) and
2 Air admittance valves (AAV).

The combination of the PAPA and AAVs protects water trap seals in the fixture drain pipes of skyscraper buildings, within a fraction of a second and efficiently throughout the system; this prevents siphonage and blowing of traps, thus overcoming the limitations of the secondary

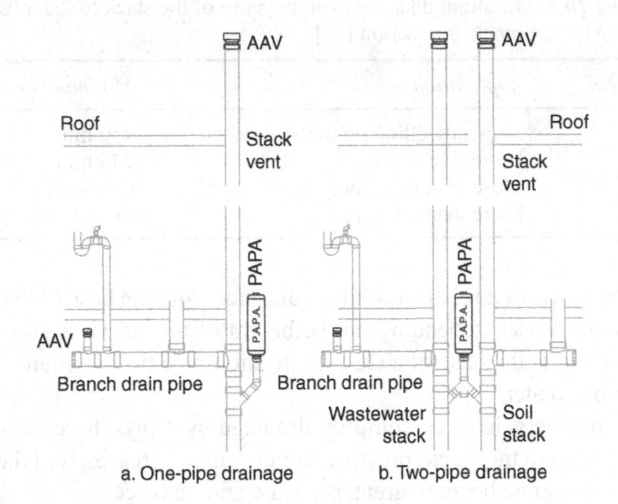

Figure 10.23 Active ventilation using Positive Air Pressure Attenuator (PAPA) and Air Admittance Valves (AAVs).

ventilation stacks. Active ventilation using a positive air pressure attenuator (PAPA) and air admittance valves (AAVs) in one- and two-pipe systems is shown in Figure 10.23.

10.12.1 Positive air pressure attenuator

A positive air pressure attenuator (PAPA) effectively deals with positive pressure transients in the drainage stacks, thereby protecting the trap seals. Use of PAPA will eliminate the need for relief vent piping throughout the system, as both the negative and positive transients are taken care of by this element. PAPA can reduce the magnitude of a positive air pressure transient by up to 90 percent.

When the transient pressure wave is generated, a large proportion of it enters the PAPA, actuating the isoprene membrane in the device, as shown in Figure 10.24. The membrane expands too quickly (in about 0.2 seconds), absorbing all the energy, and the speed of the wave is drastically reduced to a moderate 12 m/s. The remaining small volume of air goes back to the drain system, thus helping to maintain the trap seal of the fixtures. The small PAPA device has a capacity of 3.785 litres of air and can be installed as a standalone fitting, either horizontally or vertically, which is completely maintenance-free. These can also be installed in series, with a maximum up to four units in series (one on another) to ensure additional protection.

The pressure profile in the stacks of very tall buildings is found to be constantly changing; the location and time of pressure transient is variable. So, the PAPA units should be distributed throughout the stack length. The base of a stack or points at offset are the most likely places for a full blockage or surcharge to occur. So, installation of PAPA is a must near these locations. Depending upon the flow discharge to the stacks, PAPA needs to be installed in third to tenth floor [12] intervals, on an average in addition to the locations at offsets and near the base.

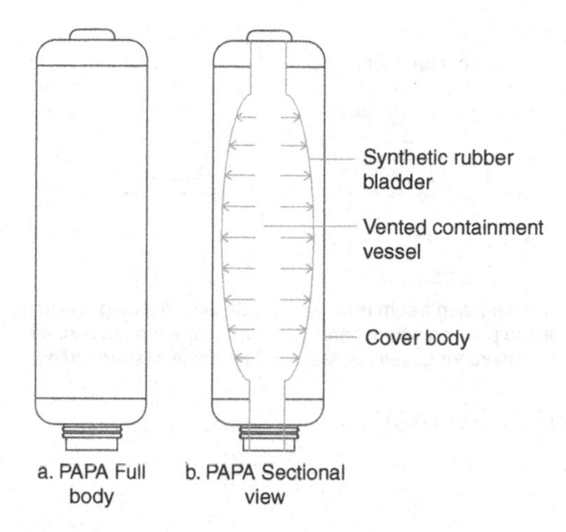

Figure 10.24 Positive air pressure attenuator (PAPA).

10.12.2 Air admittance valves

Air admittance valves (AAVs) are installed at the "point of need" near the water seal traps that require protection against any vacuum creation. AAVs are mostly used to vent negative pressure, particularly with regard to branch venting, which basically performs as a vacuum valve, as shown in Figure 10.25. AAVs have a large capacity for air admittance and a high discharge load. AAVs are installed indoors, at the end of branch drain pipes or near the fixture trap.

10.13 Sizing sanitary drainage piping

The efficiency of a plumbing system primarily depends upon appropriate sizing of piping, and so, sizing of drainage piping is one of the major jobs of plumbing technology. Complete drainage piping in buildings comprise various groups of piping with respect to their functional position. The sizing approaches of these groups of piping also differ due to the different patterns of connectivity of the discharge piping, pressure development and loss. The sizing of all categories of drainage piping is to be done carefully. Both under-sizing and oversizing of drainage pipe have negative implications. So, in the following paragraphs, the design approaches of all sorts of drainage piping are discussed comprehensively.

10.13.1 Sanitary drainage pipes

Sanitary drainage piping systems in a building involve various groups of drainage piping from the start of wastewater generation to the final disposal of wastewater into building or public sewers. The groups of drainage piping are as follows, which need to be correctly sized.

1 Fixture drain pipe
2 Horizontal branch drains

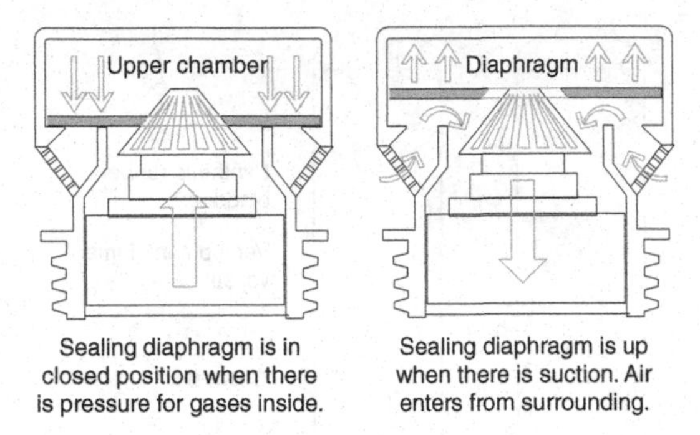

Figure 10.25 Air admittance valves (AAV).

3 Drainage stacks
4 Branch vent pipe
5 Vent stack and
6 Building drain.

10.13.2 Sanitary Drainage Fixture Unit (SDFU)

The sanitary drainage fixture unit (SDFU) is the value assigned to a particular fixture based on the volume of waste discharge, the average rate of waste discharge and the average use of that fixture. The value may be determined from the total discharge flow of the fixture in litres per minute divided by 3.5. The drainage fixture unit value is used to know the waste flowing load in a drainage pipe that will govern the computation of sizes of soil, wastewater, vent, building drain and sewer pipes. The fixture unit value assigned for various fixtures is furnished in Table 10.8.

10.13.3 Fixture drain pipe sizes

All plumbing fixtures have an outlet for allowing waste to go out. These outlets are connected to a discharge pipe to drain out the waste accumulated in the fixtures. The orifice of the fixture outlet is adequately sized to provide satisfactory flow rate for proper fixture drainage and scouring action in the connected fixture drain. In no case can the fixture drain pipe size be reduced to less than the fixture outlet orifice. Hence, the minimum size of a fixture drain pipe is assumed to be as same the orifice size of the fixture. Again, the size of a trap to be used for any fixture has been fixed, which is universally followed. The trap size for a particular fixture is the same as the outlet size of that fixture. The outlet orifice size of various fixtures is also made universally in same sizes. The sizes of outlets of various fixtures are given in Table 10.9.

10.13.4 Sizing horizontal branch drains

Horizontal branch drain pipes are those drain pipes which receive waste from one or more fixture drain pipe that, installed on the same floor, finally discharge into the corresponding drainage stacks. The wastewater load carried by a branch drain pipe is determined by

Table 10.8 Sanitary drainage fixture unit (SDFU) values for various plumbing fixtures [13].

Sl. No	*Plumbing appliances, appurtenances or fixtures*		*Minimum size trap and trap arm (mm)*	*Drainage fixture unit values (DFU)*		
				Private	*Public*	*Assembly*
1	Bathtub or combination bath/shower		38	2	2	–
2	Bidet		32	1	–	–
			38	2	–	–
3	Clothes washer, domestic, standpipe		50	3	3	3
4	Dental unit, cuspidor		32	–	1	1
5	Dishwasher, domestic, with independent drain		38	2	2	2
6	Drinking fountain or water cooler		32	0.5	0.5	1
7	Food waste disposer, commercial		50	–	3	3
8	Floor drain	Emergency	50	–	0	0
		For additional sizes	50	2	2	2
9	Shower	Single-head trap	50	2	2	2
		Multi-head, each additional	50	1	1	1
8	Lavatory	Single	32	1	1	1
		In sets	38	2	2	2
9	Wash-fountain	Small	38	–	2	2
		Large	50	–	3	3
10	Mobile home or manufactured home, trap		75	6	–	–
11	Receptors	Indirect waste	38	3	3	3
		Indirect waste	50	4	4	4
		Indirect waste	75	6	6	6
12	Sinks	Bar	38	1	–	–
		Bar of 50 mm drain	50	–	2	2
		Clinical	75	–	6	6
		Commercial with food waste	50	–	3	3
		Exam room	38	–	1	–
		Special purpose	50	2	3	3
		Special purpose	50	3	4	4
		Special purpose	75	–	6	6
		Kitchen, domestic (with or without food waste disposer, dishwasher, or both)	38	2	2	–
		Laundry (with or without discharge from a clothes washer)	38	2	2	2
		Service or mop basin	50	–	3	3
		Service or mop basin	75	–	3	3
		Service flushing rim	75	–	6	6
		Wash, each set of faucets	–	–	2	2
13	Urinal	Integral trap 1.0 GPF	50	2	2	5
		Integral trap greater than 1.0 GPF	50	2	2	6
		Exposed trap	50	2	2	5
14	Water-closet	1.6 GPF gravity tank	75	3	4	6
		1.6 GPF flushometer tank	75	3	4	6
		1.6 GPF flushometer valve	75	3	4	6
		Greater than 1.6 GPF gravity tank	75	4	6	8
		Greater than 1.6 GPF flushometer valve	75	4	6	8

Table 10.9 The sizes of outlets of various fixtures [14].

Fixture	*Trap size (mm)*
Bathtub	40
Bidet	40
Dishwasher	40
Drinking fountain	32
Floor drain	50
Kitchen sink	40
Laundry tub	40
Lavatory	32
Public toilet	integrated in the fixture
Shower	50
Toilet	integrated in the fixture
Urinal	50
Washing machine	50

Table 10.10 Maximum permissible load, in terms of drainage fixture unit (DFU) for sanitary drainage piping, corresponding to pipe diameter [15].

Sl. No	*Diameter of pipe (mm)*	*Maximum number of drainage fixture units (DFU) that may be connected*	
		Any horizontal fixture branch	*1 stack of 3 or fewer branch intervals*
1	40	3	4
2	50	6	10
3	65	12	20
4	80	20*	48*
5	100	160	240
6	125	360	540
7	150	620	960
8	200	1,400	2,200
9	250	2,500	3,800
10	300	3,900	6,000
11	375	7,000	6,000

() * No water closets permitted.

summing up the drainage fixture unit values of all the fixtures connected to the concerned branch drain pipe. With respect to this total drainage fixture unit value of a particular drain pipe, the corresponding pipe diameter to be selected can be found in Table 10.10.

10.13.5 Sizing drainage stacks

The capacity of a drainage stack is determined by the pipe size, the waste load entering at any point within a branch interval and the load entering the entire stack. The waste load is expressed in drainage fixture units, as usual.

A branch interval is a vertical length of a stack, at least 2.4 m high (generally one storey in height), within which a horizontal branch drain pipe from a storey or floor of a building is connected to the stack. Figure 10.26 illustrates how the branch intervals are numbered on a stack. Table 10.11 provides the total permissible loading expressed in drainage fixture units

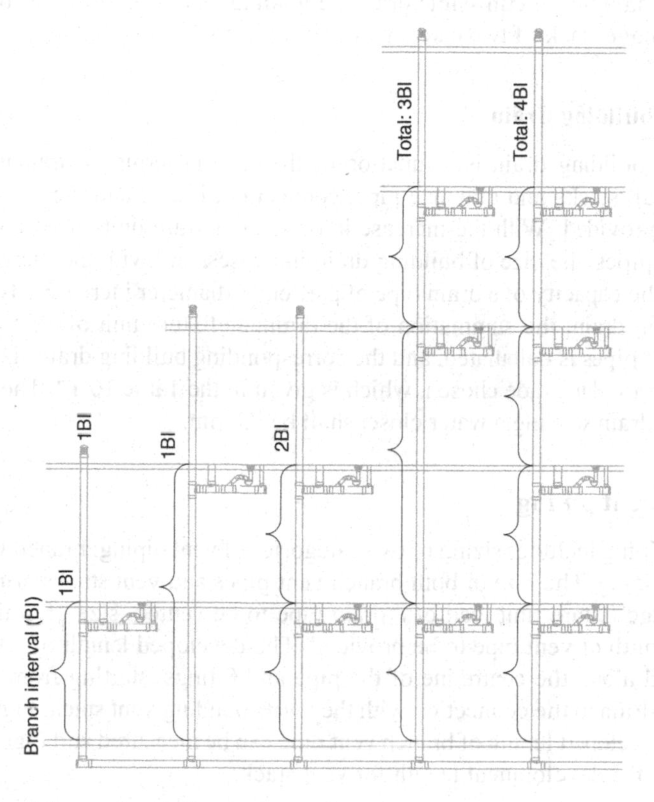

Figure 10.26 Branch intervals numbered on a stack.

Table 10.11 Maximum permissible load, in terms of drainage fixture unit, for soil and wastewater stacks [15].

Sl. No	*Diameter of pipe (mm)*	*Maximum number of drainage fixture units (DFU) that may be connected*		
		1 stack of 3 or fewer branch intervals	*Stacks exceeding 3 branch intervals*	
			Total for stack	*Total at 1 branch interval*
1	40	4	8	2
2	50	10	24	6
3	65	20	42	9
4	80	48[a]	72[a]	20[a]
5	100	240	500	90
6	125	540	1,100	200
7	150	960	1,900	350
8	200	2,200	3,600	600
9	250	3,800	5,600	1,000
10	300	6,000	8,400	1,500
11	375	6,000	8,400	1,500

with respect to maximum permissible drainage loading from one storey or branch interval, for sizing a drainage stack of two categories.

10.13.6 Sizing building drain

The sizing of a building drain is a function of the load in terms of drainage fixture unit, discharge from all stacks and drainage pipe, connection, if any, and the grade of the building drain to be provided. With the increase in drainage fixture units from connected stacks or branch drain pipes, the size of building drain increases, and with the increase in grade of building drain, the capacity of a drain pipe of particular diameter increases. To determine the size of a building drain, the summation of the drainage fixture unit of all connected stacks and branch drain pipes is calculated, and the corresponding building drain size is found with respect to the particular grade chosen, which is given in the Table 10.12. The minimum size of any building drain serving a water closet shall be 75 mm.

10.14 Sizing vent piping

Sizing of vent piping includes sizing of two categories of vent piping: branch vent pipe sizing and vent stack sizing. The size of both branch vent pipes and vent stacks primarily depends upon the drainage fixture unit load of a drain pipe to be vented, size of drain pipe and the development length of vent pipe to be provided. The developed length of a vent pipe is the length measured along the centreline of the pipe and fittings, starting from the connection with the fixture drain to the connection with the corresponding vent stack. Figure 10.27 illustrates how the developed length of branch vent pipe can be measured and Figure 10.28 shows measurement of the development length for vent stack.

10.14.1 Sizing branch vent pipes

The size of a branch vent pipe, including individual relief and circuit vent pipes, primarily depends upon the developed length of the vent pipe to be provided with respect to the drainage fixture unit load of the drain pipe to be vented. The permissible developed length of a particular diameter vent pipe decreases with the increase in drainage fixture unit, to be

Table 10.12 Maximum permissible load, in drainage fixture units, for building drain at particular grade or slope [16].

Sl. No	*Diameter of pipe (mm)*	*Maximum hydraulic load in drainage fixture units (DFU)*					
		Slope					
		1 in 400	*1 in 200*	*1 in 133*	*1 in 100*	*1 in 50*	*1 in 25*
1	75	–	–	–	–	27	36
2	100	–	–	–	180	240	300
3	125	–	–	380	390	480	670
4	150	–	–	600	700	840	1300
5	200	–	1400	1500	1600	2250	3370
6	250	–	2500	2700	3000	4500	6500
7	300	2240	3900	4500	5400	8300	13000
8	375	4800	7000	9300	10400	16300	22500

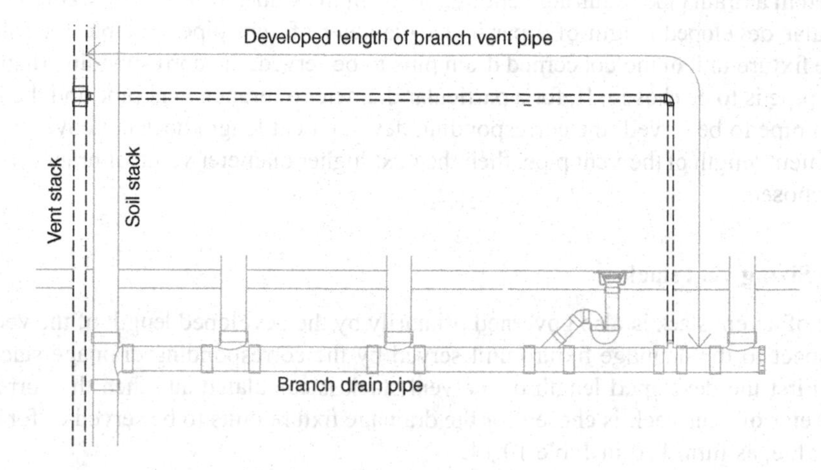

Figure 10.27 Measurement of the developed length of vent pipe.

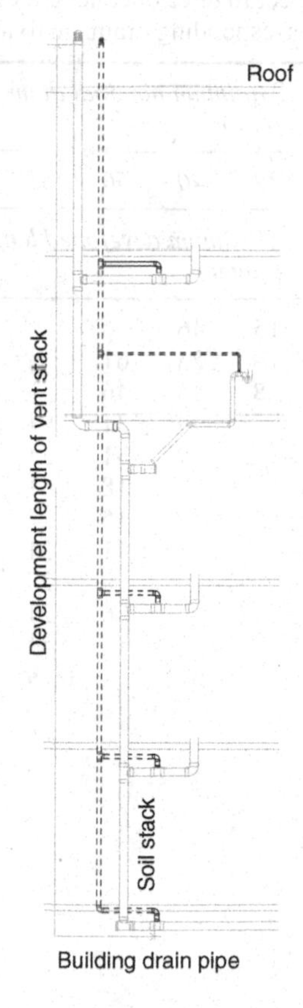

Figure 10.28 Measurement of the developed length of vent stack.

addressed in a drain pipe requiring venting. So, from the values furnished in Table 10.13, for a particular developed length of a particular diameter of vent pipe, suitable for calculated drainage fixture unit of the concerned drain pipe to be served, the corresponding diameter of the vent pipe is to be chosen. If, for a particular diameter of branch vent pipe and the DFU of the drain pipe to be served, the corresponding development length does not cover the actual development length of the vent pipe, then the next higher diameter value of branch vent pipe is to be chosen.

10.14.2 Sizing vent stack

The size of a vent stack is also governed primarily by the developed length of the vent stack with respect to the drainage fixture unit served by the corresponding drainage stack to be vented. First the developed length of the vent stack is calculated and then the corresponding diameter of vent stack is chosen for the drainage fixture units to be served or for its next higher value, as furnished in Table 10.14.

Table 10.13 Size of vent pipe with respect to developed length of individual, relief, circuit and branch vent and vent stack and corresponding drainage fixture units served [15].

Size of soil or waste stack, (mm)	*Maximum number of fixture units (DFU) served*	*Minimum nominal diameter of ventilating pipe (mm)*								
		32	*40*	*50*	*65*	*80*	*100*	*125*	*150*	*200*
		Maximum developed length of ventilating pipe (meter)								
40	8	15	46							
50	12	9	23	61						
	20	8	15	46						
65	42		9	31	91					
80	10		9	31	31	183				
	30			18	61	152				
	60			15	24	122				
100	100			11	31	79	305			
	200			9	27	76	274			
	500			6	21	55	213			
125	200				11	24	107	305		
	500				9	27	76	274		
	1100				6	21	55	213		
150	350				8	15	61	122	396	
	620				5	9	38	91	335	
	960					7	31	76	305	
	1900					6	21	61	213	
200	600						15	46	152	396
	1400						12	31	122	366
	2200						9	24	107	335
	3600						8	18	76	244
250	1000							23	38	305
	2500							15	31	152
	3800							9	24	107
	5600							8	18	76

Table 10.14 The size of vent stack based on the developed length of the vent stack with respect to the drainage fixture unit served by the corresponding drainage stack to be vented [15].

Sl. No	*Size of soil waste stack (mm)*	*Drainage fixture units connected*	*Diameter of required vent stack (mm)*						
			32	*38*	*50*	*63*	*75*	*100*	*125*
1	32	2	9						
2	38	8	15	45					
		10	9	30					
3	50	12	9	22.5	60				
		20	7.8	15	45				
4	63	42		9	30	90			
5	75	10		9	30	60	180		
		30			18	60	150		
		60			15	24	120		
6	100	100			10.5	30	78	300	
		200			9	2.7	75	270	
		500			6	21	54	210	
7	125	200				10.5	24	105	300
		500				9	21	90	270
		1100				6	15	60	210

10.15 Sizing sump pit and vent

10.15.1 Sizing sump pit

To determine the size of a sump pit or a wet-well, the most important factor is that the drawdown of the required size of pump is known. The drawdown is measured in millimetres or meters from the invert of the sewage or wastewater inlet port to the top of the pump suction submergence level determined by Formula 10.1. If a sewage pump of 4 cycles per hour is used, then the pump runs for 10 minutes and remain stopped for about 5 minutes between stop to next start, in an hour. For sewage pumping with 50 kW (67 bhp) or smaller pumps [17], the wet-well should be of sufficient size to allow for a minimum running time of 10 minutes per cycle for each pump.

From the flow capacity of the pump, the total volume of sewage in litres is calculated for 10 minutes.

The minimum base dimension of pit is 600 mm for simplex pumping, while duplex pumping systems require a minimum of 900 mm. After determining the base area in sqmm or sqm, the drawdown height of the pump in 10 minutes is determined simply by dividing the volume of sewage flow per cycle by the selected base area.

Inadequate submergence of the suction inlet can result in pre-rotation of the wastewater in the sump or wet-well, resulting in the formation of strong free-surface air core vortices that thereby cause entrance of air into the pump suction inlet. This phenomenon is termed "vortexing". Vortexing can cause a pump to be unstable for operation due to generation of vibration, pulsation and noise, ultimately resulting in severe mechanical damage. Therefore, a minimum submergence height above the suction inlet, in meters, is to be maintained, which can be determined by the formula below.

$$S = D + 2.3\left(\frac{4}{\sqrt{9.81}\ \pi}\right)\left(\frac{Q}{D^{1.5}}\right) \text{ (m)} \qquad 10.1$$

Where
Q = flow in cum/s and
D = diameter of pipe in m.

For a submersible sewage pump, the minimum submergence level is considered as the pump's top height from the floor of the pit; because of varying wastewater levels and inflow conditions, there may be occasions when the pump motor may be partially or fully non-submerged for periods longer than 15 minutes, which might cause damage to the motor due to overheating [18].

So, the total depth of a sump pit or wet-well is the summation of the invert depth of the inlet pipe, pump drawdown, suction submergence depth and depth of the suction pipe above the pit bottom, as shown in Figure 10.29.

10.15.2 Sizing sump vent pipe

Sump pits shall be well ventilated in the same manner as that of a gravity system. Building sump vent sizes for sumps furnished with sewage pumps or sewage ejectors, other than pneumatic, shall be determined in accordance with Table 10.15.

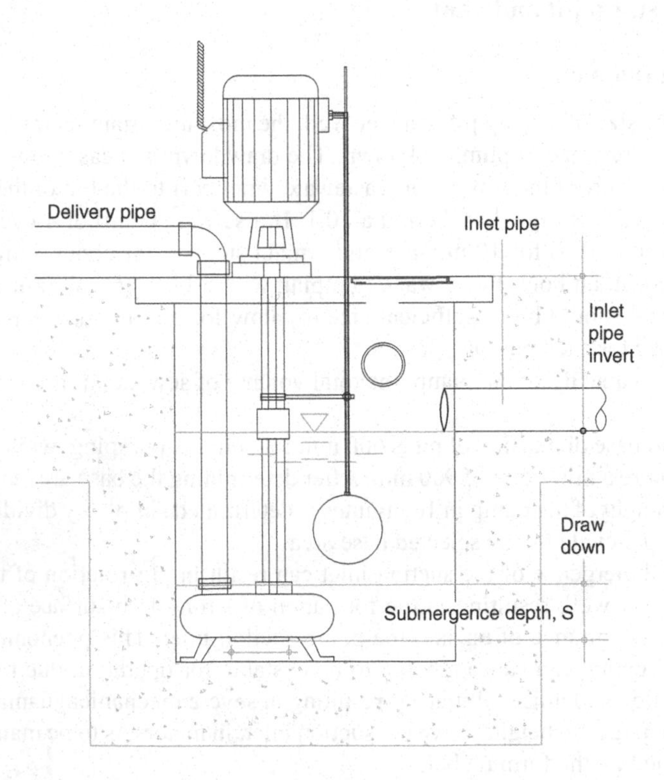

Figure 10.29 Sump pit for a simplex pumping system

Table 10.15 Size and length of sump vent pipes [19].

Sl. No	*Discharge capacity of pump (lpm)*	*Maximum developed length of vent (m)*[a]				
		Diameter of vent (mm)				
		32	*40*	*50*	*63*	*75*
1	37.85	No limit[b]	No limit	No limit	No limit	No limit
2	75.70	82.3	No limit	No limit	No limit	No limit
3	151.42	21.95	48.77	No limit	No limit	No limit
4	227.12	9.45	22.86	82.3	No limit	No limit

a Developed length plus an appropriate allowance for entrance losses and friction caused by fittings, changes in direction and diameter. Suggested allowances shall be obtained from approved sources. An allowance of 50 percent of the developed length shall be assumed if a more precise value is not available.

b Actual values greater than 152 m.

References

[1] Deolalikar, S.G. (1994) *Plumbing: Design and Practice*, Tata McGraw-Hill Publishing Company Ltd, New Delhi.

[2] Cyril, M. Harris (1991) *Practical Plumbing Engineering*, Tata McGraw-Hill Publishing Company Ltd, New York.

[3] Escritt, L.B. (1971) *Sewers and Sewage Works, Metric Calculation and Formulae*, George Allen and Unwin Ltd, London.

[4] Taylor, F., and Wood, W., eds. (1982) "Guidelines on Health Aspects of Plumbing" International Reference Centre for Community Water Supply and Sanitation (IRC), p. 59, Technical Paper Series No. 19, The Hague, IRC.

[5] Housing and Building Research Institute (HBRI) (1993) *Bangladesh National Building Code 1993*, HBRI, Dhaka.

[6] Deppmann, R.L. (2017) "Wastewater Pumps and Packaged Systems for Building Trades" Retrieved from www.deppmann.com/products/sump-sewage/#pipingsizing on 3 January 2021.

[7] North Carolina Plumbing Code (2012) UP Code Sec 9, Vent 912.4.3.

[8] Carson Dunlop Associates (2008) CarsonDunlop.com 2008 Found in "Plumbing Vent Diagram: How to Properly Vent Your Pipes" by The Plumbing Info Team Retrieved from https://theplumbinginfo.com/plumbing-vent-diagram/ on 17 June, 2021.

[9] American Society of Plumbing Engineers (ASPE) (2012) "Vent System" CEU 189, Read, Learn, Earn, Report, Continuing Education from ASPE, Illinois.

[10] Standards Australia Limited and Standards New Zealand (2003) *Plumbing and Drainage, Part 2: Sanitary Plumbing and Drainage*, SAI Global Limited, Sydney, and Wellington. NZ Code, p. 82.

[11] Building.co.uk (2015) "Introduction to Sanitary Pipe Work Design" CPD 13 2015; Sponsored by Marley Plumbing & Drainage, Retrieved from www.building.co.uk/cpd/cpd-13-2015-introduction-to-sanitary-pipework-design/5076509.article on 28 December 2020.

[12] *IPS Corporation Studor Engineered Products Technical Manual*, 8th Ed, p. 33, Retrieved from www.ipscorporation.com/pdf/professionals/StudorSpecPackage.pdf on 28 December 2020.

[13] International Association of Plumbing and Mechanical Officials, "California Plumbing Code, 2016" Retrieved from https://up.codes/viewer/california/ca-plumbing-code-2016/chapter/7/sanitary-drainage#702.0 on 28 December 2020.

[14] Engineering ToolBox (2008) "Fixtures and Trap Sizes" Retrieved from www.engineeringtoolbox.com/trap-sizes-d_1111.html on 28 December 2020.

[15] Environmental Agency – Abu Dhabi (2009) *Uniform Plumbing Code of Abu Dhabi Emirate an Environmental Guide for Water Supply and Sanitation*, Abu Dhabi, p. 358.

[16] British Columbia Plumbing Code (BCPC) (2018) "Drainage System Section 2.4, Part 2 – Plumbing Systems, Division B: Acceptable Solutions" Retrieved from https://free.bcpublications.ca/civix/document/id/public/bcpc2018/bcpc_2018dbp2s24 on 15 December 2020.

[17] Queen's Printer for Ontario (2019) "Sewage Pumping Stations" Chapter 7, Sec 7.2.7, Retrieved from www.ontario.ca/document/design-guidelines-sewage-works/sewage-pumping-stations on 30 December 2020.

[18] Page, D.W. (UD) *Submersible Non-Clog Pumps Minimum Submergence Considerations*, Yeomans Chicago Corporation, Retrieved from www.yccpump.com/assets/submergence.pdf on 31 December 2020.

[19] Up.codes (2014) "Vents" Chapter 31, Retrieved from https://up.codes/viewer/florida/fl-residential-code-2014/chapter/31/vents#P3113.4.1 on 3 January 2021.

11 Rainwater drainage system

11.1 Introduction

Rainwater is the root source of all water on the earth, for both surface and groundwater, which are mostly used as the primary sources of water supply for buildings. Rain that falls on a building and its premises might be a potential source of water for buildings where water is scarce. When the potential of such a natural resource is overlooked, then it can turn into a hazard in the form of water logging and ponding of rainwater because it becomes difficult to drain it. Rain falling on any surface or ground open to the sky flows downward until it reaches any flowing natural or man-made stream, if not interrupted or blocked while flowing. While flowing over or remaining stagnant on natural ground, some rainwater percolates into the ground, which ultimately recharges the groundwater. In this chapter, discussion will be limited to covering how rainwater can be drained from a building directly into the building or public storm sewer; in addition, some techniques for minimizing rainwater or storm water load for drains or sewers are discussed that give some idea of recharging the ground while draining out.

11.2 Rainwater characteristics

To develop efficient rainwater drainage systems for a building, various information about rain, rainfall etc. must be known. Without this information or knowledge, development of rainwater drainage not only will cause economic loss but may invite catastrophic accident as well.

11.2.1 Rainwater

Raindrop produced in the atmosphere is relatively free from impurities. Pollutants or contaminants found in rainwater may be due to absorbing airborne chemicals and materials while falling. Surface material where rain falls may also cause contamination of rainwater when it flows over. Rainwater absorbs chemicals from these materials. Wind-blown suspended particles, debris and dusts, fecal matters from birds, animals, insects etc., dead rodents or carcasses, organic matters and contaminated littering, present on the ground surface, are also a major source of contamination and pollution of rainwater or storm water. While falling, absorption of various gases like carbon dioxide, sulphur dioxide, nitrous oxide etc. in the atmosphere makes rainwater acidic.

11.2.2 Rainfall intensity

In designing drainage systems for rain or storm water, the maximum rainfall intensity data for the locality of the building shall be studied to arrive at the required design parameters.

DOI: 10.1201/9781003172239-11

Table 11.1 Hourly maximum rainfall intensity at some districts of Bangladesh [1].

Sl. No	*District*	*Return period (year)*	*Hourly maximum (mm/hour)*
1	Barisal	10	62.71
		25	74.21
		50	82.74
2	Bogra	10	74.59
		25	90.93
		50	103.05
3	Chitagong	10	94.50
		25	113.65
		50	127.86
4	Cox's Bazar	10	80.45
		25	97.96
		50	110.96
5	Dhaka	10	79.41
		25	93.86
		50	104.58
6	Jessore	10	84.94
		25	106.14
		50	121.88
7	Sylhet	10	84.24
		25	100.03
		50	111.71

Annual, monthly or even daily rainfall intensity data is not very effective for calculation of runoff quantities for vertical or horizontal conveying media for rainwater. It is the maximum intensity of rainfall in an hour which is the most important data for drainage system designers. When rain falls over a longer period, it is recorded for every hour separately. Peak intensity per hour over the total falling period is then calculated for design purposes. The hourly intensity of rainfall is very much dependent on local conditions. So, for designing drainage structures, local hourly rainfall intensity must be known. The hourly rainfall intensity for different part of Bangladesh, the 10th ranking country in receiving the highest annual rainfall (about 2666 mm/yr) [2] is shown in Table 11.1.

11.3 Scopes of rainwater drainage

Rainwater drainage may be defined as the driving out of rainwater, fully or partially, from the building or its premises to an approved disposal point like a storm sewer, combined sewer or any natural water body nearby. Rainwater drainage from a building implies collection of rainwater that falls on a building or its premises and draining the excess rainwater. In this system, rainwater is intercepted and diverted, and a portion may be infiltrated into the ground before draining to minimize the volume of storm or rainwater in drainage. So, from a broader perspective, rainwater drainage is applied in two major sectors, as mentioned below.

1 Infiltrating into the ground and
2 Draining excess rainwater.

11.4 Functional techniques of rainwater drainage

In order to develop a complete drainage system for rainwater, the following techniques are to be adopted.

1 Collecting rainwater from potential catchments
2 Conveying rainwater to desired location
3 Infiltrating into the ground and
4 Draining excess rainwater.

A schematic diagram showing all major components of rainwater drainage in a building and its premises is shown in Figure 11.1.

11.5 Collecting rainwater

Building elements which remain fully or partially exposed to the open sky are potential catchments from where rainwater can be collected or needs to be drained out. These catchments are as follows.

1 Roof
2 Verandas and balconies
3 Sun shades and cornices

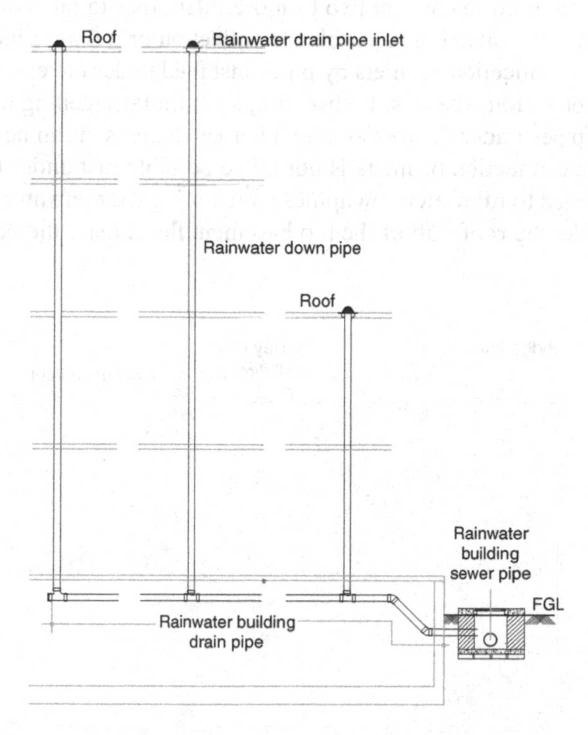

Figure 11.1 Schematic diagram showing all major components of rainwater drainage in a building and its premise.

4 Car porch and
5 Part of any side walls.

Most rainwater is collected and drained as well, from roof for being very large catchments, in comparison to other building elements serving as catchment. In case of rainwater accumulation, other building elements should be accounted for. Ways of collecting rainwater from the flat roof of a building are discussed below.

11.5.1 Rainwater collection from flat surface

Rainwater falling on any flat roof surface is collected by providing a good number of appropriately sized inlets on rainwater down pipes (RDP) placed at suitable locations, mostly around the periphery of the roof and preferably at regular intervals. If the roof is not provided with any parapet wall, then to prevent overflowing of rainwater, a curb should be made all around at the periphery of the roof. To direct the flow of rainwater towards the inlets, the flat roof shall be provided with another finishing layer, making its surface slope very gently towards the inlets. Multiple sloped surfaces are made in the same way to direct the rainwater towards inlets provided for each sloped surface. Each sloped surface for a particular rainwater inlet is configured depending upon the position of that particular inlet. The number of inlets provided is based on the area of the roof served, configuration of the roof, etc. The configuration of sloped surfaces made for a number of rainwater-drain pipe inlets in a rectangular roof is shown in Figure 11.2.

Sometimes, it is required to connect two or more inlet pipes to one collection pipe, leading to either a rainwater collection reservoir for collection or a drain pipe for draining out. This system needs a connection of inlets by pipes installed under the catchment, which may not be acceptable for various reasons. In this case, an extra false ceiling may be provided to cover the exposed pipes under the roof or any other catchments, as an aesthetic and protective measure. If the connection of inlets is not made possible just under the roof slab, then the inlets are connected to rainwater downpipes which bring the rainwater down to a suitable level, generally under the roof slab of the top basement floor; here, the bottom of rainwater

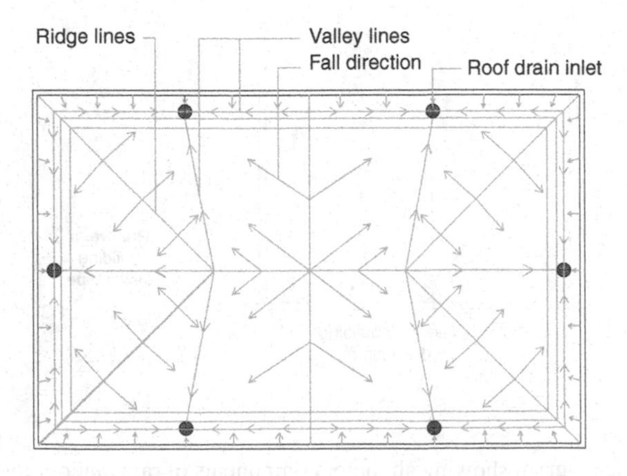

Figure 11.2 Roof drain inlets and sloping of surface on a flat roof.

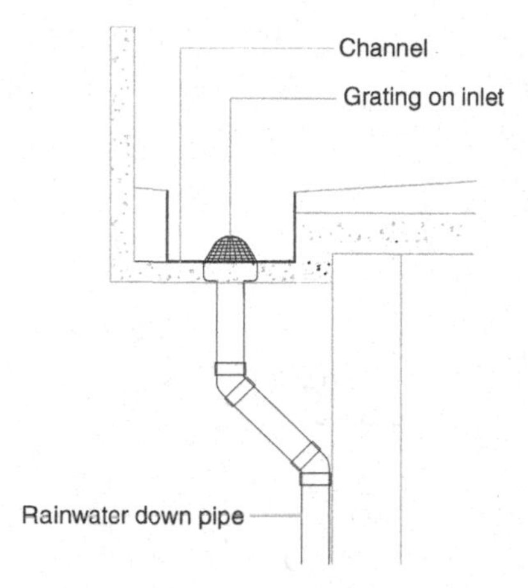

Figure 11.3 Roof draining channel around a flat roof.

downpipes are connected by a drain pipe called a building storm drain pipe, which conveys all of the rainwater outside the building, generally to a building storm or combined sewer, as illustrated in Figure 11.1.

Another way of collecting rainwater from flat surfaces is by creating channels around the catchment, as shown in Figure 11.3. The channel bed shall be sloped towards one or two collecting points, depending on drainage planning. At the collecting points, there shall be two inlets, but collection shall be made by one pipe receiving rainwater from two inlets and leading to the building or public sewer. During design of the channel, the depth and width of the channel should be determined properly.

11.5.2 Rainwater collection from sloped surface

The most effective way of collecting rainwater from any sloped surface is to provide a gutter at the periphery of the sloped surfaces to be used as catchment. The gutter shall be sloped towards a collecting point. At the collecting point there shall be two inlets but collection shall be made by one pipe receiving rainwater from two inlets and leading to the ground for drainage. During design of the roof, the size of the gutter should be calculated properly.

11.6 Determining catchments

11.6.1 Catchments of flat surface

For a flat roof surface, the effective catchment area is its horizontal plan area plus 50 percent of the one adjoining vertical wall, which contributes to rainwater accumulation on the concerned catchment.

Let us consider a building having a flat roof at different levels as shown in Figure 11.4. In level 3 the area ABCD is the catchments contributing to the rainwater down pipe RDP1.

In level 2, horizontal area FEGH and 50 percent of the adjacent vertical wall surface area AFHM projecting above are the catchments contributing rainwater towards rainwater downpipe RDP2.

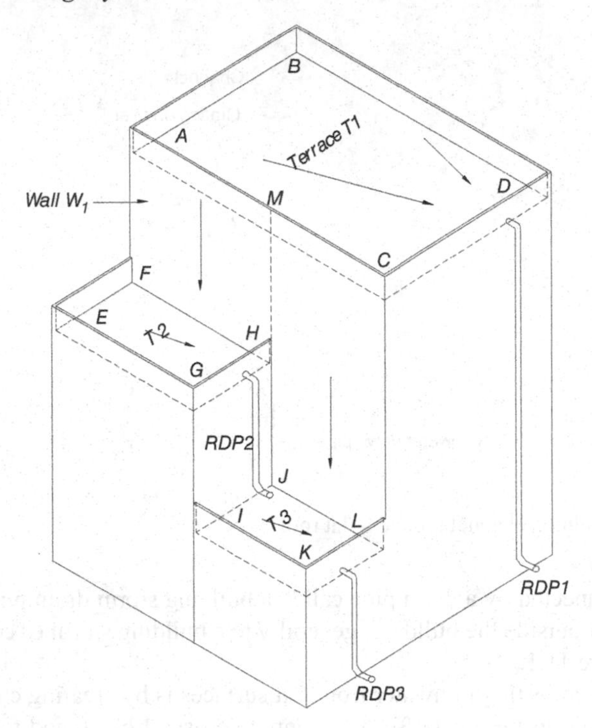

Figure. 11.4 Catchments of flat roof surfaces

For rainwater downpipe RDP3, the catchments will be the terrace at levels 2 and 3 and 50 percent of the adjacent contributory wall AFHM and MJLC.

11.6.2 Catchments of inclined surface

In determining the total effective catchment area of the inclined roof on a building, shown in Figure 11.5, the following areas should be calculated.

1 The horizontal plan area
2 50 percent of the vertical elevation area and
3 50 percent of the adjacent wall area.

Let us consider an inclined roof ABCD of Figure 11.5. Its horizontal plan area is ABC'D'; the vertical elevation area is CDC'D'; and the adjoining wall ADE contributes to the accumulation of rainwater on this roof. The total effective catchments for the roof of the building will thus be the plan area ABC'D' plus half of the elevation area CDC'D' plus half of the adjoining wall ADE.

Example:

Let us consider an inclined roof ABCD of Figure 11.5 with a length of 25 m, a width of 10 m and a pitch height of 5 m situated just adjacent to a side wall, which will also contribute

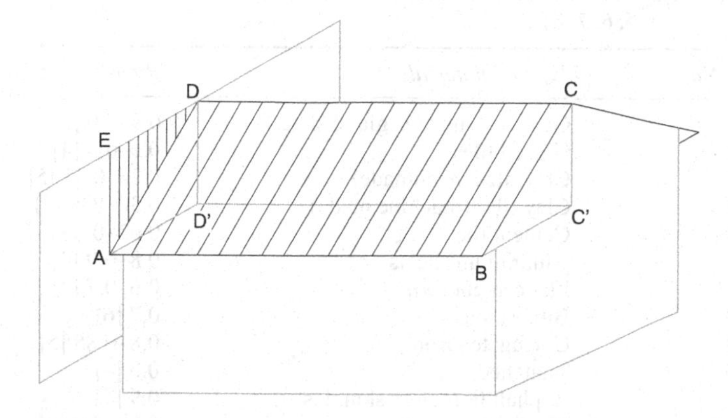

Figure 11.5 Catchment areas of sloped roof surfaces.

to generating rainwater for collection. Then the total effective catchment area can be calculated as follows.

Catchment area = Roof plan ABC'D' + ½ beveled surface CDC'D' + ½ side wall ADE
= (25 × 10) + (5 × 25) / 2 + (10 × 5) / 2
= 250 + 62.50 + 12.50
= 325 sqm.

11.7 Rainwater generation

All the rainwater that falls on any surface can hardly be collected, as the surface itself absorbs some, depending on its nature. The generation of rainwater from a catchment is therefore usually represented by a runoff coefficient (R_C). The runoff coefficient for any catchment of particular material is the ratio of the volume of rainwater that runs off a surface to the volume of rainfall that falls on the catchment area. A runoff coefficient of 0.8 of the material of a catchment means that 80 percent of the rainfall can be collected from that catchment. So, the higher the runoff coefficient, the more rainwater can be collected. An impermeable catchment will yield the highest volume of runoff. Runoff coefficients for various roofing materials are furnished in Table 11.2.

Example: Let us consider a catchment made of certain material whose runoff coefficient is 0.9. If the mean hourly rainfall is 100 mm and the catchment area is 100 sqm, then the rainwater that could accumulate from the catchment would be as follows.

= Catchment area × hourly rainfall intensity × runoff coefficient.
= 100 sqm × 100 mm × 0.9 = 9 cum or 9,000 litres per hour.

The surface rainwater runoff is considered to be some percentage of rainfall. For small areas less than 15 hectares, a rational method is used, as given below.

$$Q = 2.78CiA \qquad 11.1$$

Table 11.2 Runoff coefficient for flat catchments of various materials [3, 4, 5, 6, 7, 8].

Sl. No	*Type of materials*	*Runoff coefficient*
1	GI sheets, metals, glass, slate	0.9–1.0 [3]
2	Glazed tiles	0.6–0.9 [4]
3	Clay tiles (handmade)	0.24–0.31 [5]
4	Clay tiles (machine made)	0.30–039 [5]
5	Cement tile	0.62–0.69 [5]
6	Aluminium sheets	0.8–0.9 [4]
7	Flat cement roof	0.6–0.7 [4]
8	Bituminous	0.7 [6]
9	Corrugated iron	0.8–0.85 [5]
10	Thatched	0.2 [4]
11	Asphalt fibreglass shingles	0.9 [7]
12	Plastic sheets	0.80–0.9 [8]
13	Terracotta	0.65 [7]
14	Wood	0.65 [7]

Where
Q = runoff in litres per second (lps)
C = runoff coefficient (see Table 11.2)
i = average rainfall intensity (mm/hr) and
A = area of catchment (hectares).

11.8 Conveying rainwater

In order to avoid the accumulation of water falling on various catchments of a building, rainwater is conveyed to the desired location through various means, depending upon the type of catchment. Paramount ways of draining rainwater from various types of catchments of buildings are the following.

1 Gutter and
2 Roof drains.

11.8.1 Gutter

In inclined or pitched roofs, rainwater falling on a catchment is conveyed to desired locations by a gutter at the bottom edge of the roof surface, as shown in Figure 11.6. A gutter is a narrow channel that collects rainwater generally from any inclined roof of a building or any inclined surface and diverts it to the inlet of a collection pipe. Depending on the position of the gutters and the surrounding conditions, gutters are classified as follows.

1 External gutter
2 Parapet or boundary wall gutter and
3 Valley gutter.

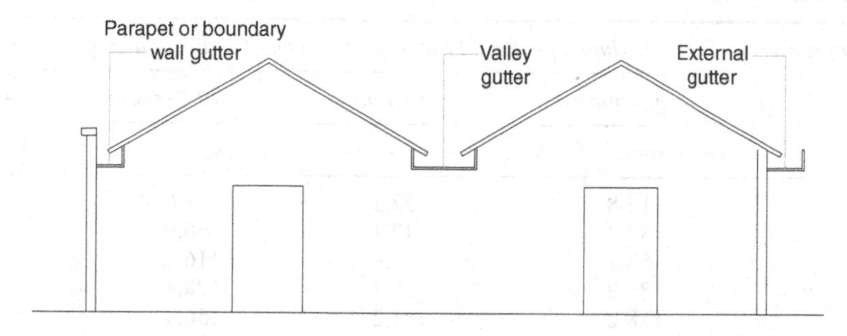

Figure 11.6 Various gutters.

A ridge is a horizontal line formed along the elevated end by the juncture of two sloping planes. In a pitch roof, the ridge line is formed by the sloping surfaces at the top of the roof. From the ridge at the top of the inclined catchment, rainwater flows downward to the gutters below. The inclination of the gutter directs the rainwater to the inlets of the collection pipes.

Gutter design: To determine the shape and size of a gutter correctly, it is necessary to calculate the rainwater discharge rate from the roof. This needs to assess the rainfall rate and the effective catchment area from which the rainwater should be collected.

The gutter has to be designed to provide sufficient capacity for the predicted discharge rate. The following assumptions are generally considered in the design of gutters.

1 The rule of thumb is that a gutter of 1sqcm cross-section is required for a 1sqm roof surface [9]
2 Gutters shall fall towards the collection pipe. The slope of the gutter should be less than 1: 350 [10]. A steeper slope makes for a greater flow rate. By steepening the slope of the gutter from 1:100 to 3:100, potential rainwater flow can be increased by 10 to 20 percent [9].
3 The gutter should have a uniform cross-section and should be large enough to ensure free discharge without spillage of rainwater.
4 The dimension from a stop end to an outlet should be less than 50 times the maximum water depth [10].
5 The distance between outlets should be less than 100 times the maximum water depth [10].

The depth of flowing water in the gutter is not constant but rather varies: the depth is the maximum at the upstream end and the minimum depth, also called the "critical depth", is at the outlet. The depth of water flow is dependent on the shape and slope of the gutter. For a rectangular-shaped gutter section, the maximum depth of flowing water is equal to twice the depth at the outlet [10]. In addition to the depth of water flow, all the gutters should include an allowance in depth as a freeboard to prevent splashing and to allow rainwater flowing below the spillover level of the gutter. A minimum 50 mm freeboard is often considered a good practice [10].

The size of a semi-circular gutter should be based on the maximum projected roof area according to Table 11.3.

Table 11.3 Size of semi-circular gutter* based on the maximum projected roof area [11].

Diameter of gutter	*Maximum projected roof area for semi-circular gutter of various slopes*			
	5.2 mm/m	*10.4 mm/m*	*20.8 mm/m*	*41.7 mm/m*
mm	*sqm*	*sqm*	*sqm*	*sqm*
76	15.8	22.3	31.6	44.6
102	33.4	47.4	66.9	94.8
127	58.1	81.8	116.1	164.4
152	89.2	126.3	178.4	257.3
178	128.2	181.2	256.4	362.3
203	184.9	260.1	369.7	520.2
254	334.4	473.8	668.9	929.0

*Based upon a maximum rainfall of 102 mm/hour for 1 hour duration. The value for drainage area shall be subject to rainfall in mm/hour of local conditions.

11.9 Roof drain inlets

With flat surface rainwater, drainage is generally accomplished by providing a minimum of two of rainwater inlets as required. Rainwater inlets are to be located in suitable positions for quick and economic drainage. Inlets need to be covered by appropriate gratings for protection of the drainage system. Necessary measures are to be taken in installing inlets and gratings for complete drainage. Inlets are connected to rainwater downpipes to convey rainwater to the building storm drain or sewer.

11.9.1 Positioning RDP inlets

After finding the number of rainwater downpipes, its positioning shall be well planned in accordance with the configuration of the roof plan and position of other catchment areas of the building. Where possible, rainwater downpipes should be proportionately distributed along all the roof-sides of buildings. Rainwater downpipes should not be installed too far away from the ridge, so that at the ridge the thickness of the finishing layer on the roof become too thick.

Following are the guidelines for positioning roof drain inlets.

1 On all independent roofs, provide minimum two drain inlets
2 One drain inlet can be provided for a maximum area of 2320 sqm when a gravitational drainage system is chosen
3 In flat roofs, the maximum distance from edge line to drain shall not be more than 15 m
4 In sloped roofs, the maximum distance from the end of the valley to the drain inlet shall not be more than 15 m
5 The maximum distance that can be allowed between roof-drains is 60 m and
6 The minimum diameter of RDP shall not be less than 50 mm.

11.9.2 Covering RDP inlets

Rainwater drain pipe systems have an inlet at the top at roof-surface level which should be kept covered by a grating, the type dependent upon the type of drainage. In gravitational flow drainage, a simple dome-shaped grating is provided. The height of the dome strainer shall not be less than 100 mm from the finished roof surface. The openings of the dome-shaped grating shall have an area greater than 1.5 times the cross-sectional area of the drain pipe. In unavoidable cases, if the flat strainer is to be provided, then the openings of the flat grating shall have an area greater than 2–3 times the cross-sectional area of the drain pipe. It is also recommended that the inlet of any vertical roof drain pipe be one size bigger than the selected size of the drain pipe.

11.9.3 Sump around RDP inlets

Roof sumps for rainwater inlets should be created to avoid ponding of water around the inlet after the rain is over. There are many different ways to create roof sumps, as illustrated in Figure 11.7. In a depressed sump there is a flat bottom, so that it may hold minimal water and stain the very small sump area over time. For a funnel-shaped sump, the ideal slope of the side is a 7.5:1 slope.

11.10 Rainwater drain pipe

For multi-storey buildings, free fall of rainwater from the roof is not recommended. Free fall of rainwater from the roof of a single-storey building may be allowed where there is no danger of soil erosion. Good practice is therefore to bring rainwater down from the roofs of multi-storey buildings to the drainage infrastructures in the ground through installing rainwater drains or downpipes.

Rainwater drain pipes are those pipes which lead rainwater down from a roof or any other elevated flat surface to a surface at a lower level, generally to the ground. Rainwater drain pipes are also termed rainwater downpipes (RDP) or leaders. Rainwater drain pipe systems have inlets at the top of the rainwater down pipes, which are kept covered by a grating, the type dependent upon the type of drainage; the other end is kept open to discharge on an open surface or any drain for finally discharging out of the building or premises. No wastewater or soil drain pipe can be connected to the rainwater drain or downpipe, even it discharges into the combined drain or sewer.

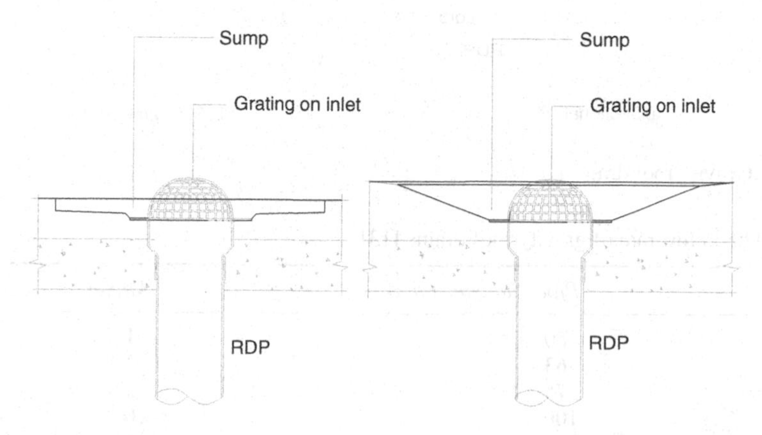

Figure 11.7 Sump around rainwater drain pipe inlet.

Virtually two types of drainage are followed, as mentioned below, for which two types of inlet are designed.

1 Gravity system and
2 Siphonic system.

11.10.1 Gravity roof drainage system

This is the conventional roof drainage system, which comprises an open inlet, generally kept covered by a grating, connected to a vertical rainwater down pipe or leader that is designed to operate at atmospheric pressure, with flow of air along with rainwater in the system. The size of the roof drains and the allowable depth of the rainwater above the inlets usually determine the discharge capacity of draining pipes. The flow into the inlet is a weir type flow, and rainwater flows down along the side wall of the pipe while the central core remains full of air, as shown in Figure 11.8, and in the horizontal drain portion, the flow is partial full flow with the remaining space filled by air. In this system, no full flow occurs in any portion of the total drainage piping.

Disadvantages:

1 In a gravity flow downpipe, excessive flow is not expected; standards or codes limit rainwater to occupy approximately 30 percent, with air taking up other 70 percent of pipe capacity [12]

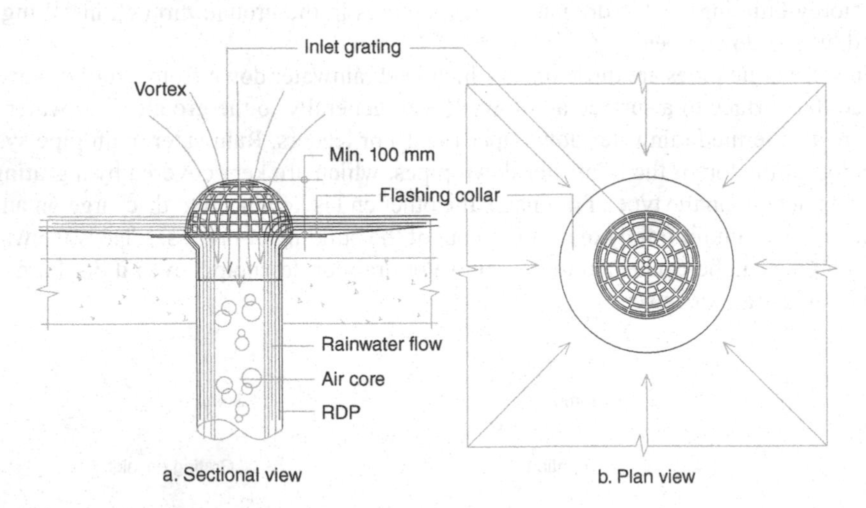

Figure 11.8 Gravity roof drain

Table 11.4 The inflow rate of gravity roof drains [13].

Sl. No	*Pipe diameter (mm)*	*Maximum inflow rate (lps)*
1	50	1.7
2	63	3.1
3	75	5.0
4	100	10.7
5	150	31.6
6	200	68.0

2 Introducing a greater volume of air than water in the draining system results in inefficiencies and excess costs in material and labour
3 Excessive flow in a gravity flow drain pipe can cause excessive vibration, pipe wear, noise and pipe implosions [13]; to avoid excessive flow in gravity roof drains, the inflow rate is limited, as shown in Table 11.4.

11.10.2 Siphonic roof drainage system

A siphonic roof drainage system consists of a series of specially designed roof drains connected to a horizontal drain pipe and, finally, connects through a single vertical drain pipe or leader, discharging at or below ground level. The piping system is designed in such a way that outside air cannot enter, and inside air is moved outward from the piping. For this purpose, a specially designed siphonic inlet is placed at the inlet of drain pipes, as shown in Figure 11.9. In specially designed siphonic inlets, an air baffle is provided to prevent the entry of air into the piping system. The air baffle prevents formation of vortices, thus enabling rainwater to enter the drain pipe stably. As air is kept from entering into the pipe, siphonage is formed to make drainage continuous and faster.

Siphonic systems may be used on any building over approximately 4.5 m in height above ground. As the drainage piping remains water filled (primed) to exclude air from the pipe, it causes the flow of drainage under pressure. The greatly increased driving head of water in the vertical portion of the pipe, with the increase in building height, causes a large amount of discharge through a substantially reduced size of pipe, in comparison to the required pipe size for the same amount of discharge through conventional gravity drainage systems.

In this system, there is a chance of developing negative pressure. This negative or suction pressure in the system should not exceed 8.0 m in water head to avoid cavitation. The minimum velocity of flow should be 1.0 mps for self-cleansing purposes.

Following are the advantages of siphonic roof drainage systems in comparison to gravity roof drain systems.

1 Fewer drain inlets are required, and thus the number of roof penetrations and length of pipe run is minimized

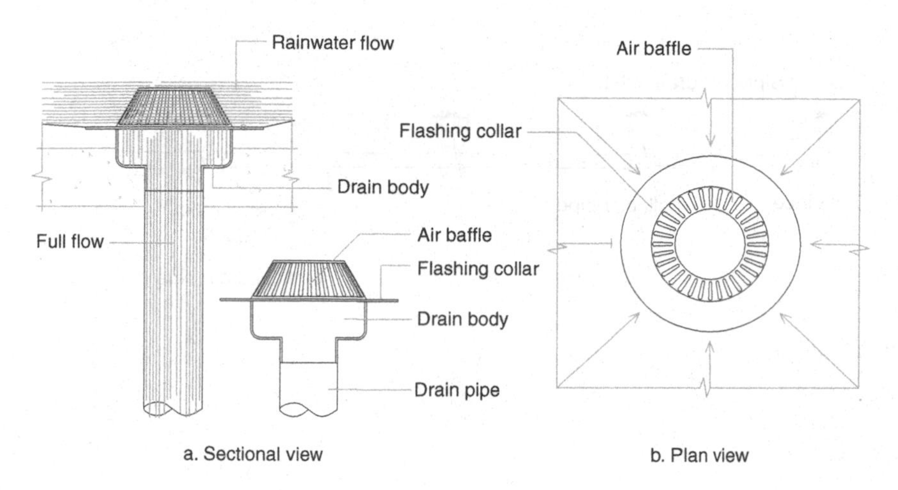

Figure 11.9 Siphonic roof drain

2 In very large and wide buildings, the need for rows of inside vertical rainwater pipes can be eliminated together with the underfloor drainage to serve those
3 The collecting pipe can be installed horizontally, at no fall, just below the roof structure, freeing more space within the building and allowing greater freedom for installing various other service pipes
4 Full flow at comparatively high velocity results in reduced pipe size and fewer pipe requirements and
5 In a single-storey building, a siphonic drain pipe can drain 10 to 15 times more than can be done under gravity flow [14].

Following are the disadvantages as well.

1 Siphonic systems require sufficient and consistent rainfall intensity to build up consistent levels of rainwater flow into the drain inlet, to instigate the siphoning process
2 Air traps in the piping may result in system failure due to lower intensity rainfall events than the designed intensity of rainfall
3 Inconsistent rainfall intensities may result in variations in incoming water level and system pressures which can lead air being drawn into the system that might result in generation of noise and vibration, so the pipe needs to be firmly secured
4 Development of cavitation may cause damage to the pipes and
5 If the entrance of air in the pipe reaches more than 40 percent of the volume of the pipe, the siphonic action will be disrupted [15].

In a siphonic roof drainage system, the calculation for the exact pipe sizes requires a full understanding of the forces acting within the system, the type of material used for the drainage piping, the maximum probable rain falls, the catchment area to be drained and the height of the building.

In this drainage system, eccentric reducers are used in the horizontal portion of drainage piping whenever there is a requirement for a change in pipe diameter, as shown in Figure 11.10. The top of the horizontal pipe is kept as a flat surface along the pipe run, thereby eliminating the chance of any air pockets.

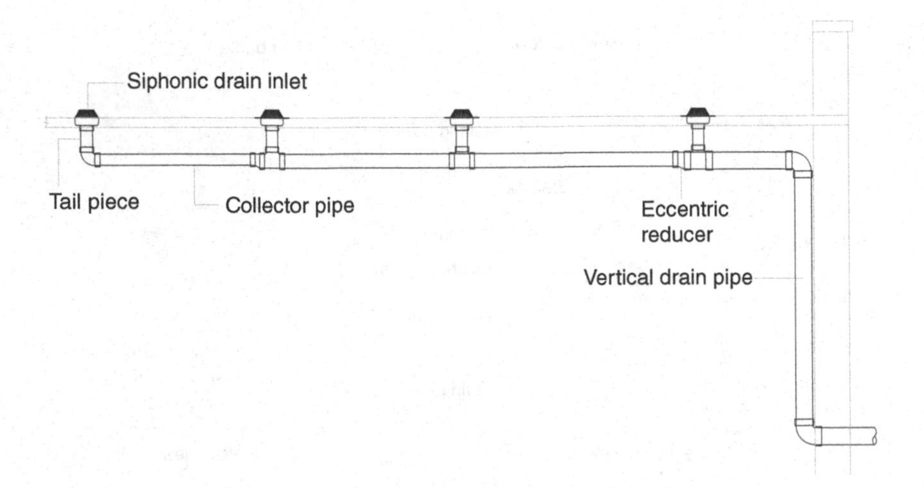

Figure 11.10 Siphonic roof drainage piping system

No more than 4645 sqm of roof area should be drained by one common collector pipe to avoid managing use of large diameter pipes and the heavy load thereof. Horizontal collector pipes should be spaced no more than 20 m apart, and stacks should be located no more than 20 times the building height away from the furthest drain.

11.11 Discharging of rainwater down pipes

The discharge from the outlet of rainwater downpipes is to be done in approved manner. The discharge of rainwater down pipes can be disposed of in two ways, as mentioned below.

1 Disposing on natural ground and
2 Disposing into the building or public sewers.

11.11.1 Disposing on natural ground

After collecting rainwater from various catchments and then conveying it to the ground rainwater can be discharged on ground, which will allow it to percolate into the ground. This can be done where large green surface is found available. To avoid erosion of soil surface a confined stone bed under rainwater down pipes is to be constructed, as shown in Figure 11.11.

11.11.2 Disposing in to the building or public sewer

Most rainwater down pipes or building storm drains discharge into the building sewers where needed; otherwise, it can directly discharge rainwater into the public storm sewer found nearby. In no case can the building storm drain or rainwater downpipe be directly connected to a combined drain or sewer unless a befitting trap on the drain pipe is installed

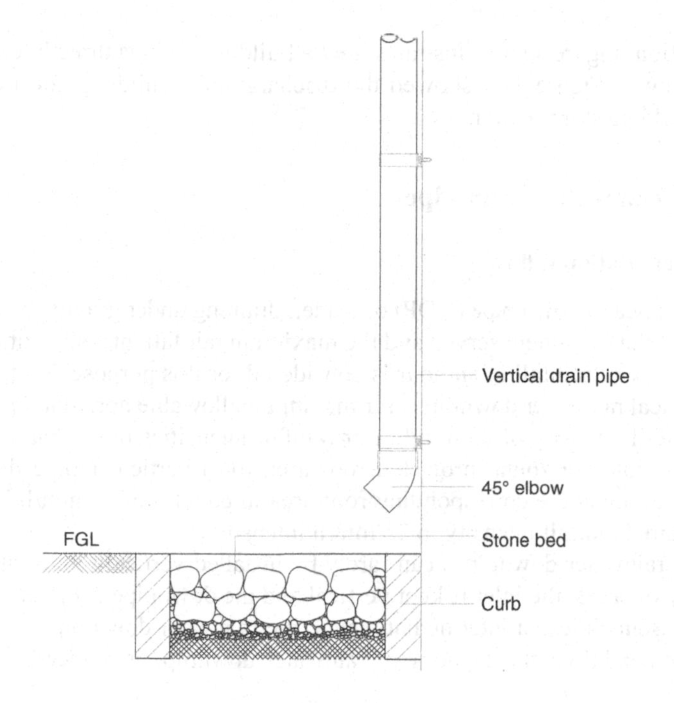

Figure 11.11 Rainwater discharged on stone bed on ground.

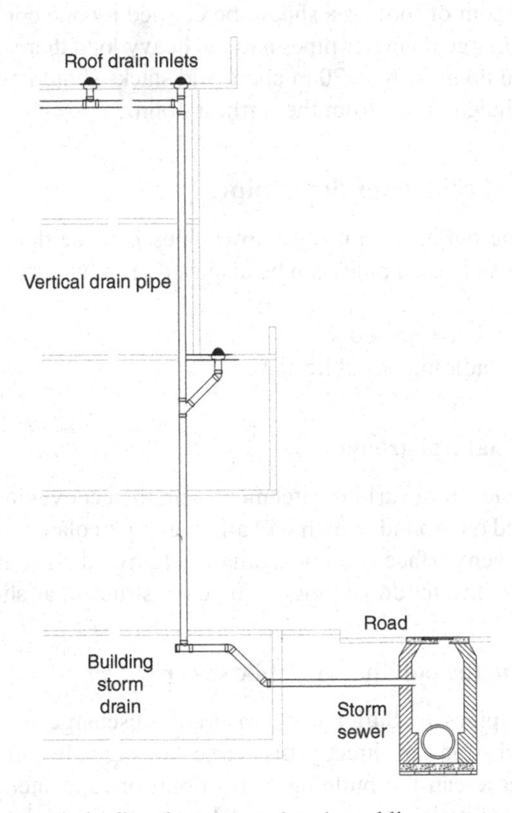

Figure 11.12 Building storm drain directly connected to the public storm sewer.

before connection. Figure 11.12 illustrates how a building drain is directly connected to the public storm sewer. Figure 11.1 showed the discharge of a building storm drain which is made into a building storm sewer.

11.12 Sizing rainwater drain pipes

11.12.1 For gravitational flow

The size of a rainwater down pipe (RDP) or leader, draining under gravity, basically depends upon the size of the catchment served and the maximum rainfall intensity anticipated on that area. The intensity of rainfall in an hour is considered for this purpose. Table 11.5 provides the size of vertical rainwater downpipes for maximum allowable horizontal projected catchments, for rainfall intensity of 25 mm/h. For rainfall intensities other than those listed, the maximum allowable horizontal projected roof area for a particular pipe diameter can be determined by dividing the corresponding roof area in column of 25 mm/h intensity by the ratio of the desired rainfall intensity to 25 mm/h intensity.

In building, rainwater downpipes can rarely be installed vertically throughout its length. In the majority of cases, the inlet is kept vertical and the downpipe is offset at either the top or the bottom. Sometimes, at intermediate level, the rainwater down pipe needs to be made offset. In these conditions the capacity of rainwater downpipe is reduced. So, it is wise to

Table 11.5 Sizing roof drain, leader or vertical rainwater down pipes [16].

Sl. No	*Size of drain, leader or pipe in mm*	*Flow lps*	*Maximum allowable horizontal projected roof areas in sqm, at various rainfall intensities in mm/h*					
			25	*50*	*75*	*100*	*125*	*150*
1	50	1.5	202	101	67	51	40	34
2	80	4.2	600	300	200	150	120	100
3	100	9.1	1286	643	429	321	257	214
4	125	16.5	2334	1167	778	583	467	389
5	150	26.8	3790	1895	1263	948	758	632
6	200	57.6	8175	4088	2725	2044	1635	1363

Notes:

1 The sizing data for vertical conductors, leaders and drains are based on the pipe flowing 7/24 full.

2 Vertical pipes may be round, square or rectangular of equivalent cross-sectional areas. The ratio of width to depth of rectangular RDP shall be within 3:1.

Table 11.6 Diameter of horizontal rainwater drain pipe in mm based on roof area (sqm) and intensity of rain (mm/hr). Slope of horizontal portion of pipe is 10.4 mm per meter [17].

Size of RDP (mm)	*Flow rate (lps)*	*Maximum allowable catchment areas in sqm. at rainfall rates, in mm/h, of*					
		25.40	*51.00*	*76.00*	*102.00*	*127.00*	*152.00*
76	2.1	305	153	102	76	61	51
102	4.9	700	350	233	175	140	116
127	8.8	1,241	621	414	310	248	207
152	14.0	1,988	994	663	497	394	331
203	30.2	4,273	2,137	1,424	1,068	855	713
254	54.3	7,692	3,846	2,564	1,923	1,540	1,282
305	87.3	12,375	6,187	4,125	3,094	2,476	2,067
315	156.0	22,110	11,055	7,370	5,528	4,422	3,683

Note: Pipe flowing full.

consider the capacity of the rainwater downpipe, with both horizontal and vertical alignment, as that of a horizontal rainwater-conveying pipe.

The diameter of a horizontal rainwater-conveying pipe in millimetres shall be based on the maximum projected roof area (sqm) and intensity of rainfall (mm/hr) according to Table 11.6. The slope of the horizontal portion of a pipe is considered to be 100:1.

11.12.2 For siphonic flow

The size of a rainwater downpipe depends upon the rainfall intensity and the projected catchment to be served, and the number of inlets of rainwater depends upon the maximum projected catchment areas to be served by a single rainwater inlet of a particular size. In this system, more than one inlet is served by one downpipe or leader. So, the size of a rainwater downpipe or leader is dependent on the number of inlets connected and thereby the summation of projected catchments served by each inlet. The size of a rainwater downpipe or leader can be chosen from Table 11.7 for catchment areas in sqm receiving rainfall of varied intensity.

Table 11.7 Size of drain pipe having siphonic flow for catchments in sqm receiving rainfall of varied intensity [18].

Sl. No	*Size of drain, leader or pipe in mm*	*For rainfall intensities in mm/h*							
		25	*38*	*50*	*63*	*75*	*100*	*125*	*150*
		*Maximum allowable horizontal projected roof areas in sqm**							
1	50	1320	880	660	528	440	330	264	220
2	80	3405	2270	1702	1362	1135	851	681	568
3	100	6300	4200	3150	2520	2100	1575	1260	1050

*Theoretical value based on ideal condition of test in the University of Munich (without safety factor).
Converted to metric system from source: Verdecchia, 2014. Conversion factors: 1 inch = 25 mm; 1sqin = 645 sqmm and 1 sqft = 0.093 sqm

References

[1] Un-named (UD) *Dissertation of Post Graduate Diploma in Civil Engineering*, Under Joint Program of Asian Institute of Technology (AIT) Bangkok, Thailand and Bangladesh University of Engineering and Technology (BUET), Dhaka, p. 26.

[2] Indexmundi.com "Average Precipitation in Depth (mm per year) – Country Ranking" Retrieved from www.indexmundi.com/facts/indicators/AG.LND.PRCP.MM/rankings on 28 December 2020.

[3] Indian Institute of Technology (2006) "Water and Wastewater Engineering" Retrieved from http://nptel.ac.in/courses/105104102/Lecture%206.htm on 25 January 2016.

[4] HarvestH2o, "Potable Rainwater: Filtration and Purification" by Doug Pushard, Retrieved from www.harvesth2o.com/filtration_purification.shtml#.VgpN2dKqqko on 29 September 2015.

[5] The Caribbean Environmental Health Institute (2009) "Rainwater: Catch It While You Can" *A Handbook on Rainwater Harvesting in the Carribean*, p. 26.

[6] Pfafflin, James R., and Ziegler, Edward N. (2006) *Encyclopedia of Environmental Science and Engineering*, 5th Ed, vol. 2 M-Z, CRC Press, Taylor and Francis Group, London, p. 1191.

[7] Metcalf and Eddy (1991) *Wastewater Engineering: Treatment, Disposal, and Reuse*, 3rd Ed. Cited in "Technical Memorandum" Contributed by Bruce Tiffany, Retrieved from www.kingcounty.gov/~/media/services/environment/wastewater/industrialwaste/docs/TechAssistance/CDW_SedTank_Tech_Memo1111.ashx?la=en on 06 June 2016.

[8] Caltrans (2001) Field Guide to Construction Site Dewatering – Appendix B: Sediment Treatment Options. California State Department of Transportation (Caltrans) – Construction Division. Publication No. CTSW-RT-01–010. October 2001. Cited in "Technical Memorandum" Contributed by Bruce Tiffany, Retrieved from www.kingcounty.gov/~/media/services/environment/wastewater/industrialwaste/docs/TechAssistance/CDW_SedTank_Tech_Memo1111.ashx?la=en on 06 June 2016.

[9] Worm, J., and van Hattum, T. (2006) *Rainwater Harvesting for Domestic Use*, Agromisa Foundation and CTA, Wageningen, The Netherlands.

[10] Kingspan (2015) Sandwitch Panel India, Building Design, Retrieved from http://panels.kingspan.in/Roof-Drainage-%7C-Roof-Drains-%7C-Gutter-Layout-%7C-India – 13266.html on 22 September 2015.

[11] IAPMO Plumbing Code and Standards, India (2007) *Uniform Plumbing Code- India 2008*, IAPMO Plumbing Code and Standards Pvt Ltd, New Delhi.

[12] Hydromax Ltd (2011) "How Does Hydromax Siphonic Drainage Work" Retrieved from www.hydromax.com/144_HowdoesHydroMaxSiphonicDrainageWork.html on 28 December 2020.

[13] Metal Gutter Manufacturers Association (MGMA) (2012) "Rainwater Drainage Design, BS EN12056:3–2000" Information Sheet No 03, Retrieved from https://mgma.co.uk/wp-content/uploads/2016/04/MGMA-GD03-rainwater-drainage-design.pdf on 28 December 2020.

[14] www.polypipe.com (2009) "Siphonic Rainwater Drainage" p. 4, Retrieved from www.polypipe.com/sites/default/files/Siphonic_Hydromax_Product_Guide.pdf on 22 December 2020.

[15] Snoad, Peter (UD) "Syphonic Roof Drainage – How Does It Work?" Retrieved from www.cibse.org/getmedia/699c149b-f37e-4634-9607-58248550b3c8/SYPHONIC-RAINWATER-SYSTEMS.pdf.aspx on 22 December 2020.

[16] California Plumbing Code (2007) "Traps and Interceptors" Chapter 10, p. 165.

[17] IAPMO Plumbing Code and Standards, India (2007) *Uniform Plumbing Code- India 2008*, IAPMO Plumbing Code and Standards Pvt Ltd, New Delhi, p. 170.

[18] Verdecchia, William (2014) "Introducing Siphonic Roof Drainage: Common in Europe, Now Gaining Traction Stateside" Retrieved from www.constructionspecifier.com/introducing-siphonic-roof-drainage-common-in-europe-now-gaining-traction-stateside/ on 28 December 2020.

12 Building sewer systems

12.1 Introduction

When a building is found considerably far away from the location of the final disposal points for waste disposal, then a building drainage system is to be developed outside the building concern, and within the building premises, to convey the waste from the building to the final disposal points. When a building exists just on the roadside, the development of a building sewer system might not be needed. So, in many plumbing books this part is not discussed. Building sewer systems consist of various components; all of these should be well planned, designed, constructed and maintained, to develop an efficient drainage system holistically. In this chapter, all the aspects of a building sewer system are discussed exclusively.

12.2 Building site drainage

Building site drainage systems involves various infrastructural development, inside the building premise or site, that continue on outside the buildings, for transporting sewage and storm water from the building and site to a basically primary treatment unit or an approved disposal system. The objective of developing this drainage system is to provide efficient and sanitary conveyance of human excreta, ablutionary wastewater, laundry, kitchen wastewater etc. and storm water to the approved disposal points. Building sewage disposal might be accomplished through a primary sewage treatment unit built inside the building premises, or on-site management of sewage; or into the sewer outside of the building premises, which convey wastewater to a treatment plant elsewhere, or off-site sewage management, as shown in Figure 12.1.

12.2.1 Type of building site drainage

In a building complex, the building drainage system can be developed in two ways, based on the level of conveying media into the ground.

1 Subsurface or building sewer systems and
2 Surface drainage system.

12.3 Building sewer systems

A building's sewer is that portion of the drainage system which starts from the outer face of a building and terminates at the main sewer at the street or primary treatment units within

DOI: 10.1201/9781003172239-12

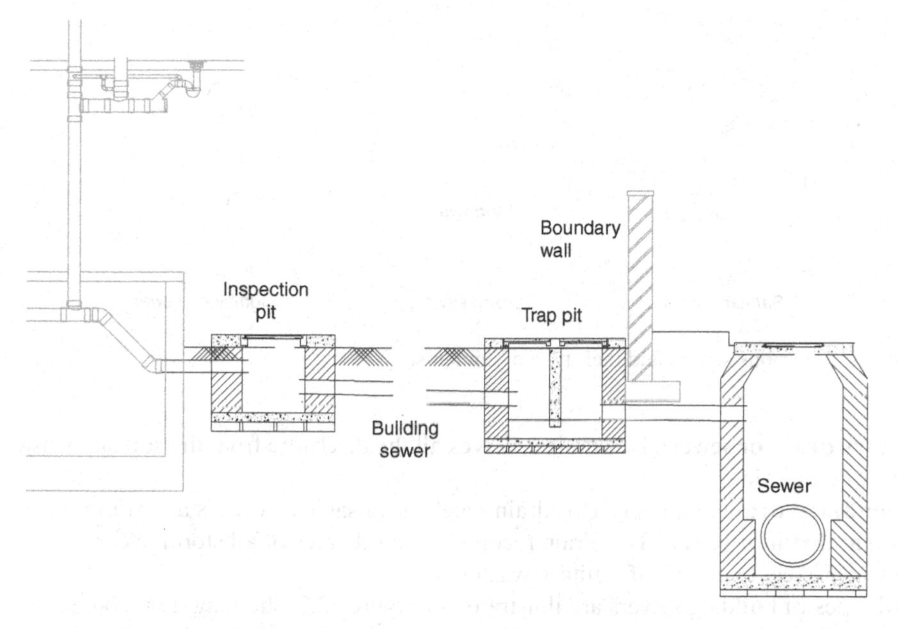

Figure 12.1 Components of building sewer systems of a building for off-site sewage management

the premises. Pipes are generally used as the conveying media in subsurface building sewer systems, laid underground.

Building sewers do not need to be built when a building drain end can be connected to a public sewer found close to the building. Otherwise, building sewers are to be constructed in the following conditions.

1 Building is located considerably away from the nearby public sewer
2 On-site treatment units are to be incorporated far away and
3 The stacks or building drain are to be terminated on the ground and around the periphery of the building.

For underground construction, not only the size of the pipe but also its material, depth and strength need to be determined, considering the site conditions and anticipated surcharge loading. Accordingly, an appropriate pipe is to be chosen for building the sewer.

Building sewer systems incorporate another structure, termed an inspection pit or manhole. At the end, before connecting to the public sewer, special measures are to be taken depending upon the type of sewer.

12.3.1 Building sewer categories

Three categories of building sewers can be developed, as mentioned below, depending upon the type of wastewater handled.

1 Sanitary drain or sewer
2 Combined drain or sewer and
3 Storm drain or sewer.

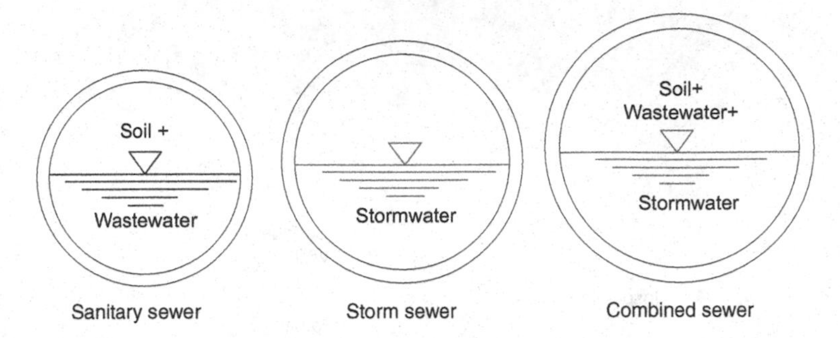

Figure 12.2 Various sewers and their type of contents.

Sanitary drain or sewer: This drain receives all the discharge from the soil and wastewater stacks of a building.

Combined drain or sewer: This drain receives all sanitary wastes and storm water.

Storm drain or sewer: This drain receives all discharges of all storm, clear water or surface water except any sort of sanitary wastes.

All types of building sewers are illustrated in Figure 12.2, showing their content type.

Generally, public sewer systems are developed by incorporating sanitary sewers and storm sewers. Where there is very limited scope of treatment and the drainage system is to be run for a comparatively small area for a short period, a combined sewer system may be adopted.

12.3.2 Building sewer components

The components of the building sewer system of a building, as shown in Figure 12.1, are as follows.

1 Building sewer piping
2 Inspection pit or manhole and
3 Connection chamber for off-site sewage management.

12.3.3 Draining methodology

Broadly, there are two basic methodologies adopted for draining sanitary waste and storm water in a building site or premise. Depending upon the site condition around the building, the adoption of the system of drainage is chosen. The complete drainage system may be a combination of both systems. Considering the position of disposal point, the drainage methodologies may be termed as follows.

1 Gravitational system and
2 Pumping system.

12.4 Building sewer piping

Building sewers are generally developed by installing circular pipes of diverse material, mostly underground and in a particular grade, falling and ultimately directing towards the final discharge point.

12.4.1 Subsurface pipe drainage

Subsurface drainage systems for sanitary wastewater and storm or rainwater are for directing all these wastewaters by natural or artificial means, from various building locations and their surrounding land, generally by using a system of pipes placed below the ground surface level. Underground drainage pipes must satisfy two conditions: hydraulic and structural. Both are equally important. Hydraulically, these pipes must provide adequate passage for the estimated wastewater, storm or rainwater for which they are designed, and structurally, they must sustain the surrounding ground, supporting the overburdened weight of the ground and moving load, if any, on it.

12.4.2 Positioning of sewers

The building sewer depth, top of pipe, should be a minimum of 150 mm below the frost depth of the ground; where there is no frost, the depth of pipe should be 300 mm minimum below the finished ground.

Various other piping of different content might be found around the alignment of a building sewer, and water supply pipe in particular. A building sewer pipe must be positioned to maintain a safe distance from the water supplying pipe, as illustrated in Figure 12.3. Sewer pipes must be laid below the nearby water supply line.

12.4.3 Grading for sewers

Like any drain pipe the efficiency of a sewer pipe also depends upon the velocity of flow. Velocity of wastewater flow in a sewer increases with the increase in grade of a sewer pipe.

Pipes with a velocity of flow of more than 146 mm/min have a tendency for waste separation. Water flows faster, leaving the suspended waste on the pipe bed. When velocity of flow is less than 36 mm/min, then the suspended materials settle down at the bottom of pipe while flowing very slowly. So, in sewers or drain pipes, a minimum velocity of flow of 36 mm/min is to be maintained. This velocity of flow is termed self-cleansing velocity. These phenomena of wastewater, flowing through a sewer pipe, are illustrated in Figure 12.4.

The grade of pipe should be as minimum as possible, but flow should be at a velocity just above self-cleansing velocity. To maintain self-cleansing velocity, the minimum gradient of sewers needed for different diameters of pipe is shown in Table 12.1.

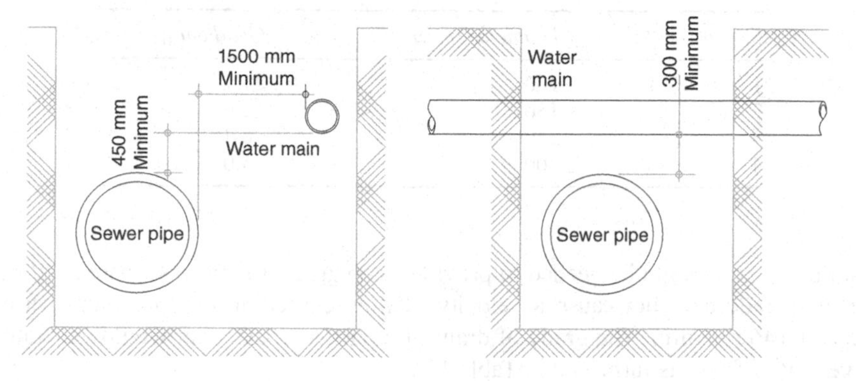

Figure 12.3 Minimum clear distance between underground sewer and water supply pipe.

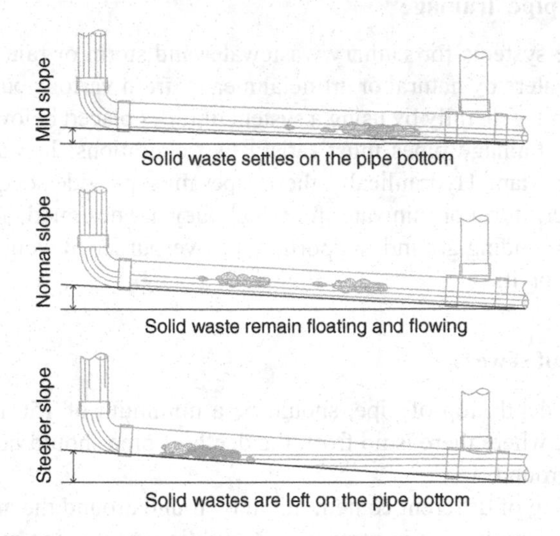

Figure 12.4 Effect of slope of sewer in drainage.

Table 12.1 Approximate minimum gradients for drains and sewer [1].

Sl. No	*Diameter of pipe, mm*	*Gradient length/fall*
1	100	105
2	150	174
3	200	251
4	225	288
5	300	417
6	375	550
7	450	692
8	525	832
9	600	1000
10	675	1148
11	750	1318

Table 12.2 Approximate maximum gradients for drains and sewer [2].

Sl. No	*Diameter (mm)*	*Gradient (percent)*
1	100	21.7
2	150	12.6
3	250	6.4
4	300	5.0

In some cases, it might be needed to provide steep grade. In that case, care must be taken so that the grade does not cause a velocity of flow exceeding the maximum permissible velocity of 146 mm/min. The grade of drain of a particular size that would generate maximum velocity of flow is furnished in Table 12.2.

12.5 Inspection pits and manholes

Inspection pits and manholes are chambers built on building sewer piping systems so as to make every length of drain accessible for inspection, removal of debris and maintenance. Inspection chambers are square, rectangular or circular in shape, with a minimum of 450 mm clear inner dimension and up to 1 m clear depth built underground. Inspection chambers do not provide man-entry access. Chambers built for man-entry access are termed manholes, made deeper than inspection chambers, depending on the depth at which the sewer is laid and the size of the sewer as well.

Inspection pits might be precast in a factory or built on-site with clay bricks or concrete blocks, but manholes are generally built on-site with clay bricks or concrete blocks. In sanitary and combined sewer systems, a concrete bed at the bottom of inspection pits, called benching, is created, which is slightly sloped towards the half-round open channels made in continuation of the sewer line. But in storm sewers, no benching of concrete bed at bottom of the chamber is to be made; instead, a porous flat bed of aggregates or stones is to be made, at least 75 mm below the outlet of sewer.

All the stacks and building drains discharge wastewaters into the inspection pits or chambers outside the building, at first. In addition to these inspection pits, more inspection pits are to be provided where the direction of run, size and grade of pipe changes and crossing of pipes are interrupted. The distances of the inspection chambers should not be more than 2 meters from building facade to end of building drain or any peripheral stack.

Inspection pits are mostly constructed independently for a particular sewer. But, when two types of sewer run parallel and comparatively close to each other, a combined inspection pit with a separating wall inside can be built for economy and space-saving. For different types of building drainage systems, the configurations of the starting inspection pits constructed at the start of building sewers are shown in Figures 12.5, 12.6 and 12.7.

12.6 Gravitational drainage

Drainage systems are mostly developed by gravitational systems that work under gravitational force, and as such is an economical system. In flat terrain where the point or level of

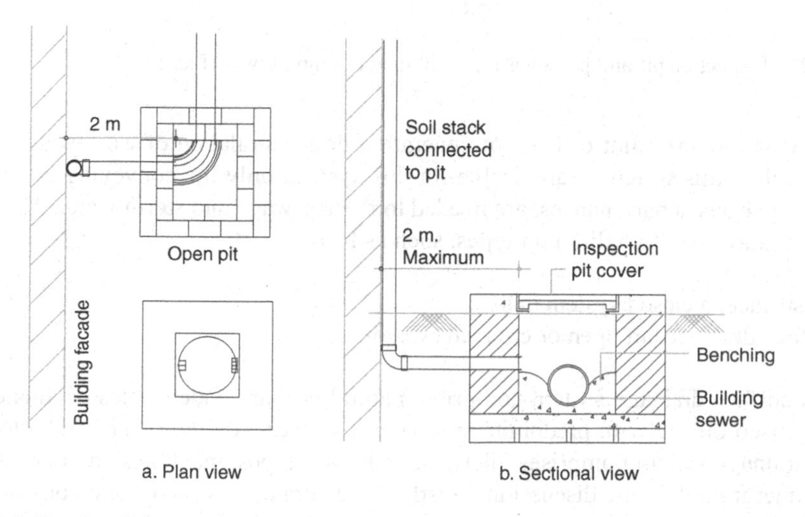

Figure 12.5 Single soil inspection pit and Pit-cover with manhole cover.

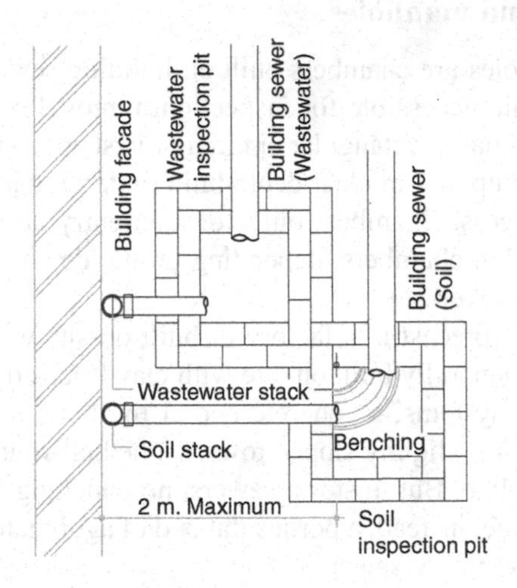

Figure 12.6 Combined (also start) inspection pits for two-pipe drainage system (Plan).

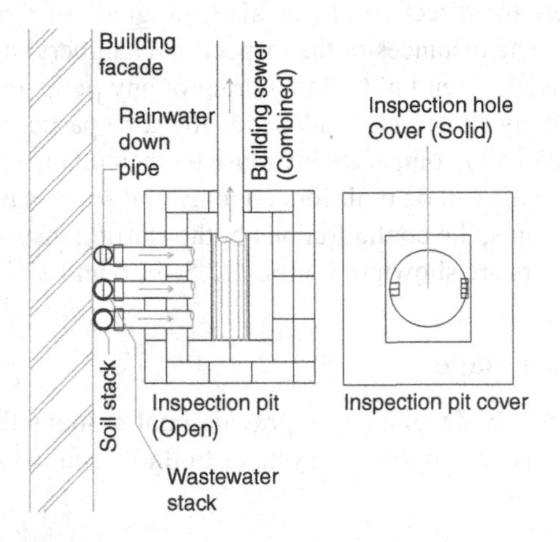

Figure 12.7 Inspection pit and pit cover for combined building sewer (Plan).

disposal is below the point or level of generation or accumulation of wastewater, storm or rainwater, then this system is applicable. In this system, only the conveying media, and if necessary, various appurtenances, are needed to drain sewage and storm water. The conveying system may be of two different types, such as below.

1 Subsurface, a closed system and
2 Surface drainage, an open or covered system.

An open surface drainage system comprises natural or man-made swales or ditches and is generally used on site with predominantly natural surfaces; on the other hand, closed subsurface drainage system comprises inlets, catch basins, pipes, manholes and outlets, mostly placed underground. In the discussion, mostly closed drainage systems and constructed surface drains will be covered.

12.6.1 Gravitational sanitary building sewer

Gravitational drainage of sanitary sewage is primarily dependent on the sewage load in terms of sanitary drainage fixture units and the grade or fall of the sewer. Sanitary drainage fixture units were discussed in previous chapters. The drainage fixture unit to be served by a segment of the sanitary sewer to be designed is the cumulative sewage load received from the preceding sewer and additionally, accumulated from both soil and wastewater stacks. The gravitational flow capacity also increases due to an increase in the fall of a sewer pipe.

12.6.2 Gravitational storm building sewer

Storm building sewers may receive rainwater from the catchments of a building through connected rainwater downpipes. While draining, it may receive storm water from the surrounding surface if allowed. If storm water is allowed, then the load capacity of storm water from the contributory area shall be calculated.

Storm water generation: From a small site around a building, the generation of storm water from a contributory area is primarily governed by the runoff coefficient (R_C) of the site. The runoff coefficients for various green surface areas are furnished in Table 12.3.

The rational formula used in computing the peak runoff from a catchment area is as follows.

$$Q_r = \frac{1}{360} C \times i \times A \qquad 12.1$$

Where
Q_r = peak runoff at the point of design (cum/s)
C = runoff coefficient
i = average rainfall intensity (mm/hr) and
A = catchment area (hectares).

Discharge capacity of storm sewer or drain: Storm drains and sewers are to be designed for steady uniform flow conditions where a one-dimensional method of analysis is considered. A drain should be designed in such a way that it has adequate discharge capacities (Q_c)

Table 12.3 Runoff coefficient for land surfaces of various soil conditions [3].

Sl. No	*Soil and surface texture*		R_C
1	Clay	Bare	0.7
		Light vegetation	0.6
		Dense vegetation	0.5
2	Loam	Bare	0.6
		Light vegetation	0.45
		Dense vegetation	0.35
3	Sand	Bare	0.5
		Light vegetation	0.4
		Dense vegetation	0.3
4	Gravel	Bare	0.65
		Light vegetation	0.5
		Dense vegetation	0.4
5	Grass land		0.35

Table 12.4 The roughness coefficient of various finished materials of surface drains [4].

Boundary condition	*Roughness coefficient (n)*
Unplasticized polyvinyl chloride (uPVC)	0.0125
Concrete	0.0150
Brick	0.0170
Earth	0.0270
Earth with stones and weed	0.0350
Gravel	0.0300

Note: If there are different flow surfaces within a drain section, equivalent roughness coefficient may be used.

to accommodate the estimated peak runoffs (Q_r). The size, cross-sectional geometry and the gradient of a drain are the factors influencing its discharge capacity (Q_c). The calculated discharge capacity (Q_c) of drain section must be at least equal to or preferably larger than the peak runoff (Q_r). The size of the drain is computed using Manning's formula given as below.

$$Q_c = \frac{1}{n} A \times R^{\frac{2}{3}} \times S^{\frac{1}{2}} \qquad 12.2$$

Where
Q_c = discharge capacity of drain (cum/s)
n = roughness coefficient materials of drain surfaces
A = area in square meter (sqm) from which flow occurs to the drain
R = A / P = hydraulic radius (m), where P = wetted perimeter (m) and
S = gradient of drain bed.

Roughness coefficient: The value of the roughness coefficient (n) depends on the type of material with which the surface of drain has been finished. The roughness coefficient of various surface finished material is given in Table 12.4.

12.6.3 Gravitational combined building sewer

A combined building sewer receives all of the discharge from the stacks, rainwater downpipes or all connected building drains. While draining, it may receive storm water from the surrounding surface if allowed.

12.7 Surface drain

A storm or rainwater surface drain is a channel, constructed on ground surface, to drain away storm water quickly and efficiently towards a disposal destination. In areas of heavy rainfall or low soil percolation, it may be necessary to build surface drainage systems for a building to get rid of surplus surface water from its various catchments and surrounding land.

Surface drains can be developed in two ways; one is an open surface drain and the other is a covered surface drain. Both systems have advantages and disadvantages. Open surface drains have the following disadvantages.

1 Receives various solid wastes that create nuisance and reduce the flow capacity
2 Not hygienic and may emanate a bad smell

3 Accidents may happen due to falling into the drain and
4 May act as a breeding ground for mosquitoes and other organisms.

The advantages of developing open surface drains are also many. These are as follows.

1 Less prone to blockage
2 Easy to inspect and get access to for removal of suspended and blocking wastes
3 Easy to identify the location and cause of the problem and rectify accordingly
4 Precautionary measures can be taken before any problem occurs and
5 Easy to maintain and repair.

12.8 Minimizing storm water drainage load

In storm water drainage, it is good practice to minimize the storm water load to be addressed by the drainage system. The most effective way of reducing storm water load is to infiltrate rainwater into the ground where soil characteristics permit easy infiltration of rainwater and where there is little or no scope for disposing of rainwater nearby. There are various easy ways of infiltrating rainwater into the ground. These are as follows.

1 Spreading rainwater on the ground and
2 Induced infiltration from storm drain appurtenances.

12.8.1 Spreading rainwater on ground

After collecting rainwater from various catchments and then conveying it to the ground, rainwater can be discharged on the ground to allow it to percolate into the ground. This method of discharging rainwater on the ground was already discussed in the previous chapter.

12.8.2 Induced recharge from storm drain appurtenances

While disposing of excess rainwater from a building or storm water in a building complex through drainage piping, rain or storm water can be facilitated to percolate into the ground by taking some special measures, as described below. These measures may not be considered as the recharging methodology but obviously will facilitate percolation of rainwater into the ground whatever may be the quantity is.

Recharging through drain pipes: In pipes used for draining out rainwater through any underground sewer, it is suggested that the bottom half perimeter should remain perforated or cut at about 150 mm intervals, as shown in Figure 12.8. This is done for facilitating rainwater to percolate into the ground while flowing through it. Conventionally perforated pipes are suggested in the literature, which are generally found to be factory-made. But, the proposed cut system can be done at site. Furthermore, cutting half of the perimeter of pipe will produce more open area than perforating the same area of pipe.

Recharging through inspection pit: In the inspection pits of a sanitary or combined drainage system, the bottom slab is found to be a solid slab made of reinforced cement concrete, which is also found in inspection pits in storm sewers. To facilitate intrusion of rainwater, placing brick chips or gravel, as shown in Figure 12.9, can be done instead of putting a solid slab underneath in rainwater or storm water inspection pits. This type of inspection pit can be termed "recharge inspection pits" [5].

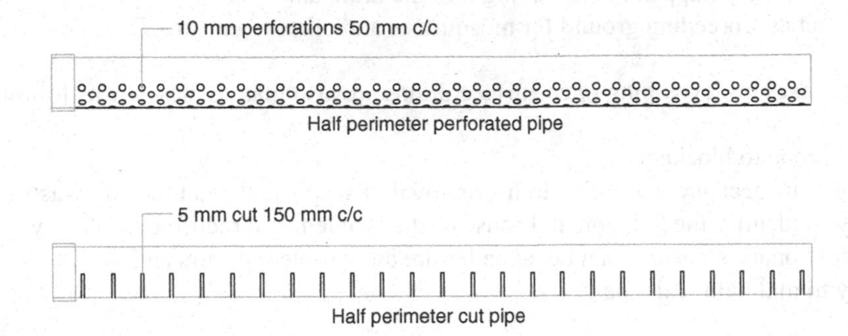

Figure 12.8 Half perimeter of rainwater drainage pipe perforated and cut.

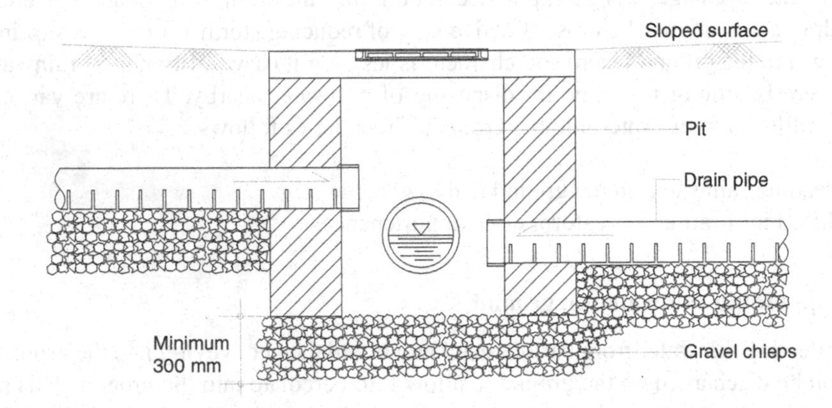

Figure 12.9 Gravel at bottom of rainwater inspection pit.

12.9 Sizing and placing of inspection pit or manhole

The size of an inspection pit or manhole chamber should be so designed that they are large enough to accommodate all of the connected pipes, allow a person to go inside (in the case of manholes) and perform an inspection or repair and maintenance of the piping system and allow access for a drain rod or water jetting hose to be inserted without physically entering the chamber.

The minimum internal dimension of an inspection chamber for a building sewer should be 0.5 meters and the minimum internal dimensions for a manhole should be 750 mm by 1000 mm for any depth less than 0.9 meters. The internal dimensions are to be enlarged depending on the size and depth of the sewer outlet in the chamber. The minimum internal dimensions of an inspection pit, depending on size and depth of outlet sewer, are given in Table 12.5.

On long straight lengths of sewer, additional inspection pits are to be provided at certain intervals as mentioned in Table 12.6.

In some special cases, as mentioned below, for pipes of a size smaller than or equal to 675 mm, the interval between manholes or inspection pits can be reduced to 60 m.

1 There is a possibility of frequent blockage in the flow
2 Opening of adjacent manhole covers at the same time may cause difficulties for traffic or pedestrian movement or
3 Pits located in narrow roads which are inaccessible to standard water-jetting units.

Table 12.5 The minimum internal sizes of the inspection pits and size of manhole covers.

Sl. No	*Invert of outlet pipe*	*Internal dimension*		*Manhole cover diameter*
		Width	*Length*	
1	0.6 m	0.5 m	0.5 m	450 mm
2	0.9 m	0.75 m	0.75 m	450 mm
3	2.1 m	0.9 m	1.2 m	550 mm
4	More than 2.1 m	Circular; minimum diameter 1.2 m		550 mm

Table 12.6 Intervals of inspection pits on long straight sewers [6].

Diameter of pipe (mm)	*Maximum intervals (m)*
Smaller than 675	80
Between 675 and 1050	100
Larger than 1050	120

12.10 Public sewer connections

No building sewer can be connected to a public sewer unless permitted by the concerned authority responsible for sewerage management. The authority makes sets of rules and regulations for this purpose; following those guidelines, building sewers are to be developed and necessary connections are to be made subject to obtaining the permit for the work.

12.10.1 Sanitary and combined building sewer connections

Building sewers convey wastewater to a disposal point from whence it is generally connected to the public sewer, or to any other sewer system that conveys the collected sewage from other buildings and establishments to the sewage treatment plant. The connection can be made simply by installing a drain pipe between the end of the disposal point of the building sewer and the nearest public sewer manhole, as shown in Figure 12.10, when the invert of the public sewer is below the invert of the building sewer's end point. The flow occurs under gravity. Building sewers are connected to the outside sewers in such a way that sewage gas and sewage cannot re-enter the building sewer. For this purpose, a trap is used for smaller diameter pipes of 100 mm to 150 mm or more, depending upon availability. The trap is termed an intercepting-trap, which is housed in a pit or chamber. For connections for larger diameter sewer pipes, an on-site constructed trap can be built up and may be termed a trap-pit.

When the invert of the building sewer's end-point at disposal location to the public sewer is below the invert of the public sewer at its connection point, then pumping of sewage would be needed at the disposal point of the building sewer.

Intercepting-trap: These are traps installed on a building sewer before connecting to a public sanitary sewer. These traps are supported by clean-out caps and housed in a pit, as shown in Figure 12.11. These pits, housing the intercepting-trap, are provided at the end of the building sewer, inside property line, in which the outlet pipe is connected to the nearest manhole of the public sanitary sewer, in the front road. These traps prevent ingress of sewer gas into the building sewer and, at the same time, provide provisions for inspecting the condition of flow to the public sewer.

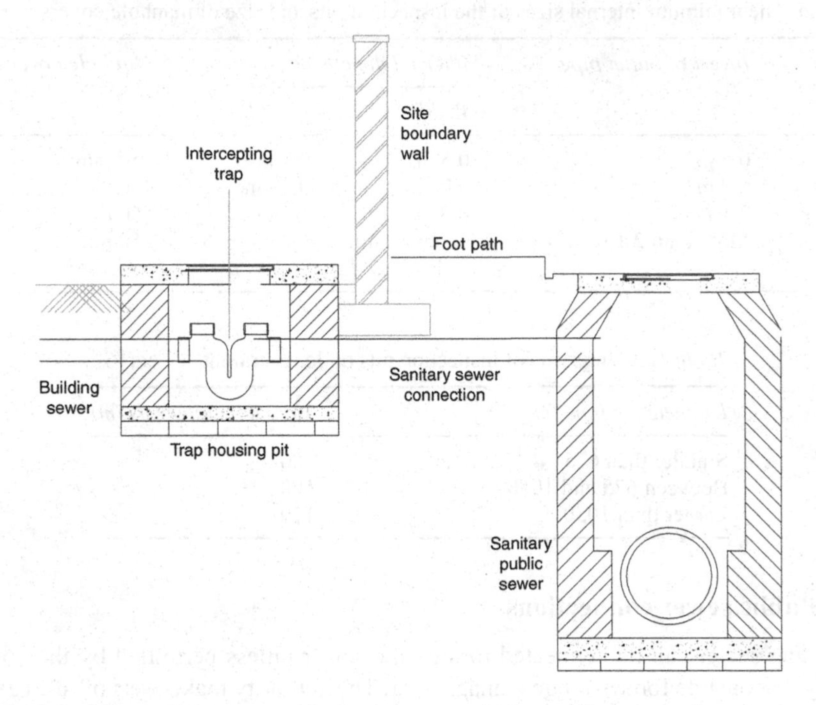

Figure 12.10 Building sanitary or combined sewer connected to public sewer using trap.

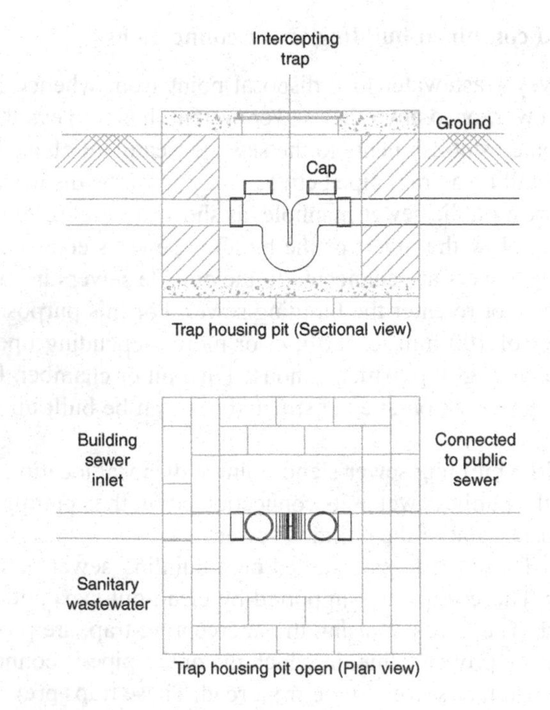

Figure 12.11 Intercepting-trap on building sewer connection pipe.

Trap-pit: These are pits constructed to function as a trap at the same time as inspection pit. In these pits, a baffle wall is provided to perform as a seal, as shown in Figure 12.12. The outlet pipe is connected to the nearest manhole on a public sanitary sewer, in the front road. The baffle wall is extended about 150 mm (6 inches) below the invert of the outlet pipe to act as a seal, thus preventing ingress of sewer gas into the building sewer. Two manhole covers are provided as provisions for inspecting the condition of flow and maintenance.

Sewage lifting: In all building premises in which the fall of the end of the building sewer is too low to permit gravity flow to the public sewer, sanitary sewage carried by such a building sewer shall be lifted by installing a swage pump and discharging into the public sewer, as illustrated in Figure 12.13.

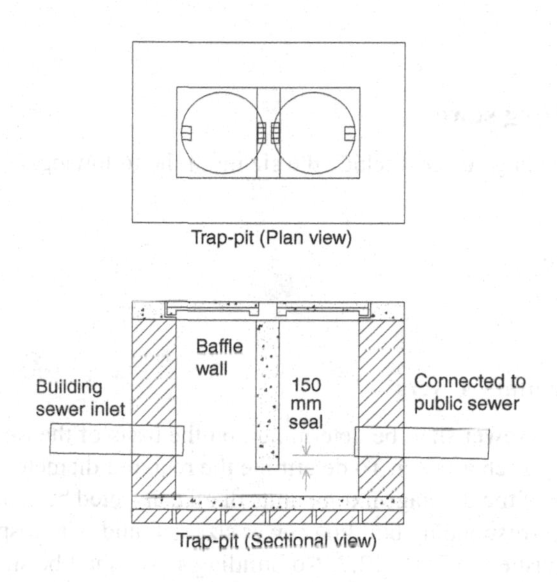

Figure 12.12 Trap-pit, plan and sectional view.

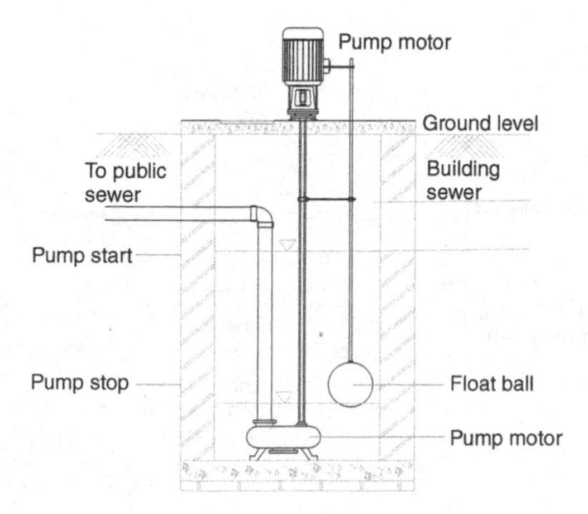

Figure 12.13 Sewage lifting to discharge into public sewer.

12.10.2 Building storm sewer connections

To connect a building storm sewer to the public storm sewer, a simple inspection pit or chamber is to be constructed at the end of the building storm sewer, inside the property line, as shown in Figure 12.14. The outlet pipe is to be connected to the nearest manhole on a public storm sewer on the road in front when the invert of the public storm sewer is below the invert of the end of building storm sewer. If a storm sewer is to be connected to a combined public sewer, then either a trap-pit or an intercepting-trap is to be installed instead, as described earlier.

While connecting the building storm sewer to the public storm sewer, in which it is found that the invert of the end of the building storm sewer is below the public storm sewer, then the disposal of storm water is to be made by installing lift pump, as mentioned earlier.

12.11 Sizing building sewers

Sizing of various building sewers include the sizing of the following.

1 Sanitary sewer
2 Storm sewer and
3 Combined sewer.

12.11.1 Sizing of sanitary sewers

The size of a building sewer shall be determined on the basis of the total number of fixture units to be drained by such a sewer. To determine the required diameter of a building sewer, the cumulative value of the drainage fixture units of all connected building drains or stacks is calculated, and the corresponding building sewer size is found with respect to the particular grade chosen, as furnished in Table 12.7. No building sewer shall be smaller than the building drain served.

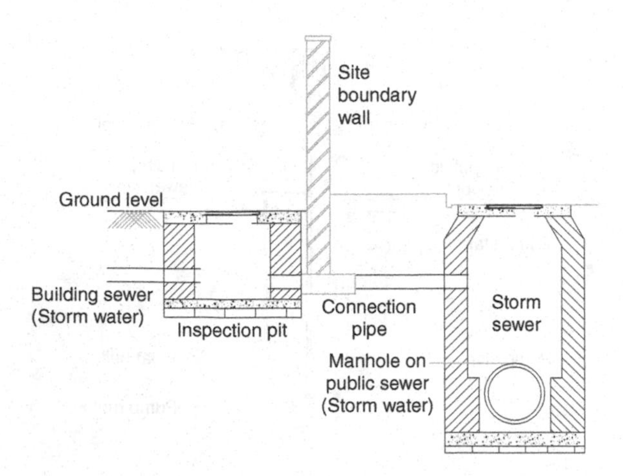

Figure 12.14 Building storm sewer connection to public storm sewer.

Table 12.7 Maximum permissible load, in drainage fixture units, for building sewer at particular grade or slope [7].

Sl. No	*Diameter of pipe (mm)*	*Maximum hydraulic load in drainage fixture units (DFU)*					
		Slope					
		1 in 400	*1 in 200*	*1 in 133*	*1 in 100*	*1 in 50*	*1 in 25*
1	150	–	–	600	700	840	1300
2	200	–	1400	1500	1600	2250	3370
3	250	–	2500	2700	3000	4500	6500
4	300	2 240	3900	4500	5400	8300	13000
5	375	4 800	7000	9300	10400	16300	22500

Table 12.8 Size of horizontal building storm drains and building storm sewer [8].

Size of pipe	*Maximum allowable horizontal projected roof areas (sqm) for and flow (l/s) in horizontal rainwater drain piping at various slopes*					
	10.4 mm/m		*20.8 mm/m*		*41.7 mm/m*	
mm	*sqm*	*l/s*	*sqm*	*l/s*	*sqm*	*l/s*
76	305	2.1	431	3.0	611	4.3
102	700	4.9	985	6.9	1400	9.8
127	1241	8.8	1754	12.4	2482	17.5
152	1988	14.0	2806	19.8	3976	28.1
203	4273	30.2	6057	42.7	8547	60.3
254	7692	54.3	10851	76.6	15390	108.6
305	12375	87.3	17465	123.2	24749	174.6
381	22110	156.0	31214	220.2	44220	312.0

Notes:
1 Table 10.5 is based upon a maximum rainfall of 25.4 mm/hour for 1 hour duration. The value for roof area of any locality shall be determined by dividing the area given in the table by the desired rainfall in mm/hour of that locality.
2 The sizing data is based on considering the pipe running in full.

12.11.2 Sizing of storm sewer

The size of a building storm drain, storm sewer or any of their horizontal branches shall be based on the maximum catchment area served, including projected roof or paved areas to be drained, in accordance with Table 12.8.

12.11.3 Sizing of combined sewer

When the combined sewer receives storm water from the surrounding land area, then the sewer pipe is to be designed considering its discharge capacity. This should be higher than the storm runoff accumulation and the flow rate corresponding to the drainage fixture units to be drained from the wastewater and soil stacks or building drains.

Storm water entering a sewer generates a load in a flow-rate like cusec or lpm, whereas stacks or building drains generate load as a drainage fixture unit (dfu). So, different load units must be converted to either a fixture unit value or a flow rate unit of cusec or lpm. The drainage fixture unit is considered equivalent to 0.47 lps of wastewater flow. On the contrary, for 100 mm/hr rainfall intensity, 0.36 sqm. roof surface area can be considered as 1 fixture unit [9]. The value for other rainfall intensity can be found by interpolation.

After determining the flow capacity of a drain pipe in cusec or lpm, the corresponding diameter of a pipe with respect to velocity of 0.6 mps can be obtained from a nomograph chart given in Figure 2.4 of Chapter 2. At the same time, the graph will give the corresponding fall for the diameter of drain pipe chosen, which is to be maintained.

12.12 Sizing surface drain

Development of effective surface drainage needs to be carefully planned and designed by taking into account the topography of the land and the peak runoff generated in the area to be served. Peak storm water runoff generation has already been discussed.

The surface drain shall be sized adequately to provide passage for conveyance of excess rain and storm water, based on climatic and soil conditions and the needs of the landscape.

$$\text{i.e. } Q_c > Q_r \qquad 12.3$$

Where
Q_c = discharge capacity of drain and
Q_r = peak runoff at the point of design.

Drain cross-section may be a variety of shapes but rectangular shapes are mostly designed, as shown in Figure 12.15.

The flow in drain Q = A.V

$$\text{So, V} = \text{Q/A} = \frac{1.486}{n} R^{2/3} S^{1/2} \qquad 12.4$$

Putting this value of V and R, it is found that

$$\text{Q} = \frac{1.486}{\text{n}} \frac{(\text{W.D})^{5/2}}{(\text{W}+2\text{D})^{2/3}} \text{S}^{1/2} \qquad 12.5$$

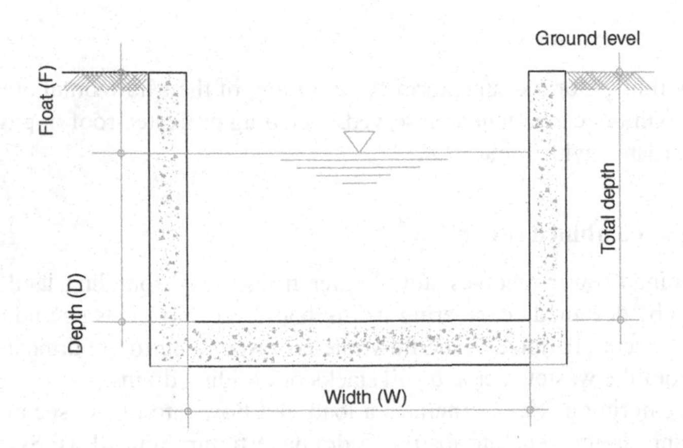

Figure 12.15 Rectangular surface drain section.

Now assuming value of W (width) and D (depth of flow), the right and left sides of the equation are satisfied. And thus, width and depth of flow are determined. After determining depth D, depth of float should be added to depth D, to determine the total depth of drain section. Float is considered 15 percent of flow depth.

12.13 Pumping drainage system

In a situation when the point or level of generation or accumulation of rain or storm water remains below the point or level of disposal, then either lifting or throwing of the storm or rainwater is required for draining. Sometimes, the storm water might need to be disposed of at a distant point from the site. In these cases, some form of energy will be necessary to dispose of or drain out the rain or storm water. Generally, a pump is used for these purposes. To ensure continuous pumping, there should be at least two pumps operating alternatively.

Depending upon the quality of the storm or rainwater, pumps of different characteristics may be required. For rainwater disposal, a normal centrifugal pump can be used, but for storm water containing heavy suspended particles or matter, sewage handling types of pump may be needed.

For facilitating storm water pumping, a detention or storage tank is to be constructed from where pumping is to be done. The storm water detention or storage tank should be of such volume that it can hold the runoff volume of a 2-hour storm generated from the site. The required volume of storm water storage may be reduced to a volume that can be discharged by pumping within a 30-minute period. But, in any case, the reduced volume of a detention tank cannot be less than 25 percent of the initial calculated storage volume, so as to allow for some storm water to remain in case of pump failure [10].

The pumping capacity shall be more than the storm runoff generation rate. To address immediate discharge of storm water ingress of intensity not less than 150 mm/hr, the pumping rate should be determined according to formula 12.7 [4].

$$P > \frac{IA}{3.6 \times 10^6} \qquad 12.6$$

Where
P = pumping rate (cum/s)
I = rainfall intensity (mm/hr) and
A = catchment area contributing to storm water generation (sqm).

References

[1] Escritt, L.B. (1971) *Sewers and Sewage Works*, Metric Calculation and Formulae, George Allen and Unwin Ltd, London, p. 20.

[2] Taylor, F., and Wood, W., eds. (1982) "Guidelines on Health Aspects of Plumbing" International Reference Centre for Community Water Supply and Sanitation (IRC) Technical Paper Series No. 19. P-59, The Hague, IRC.

[3] 911metallurgist (2015) "How to Calculate Property Run-Off" by David Michaud, Retrieved from www.911metallurgist.com/blog/how-to-calculate-rainwater-property-run-off on 29 December 2020.

[4] PUB Singapore's National Water Agency (2011) *Code of Practice on Surface Water Drainage*, 6th Ed, Retrieved from www.pub.gov.sg/general/code/Documents/HeadCOPFINALDec2011-1.pdf on 07 December 2015.

[5] Haq, Syed Azizul (2017) *Harvesting Rainwater From Buildings*, Springer International Publishing, Switzerland.

[6] Government of Hong Kong (2013) *Sewerage manual, Key Planning Issues and Gravity Collection System*, 3rd Ed, p. 40, Retrieved from file:///C:/Users/pm/Downloads/Part_1_SEWERAGE_MANUAL_Key_Planning_Issu.pdf on 17 June, 2021.

[7] British Columbia Plumbing Code (BCPC) (2018) "Drainage System Section 2.4, Part 2 – Plumbing Systems, Division B: Acceptable Solutions" Retrieved from https://free.bcpublications.ca/civix/document/id/public/bcpc2018/bcpc_2018dbp2s24 on 15 December 2020.

[8] IAPMO Plumbing Code and Standards, India (2007) *Uniform Plumbing Code- India 2008*, IAPMO Plumbing Code and Standards Pvt Ltd, New Delhi, p. 170.

[9] McGuinness, William J., Stein, Benjamin, Gay, Charles Merrick, and Fawcett, Charles De van (1964) *Mechanical and Electrical Equipment for Buildings*, 4th Ed, John Willey & Sons Inc, New York, p. 80.

[10] Lake Macquarie City Council (2013) "Handbook on Drainage Design Guidelines" p. 11, Retrieved from www.lakemac.com.au/Development/Planning-controls/Local-Planning-Controls#section-5 on 22 December 2020.

13 Special concerns

13.1 Introduction

In thc preceding chapters, the very common and most wanted concerns of plumbing, like safety, economy and durability, were discussed where required. There are various other concerns of plumbing; these are to be addressed with equal importance, where applicable. In this chapter, all of the major concerns of plumbing, other than pressure management for safety, durability etc., are discussed herein, to provide a complete understanding of plumbing. Air locking in the piping causes an interruption in flow, thereby making the system inefficient, while freezing of water in the pipes in a very cold climate might cause huge damage.

13.2 Special concerns

All the major concerns other than pressure phenomena in plumbing development are considered in this topic as special major concerns. In many cases, these special concerns are either overlooked or not properly addressed. These major special concerns of plumbing are as follows.

1. Noise
2. Vibration
3. Contamination of water
4. Air locking
5. Water hammer
6. Freezing of water
7. Corrosion
8. Explosion
9. Seismic protection and
10. Security.

13.3 Noise

Faulty plumbing, even newly installed, can generate unwanted and unpleasant noise, which causes annoyance to users and interrupts the tranquillity of the living environment in a building, particularly late at night. A noise-free plumbing system is therefore highly demanding.

Sound or noise is mainly created in a plumbing system due to flow through restrictions, discontinuities, valves or any equipment or appliances. Noise is generated in these conditions by the following mechanisms: turbulence, cavitation, water hammer, splashing and dropping due to water flow.

DOI: 10.1201/9781003172239-13

Figure 13.1 Damage of impeller due to cavitation [1].

Turbulence: It has already been mentioned that turbulent flow occurs when the Reynolds number is greater than about 4000. When the velocity of flow becomes high enough to create turbulence in flow, noise is created.

Cavitation: Cavitation is a phenomenon of forming cavities (bubbles of air) and the subsequent collapsing within the flow of water through and past a restriction in the flow. Restriction causes localized high velocity and low pressure. When pressure is reduced to vapour pressure, about 18 kPa for water at 16°C, it creates bubbles. As these bubbles move past, the restriction velocity decreases and pressure increases, helping to grow the bubbles; when they come near or into contact with any surface, this cause the bubbles to burst, ultimately creating sound. Cavitation also creates vibration. Bursting bubbles exert pressure on surfaces, resulting in a pitting of the surface on which cavitation occurs. Figure 13.1 represents a damaged pump impeller due to excessive cavitation. Cavitation also occurs at discontinuities, as might happen within bathtub spouts, shower heads, etc.

Water hammer: The phenomenon of developing water hammer in water flow was already discussed in Chapter 6. Water hammering can happen not only in flow through pipes but also during the wash and rinse cycle of a clothes washing machine or dishwasher etc.

13.3.1 Control of noise

To control the generation of noise or to bring its level down to an acceptable limit, the following measures are to be taken.

1. **Pressure management:** In a water supply system, the limit of pressure is to be maintained between 100 kPa and 500 kPa. To minimize generation of noise, pressure should be maintained at 230 kPa.
2. **Pipe transition:** A sharp transition of pipe should be avoided; instead, pipe transitions using 45° bends or wye is preferred.
3. **Isolating:** Noise is transmitted due to contact of noise-generating elements with radiating surfaces like walls, floors etc. Therefore, these elements should not be installed in

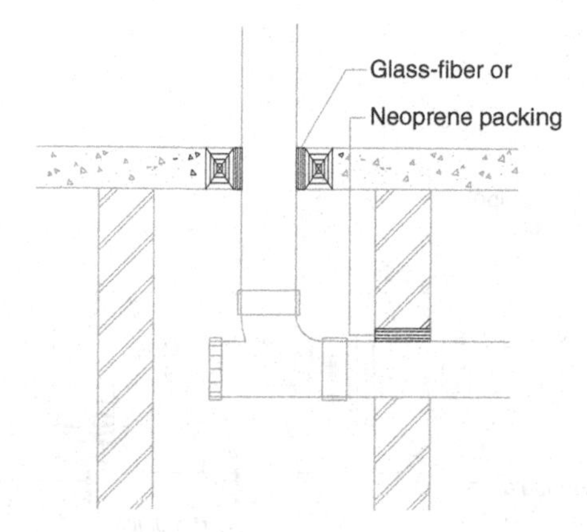

Figure 13.2 Isolation of pipe by packing.

contact with structural building elements. A pipe isolator of at least 6 mm thick, made of resilient material like neoprene or fibreglass, should be used in filling the isolating gaps, as shown in the case of piping running through structural building elements, illustrated in Figure 13.2. Vibrating elements shall also be isolated by providing vibration isolators.

4 **Jacketing:** Pipes may be jacketed by foam or fibreglass insulating material covered with a 1.5–3 mm thick aluminium metal jacket, secured by adhesive or tape, to reduce noise level to between 6 and 10 decibels (dB).
5 **Pipe shaft:** Installing a pipe inside a pipe shaft reduces transmission of noise considerably.
6 **Vibration controlling:** Vibration can be well controlled by using a water hammer arrestor on a vibrating pipe.

13.4 Vibration

Vibration is mostly caused by the reciprocating motion from rotating components within any mechanical equipment used in plumbing. Vibration in some plumbing elements is another major concern for the safety and comfort of building occupants. Pipes and some machinery, such as pumps and washing machines, may vibrate. Most commonly, vibration in pipes is attributed to the following:

1 Loose pipe installation
2 High water pressure and
3 Water hammer.

Among the machinery, pumps are the major contributor to vibration generation. A pump may cause vibration for the following reasons.

1 Cavitation
2 Misalignment of the shaft
3 Impeller imbalance and
4 Bearing failure.

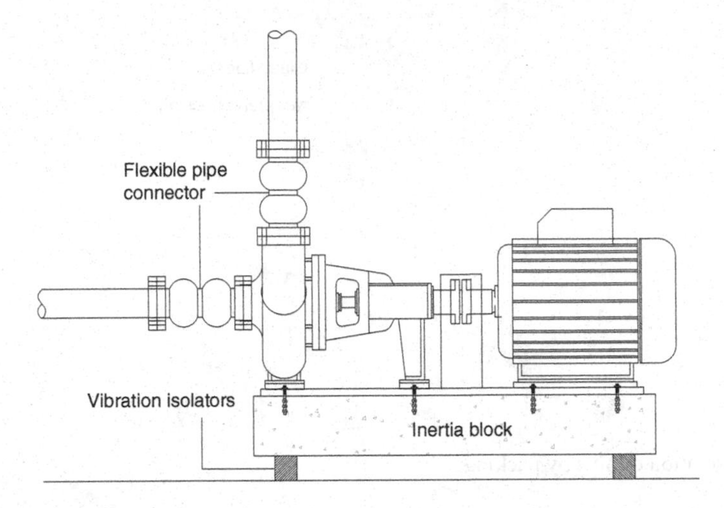

Figure 13.3 Isolation of a pump from building structure, by putting on inertial block.

It is hard to control vibration but easy to influence it by taking corrective measures, starting with spotting the problem at an early stage of occurrence. Transmission of vibration can be controlled by taking the following measures.

1 All vibrating plumbing elements should be installed, keeping those isolated from its support to reduce transmission of vibration into the structure
2 All vibrating machines or equipment shall be supported on vibration isolation mounts and placed over elastomeric pads, as shown in Figure 13.3, and
3 Pipe joints shall be made flexible and pipes shall be hung by vibration isolation hangers.

13.5 Contamination of water

In a building's water supply system, potable water in the piping might be contaminated by coming into contact with potential sources of pollution remaining outside. This phenomenon of water contamination mostly occurs due to cross-connection and temperature rise. There are three ways for cross-connection to occur: 1. back siphonage, 2. counter-pressure and 3. backflow.

Back siphonage: Back siphonage may occur by developing negative or sub-atmospheric pressure within a water distribution piping system. Back siphonage causes backflow of water. So, contaminated water can have backflow into the water distribution piping, thus contaminating water for distribution.

Counter pressure: Counter-pressure may occur when any device or equipment operates at a pressure level above that of the potable water distribution system. For example, contaminated hot water containing Legionella bacteria from the geyser, operating under higher pressure, can contaminate the potable cold water in the distribution network.

Backflow: Contaminated water can get into the potable water distribution piping if sufficient preventive measures are not taken. It may so happen, particularly in the case of open-top water retaining fixtures, that contaminated water, during using the retained water which can get into the faucets spout located just above the retained water surface, when sufficient height is not maintained in between the faucet end level and water surface level in a fixture like basin.

Temperature rise: According to the water supply regulations, no cold water supplies should be warmed above 25°C. This requirement is to ensure the possibility of the growth of bacteria in water, such as *Legionella*.

13.5.1 Prevention of contamination

The possible ways of contaminating potable water in distribution piping must be well addressed during the planning, design and installation of concerned parts of the water supply system. The effective measures to be taken are discussed below.

Air gap: An air gap in a plumbing system is the physical vertical separation between the free flowing discharge end of the potable water-supplying appliance and the overflow rim of an open or non-pressure receiving vessel. This is the effective and economical means of preventing backflow. It is necessary to provide a sufficient air gap between the outlet of any faucet or water supply pipe and open-top water retaining fixtures or containers. Air gap in a water supply system for plumbing fixtures is considered as the vertical distance between the supply fitting outlet (spout) and the highest possible water level in the receptor (when flooded), as shown in Figure 13.4. The air gap required for preventing backflow through a water supply opening (faucet or valve) under the action of atmospheric pressure and a vacuum in the water supply system depends principally on the effective opening. The common practice of keeping an air gap is at least twice the inside diameter of the inlet pipe, but never less than 25 mm. The recommended air gap for various fixtures is given in Table 13.1.

Backflow devices: To avoid contamination, cross connections must either be physically disconnected or approved backflow prevention devices to be installed, to protect the water supply system. There are four types of backflow preventing devices, as mentioned below.

1 Atmospheric vacuum breaker
2 Pressure vacuum breaker
3 Double check valve and
4 Reduced pressure principle backflow preventer.

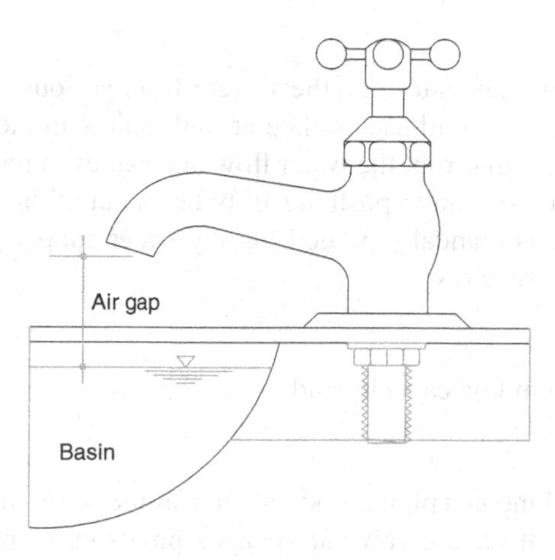

Figure 13.4 Air gap for supply fixtures.

Table 13.1 Recommended air gap for common plumbing fixtures [2].

Fixture	*Minimum air gap*	
	Away from a wall[a] (mm)	*Close to a wall (mm)*
Lavatories and other fixtures with effective openings not greater than 12 mm in diameter	25	40
Sinks, laundry trays, gooseneck back faucets and other fixtures with effective openings not greater than 19 mm in diameter	40	50
Over-rim bath fillers and other fixtures with effective openings not greater than 25 mm in diameter	50	75
Drinking water fountains, single orifice not greater than 11.11 mm in diameter or multiple orifices with a total area of 97 sqmm (area of circle 11.11 mm in diameter)	25	40
Effective openings greater than 25 mm	Two times the diameter of the effective opening	Three times the diameter of the effective opening

a Applicable where walls or obstructions are spaced from the nearest inside-edge of the spout opening a distance greater than three times the diameter of the effective opening for a single wall, or a distance greater than four times the diameter of the effective opening for two intersecting walls.

Preventing stagnation of water: If water in reservoirs or any non-flowing pipe remains stagnant for long period, then there is a possibility of growing *Pseudomonas* bacteria in stagnant water. One variety of this bacterium causes *Pseudomonas aeruginosa* infections when consumed. To avoid stagnation of water in a reservoir, the reservoir should be sized properly and the inlet and outlet or suction pipe end should be located in such a way that the maximum distance of flow occurs from inlet to outlet points in the tank.

If there are any abandoned branch pipes remaining in a water distribution system, which are not in use and remain capped, then the branch pipe must be removed from the main supply pipe.

13.6 Air locking

In water supply piping, air can enter into the system from various sources. When air enters into a water supply system, it will accumulate at high points in the piping network. These pockets of accumulated air interrupt the water flow and can even restrict the flow. If there is not enough pressure in the system to push the air bubbles out of the pipe, the air will remain locked until the pipeline is manually purged. In any water supply piping, air may enter in various ways as mentioned below.

1 Through vent pipes
2 Water tanks running in low capacity and
3 Water as it is heated.

To get relief from air locking in a piping system, various measures are taken. The most effective way is to install an air release valve at the apex points in the piping system. In case of low pressure in flow, pipes should be graded to allow air to exit from predetermined high points, thereby preventing of air locks inside the pipe from occurring.

13.7 Water hammer

Water hammer is the phenomenon of developing shockwaves in the water flow due to a sudden fluctuation in flow velocity, generally caused by a sudden stopping of the flow of water in pipes under high pressure. Water hammer is also known as a pressure surge, which develops and causes the pipe to vibrate, making a "hammering" sound in metal pipes and to some extent in plastic pipes. Violent action of water hammer can cause damage to the piping system, valves, faucets and other devices used in the system.

Fast-acting faucets and gate valves operated by lever, spring-closing valves and pumps often cause water hammer. Water hammer is related directly to the velocity of water flow – the more the velocity of water flow, the greater likelihood of developing water hammer. To avoid water hammer, pipe sizing should be done judiciously so that flow velocity is limited to below 3.0 m/sec and system pressure must not be more than 350 kPa.

13.7.1 Arresting water hammer

To reduce the impact of water hammer, the following measures should be taken in pipe installation.

1. Use water hammer arrestors at proper locations needed
2. Direct contact of pipes with structural elements should be avoided
3. Pipes should be clipped with rubber insulated clips or over the pipe insulation
4. Pipe work shall be rigidly fixed to prevent its movement
5. Relief bends or flexible sections of pipe should be provided to absorb shock
6. Fit cushioned packers where pipes pass through structural members and
7. Use water hammer arrestors on piping, which was discussed in the previous chapter.

13.8 Freezing of water in piping

If the water supply system remains unused for any reason for a long period of time during cold weather and the heating system remains turned off, the water pipes might freeze. When the water remains no-flowing, the temperature can lower quite quickly and freeze the water inside, which results in increase in the volume of ice exerting pressure on the pipe, accessories, joints etc.; these may therefore be damaged. Necessary measures must be taken to avoid the problems arising from freezing of pipes, as discussed below.

13.8.1 Precautionary measures against freezing of water

To avoid freezing of water in the piping, the following precautions are to be followed.

1. Shut off the water supply at the service pipe by closing the gate valve. Open all faucets and devoid the piping of water. Make sure all horizontal pipes are drained properly.
2. Close the controlling cocks or gate valves after the faucets have run dry.

3 Remove all water in the traps under sinks, water closets, urinals, bathtubs and lavatories by opening the clean-out plugs of traps and drain out. Use a force pump or other method to siphon the water out from traps where required. Fill all traps with a non-freezing solution such as mineral oil, windshield washing fluid etc.
4 Clean out all water from the flush tank.
5 Drain out water from hot water tanks.
6 Drain out water from the heating pipes and heater. Burner and the main water supply must be turned off.
7 Every radiator valve must be opened to release condensation.

13.9 Corrosion

Corrosion is a chemical reaction which generally occurs when one metal comes into contact with another metal in the presence of water, oxygen, moist air, soil etc. and ultimately disintegrates the metal forming rust. Corrosion gets worse once it starts and continues, reducing the strength and causing damage. Corrosion causes damage to the plumbing elements and thus reduces life and causes leakage as well. Corrosion also causes contamination of water, thereby posing health hazards, making water odorous and affecting dishes, laundry and even outdoor gardening. Lead and copper are examples of toxic metals that can leech into the water if pipes and containers are corroded. Corrosion can also damage water heaters. It can also affect plumbing fixtures, creating unsightly stains and strange odours that never seem to go away.

13.9.1 Preventing corrosion

There are ways to keep corrosion from creeping into piping or non-piping surfaces, as mentioned below.

1 **Water quality:** Water pH value should be maintained between 6.5 and 7. Pitting corrosion in copper pipe occurs when pH is above 7. Dissolved oxygen level shall be less than 7 ppb to prevent corrosion. The temperature of water is another cause of corrosion. Above 60°C, the rate of corrosion doubles for every 266K increase of temperature.
2 **Keep pipes clean:** Microbiologically induced corrosion (MIC) happens when metals are exposed to corrosive bacteria, which may be aerobic or anaerobic. *Ferrobacillus ferrooxidans* bacteria directly oxidizes iron to iron oxides and iron hydroxides, and microbial corrosion can also apply to plastics, concrete and many other materials.
3 **Apply protective coating:** Protective coatings can prevent corrosion in metal surfaces. For instance, galvanization is a layer of zinc applied to iron or steel surfaces. Painting also protects corrosion.
4 **Keep structures stable:** Stress corrosion occurs in metals due to development of stress concentration, mostly due to bending, over-tightening, hammering, tensile stressing on fittings etc.
5 **Protect against metal-to-metal contact:** Galvanic corrosion occurs when one metal of higher galvanic order acting as a cathode pulls electrons from another metal of lower galvanic order acting as anodes, in the presence of an electrolytic medium which allows electrons to flow from cathode to anode. Things like metallic U-bolts, straps and clamps in contact with metal pipes can lead to corrosion.

13.10 Explosion

Explosion is a serious concern of plumbing, though it is not a very common phenomenon. But there are incidences of explosions due to overlooking the condition of some plumbing elements of concern. There are countless examples of plumbing explosions occurring for decades.

Water heaters are one of the most well-known causes of plumbing explosions, which was discussed in the hot water supply chapter. Following are some other types of causes for plumbing explosions that must be well cared for.

Explosions might be caused by the natural gas used in plumbing systems. In particular locations, many types of pipes may run side by side, like water, natural gas, wastewater pipes etc. The explosion might occur when one purges a gas line before connecting the proper fixture. As a result, the enclosed space for working in can be filled with natural gas, which can lead to a massive explosion after being triggered accidentally by any sort of open flame.

Due to leakage of gas from a gas water heater accumulating in a closed room, this could cause an explosion when the gas comes into contact with any flame. If any sort of hissing sounds around the heater are noticed, that might be an indication of a gas leak. Depending on where the leak is located, it can also cause a blowing noise. Corrective measures should be taken immediately to avoid any accident.

A toilet explosion might be an ultimate fate of faulty plumbing, poor inspection and ignorance. Toilets have sanitary drains supported by water seal traps. There are 50 mm water seals in each trap that prevent sewage gas, containing methane and hydrogen sulphide, from entering the toilet room. If any toilet, for any reason, is not used for a long time, the water in the traps dry off, allowing easy entrance of sewage gas to fill the toilet room. Methane and hydrogen sulphide are explosive gases. The toilet room, thus filled with explosive gasses, might explode if it in any way contacts an ignition source like a lit-up cigarette, candle or a faulty electric element in the toilet sparking when switched on. If it is considered that there is a linear loss rate of 6 mm of water per day [3], then it will take only about eight days to make the traps completely devoid of water. So, a toilet unused for more than eight days might be subject to such an explosion. The impact of such an explosion might range from causing death due to burning of anyone using the toilet, to property damage, subject to the pressure of accumulated gas. To avoid occurrence of any accident by explosion, the following preventive and corrective measures shall be taken.

1 Noticing the presence of any foul gas by having smell similar to smell of rotten eggs, inside a toilet
2 Flushing toilets at an at least once a week (seven days) interval while not using any toilets for long
3 Pouring a small amount of mineral oil into the trap to just cover the trap's water surface; vegetable oil may be an alternative option if mineral oil is limited
4 Keeping windows open while toilets are not in use
5 Not entering any toilet with an open flaming light just after opening the toilet door and
6 Frequently checking the healthiness of electrical connections and elements inside the toilet.

Manholes, sanitary or combined sewers, septic or other anaerobic tanks are also exceptionally dangerous place to work inside without taking any precautionary measures, mainly due to presence of noxious sewer gases, the majority of which are explosive and life threatening.

Entering into and working in these structures must be done by trained personnel with "confined spaces" and "sewer working" safety licenses.

13.11 Seismic protection

The crust and the top of the mantle make up a thin skin on the earth's surface. This crust is not a continuous covering of the earth but is divided into many pieces, called tectonic plates, which keep slowly moving around, sliding past and bumping into each other. An earthquake is what happens when two plates of the earth suddenly slip past one another. The surface where they slip is called the fault or fault plane. The energy radiates outward from the fault in all directions in the form of seismic waves. The seismic waves shake the earth as they move through it, and when the waves reach the earth's surface, they shake the ground and anything on it, including buildings. Shaking in buildings causes horizontal forces on every structural element of buildings and all the building service elements attached to the structure. The horizontal seismic force acting on elements attached to building elements varies with weight and elevation, along with other factors, as formulized below [4].

$$F_p = \frac{Z}{2}\left(1+\frac{x}{h}\right)\frac{a_p}{R_p} I_p W_p \qquad 13.1$$

Where
Z = zone factor
x = height of point of attachment of the equipment above the foundation of the building (m)
h = height of the building (m)
a_p = amplification factor of the equipment
R_p = response modification factor
I_p = importance factor and
W_p = weight of the equipment (kN).

F_p is the design seismic tensile or shear force on anchors or bracings of a plumbing element to be restrained, in kN. This force is assumed to be equally shared by all the bolts or bracings to be used for a particular element.

In earthquake-prone areas, planning, design and construction make buildings structurally safe during a seismic event. It is observed that earthquakes damage building service elements, including plumbing elements, more than building structural elements due to not considering the force acting on these plumbing elements. So, seismic protection for ductwork and plumbing items are essential for both financial and life safety reasons. For plumbing developed in earthquake-prone areas and in buildings where all sorts of safety measures are to be ensured, plumbing items should be installed with equal importance and considering all aspects of seismicity related to plumbing.

The core purpose of seismic protection of plumbing items is basically to restrict horizontal shaking force resulting from an earthquake on these items. So, seismic braces are employed to firmly attach equipment to structural members of a building, so that they move with the structure during an earthquake. This prevents the equipment from tipping over, falling from where it is suspended or colliding with other objects.

Adoption of seismic bracing strategy is the most effective measure for seismic protection of plumbing items. Seismic braces must be able to withstand and resist the expected seismic forces in their area. The magnitude of the seismic forces depends on two factors: mass and acceleration, as the formula $F = ma$; force equals mass times acceleration. So, heavy equipment experiences a bigger force when the greater ground acceleration is generated by the earthquake.

In practice, mounting plumbing items with respect to building structural members might occur in several ways, as mentioned below.

1 Rigid mounting to the floor or a pad
2 Roof mounting
3 Wall mounting and
4 Suspension from the ceiling.

13.11.1 Seismic protection for equipment

To prevent damage of equipment or any vertical items, seismic bracing is necessary, which involves the following.

1 Bolting equipment directly to structural members
2 Using rigid bracing to fix equipment to the structure
3 Using bumpers to prevent the equipment from moving (when there is no risk of the equipment tipping over) and
4 Using a vibration isolator to dampen the effects of shaking on equipment; these are commonly found in the form of heavy-duty springs.

Figure 13.5 illustrates how to anchor pump and vertical tanks with supporting structural elements with anchor bolts and bumpers.

13.11.2 Seismic bracing for suspended pipe

During an earthquake, the hangers used for suspending plumbing items, pipes in particular, experience significant stress because of shaking, and the pipe itself may deform, collide with other objects or fall from its mounting point. Seismic restraining for horizontal pipes, ducts or shafts is therefore needed. Seismic restraints for pipe and duct are separated into two categories.

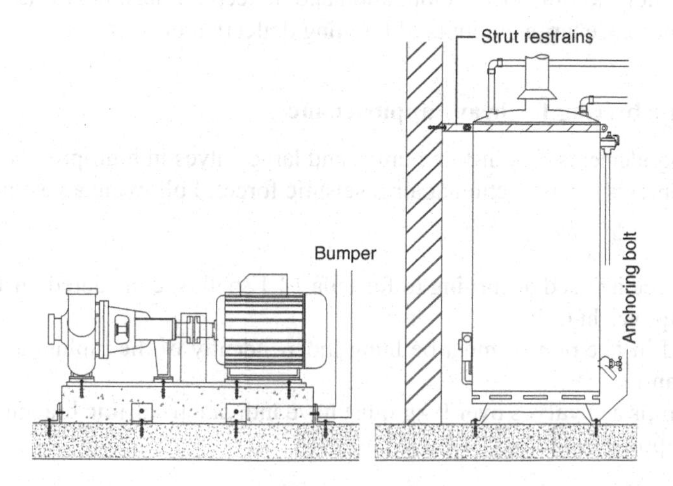

Figure 13.5 Anchoring a pump and a vertical water heater with supporting structural elements by anchor bolts and bumpers and strut restrainer.

1 **Transverse seismic restraints (T):** These act to prevent the pipe or duct from swinging side-to-side. They are normally placed perpendicular to the pipe or duct. The word lateral is often used for transverse when describing these restraints.
2 **Longitudinal seismic restraints (L):** These act to prevent the pipe or duct from swinging back-and-forth along the length of the pipe or duct. They are usually placed parallel to the pipe or duct. The word axial is also used when describing this type of restraints.

Seismic restraints for pipe and duct may be further broken down into three basic types based on the way they operate.

1 **Strut restraints (rigid braces).** These restraints carry both tension and compression loads along the axis of the strut. Only one strut is required to restrain a pipe or duct in one direction, either transverse or longitudinal.
2 **Cable restraints (only tension braces).** These restraints carry only tension loads along the axis of the cable. They are used in pairs 180° apart to restrain the pipe or duct in one direction, either transverse or longitudinal.
3 **Post restraints (omni-directional braces).** These restraints carry horizontal loads acting from any direction. One post will be required for each restraint location and can be used to restrain the pipe or duct in both the transverse and longitudinal directions.

For cable or strut restraints, the installation angle may be between 60° and 45°, as measured from the horizontal. The transverse and longitudinal seismic restraints are to be placed at or near the hanger locations for the pipe or duct. Figure 13.6 illustrates various seismic restraints for pipes.

Vertical plumbing pipes of high-rise buildings are subject to deflection due to shaking of buildings during earthquake. If the pipes are not adequately accommodated for deflection subject to building height and magnitude of earthquake, repeated stress on the piping system can cause damage to pipes and connected equipment. To accommodate total deflection, flexible grooved couplings or simple flexible connections at certain intervals provide flexibility to accommodate pipe movement in long vertical straight run pipes. There are also some special seismic fittings which are basically a combination of sleeve and ball-joint arranged in the fittings that help accommodate elongation and deflection. Seismic fittings can be used in series to accommodate large amounts of building deflection or drift.

13.11.3 Seismic bracing for heavy appurtenances

Some plumbing elements like in-line pumps and large valves in high-pressure systems need special requirements for protection against seismic forces. Following are some guidelines in these respects.

1 Seismic force induced according to formula 13.1 shall be considered, instead of considering pump weight only
2 Suspended in-line pumps must be hung independently of the piping, as shown in Figure 13.7, and
3 In line pumps and valves over 9 kg must have individual seismic bracing when the piping has importance factor (I_p) 1.0 [5].

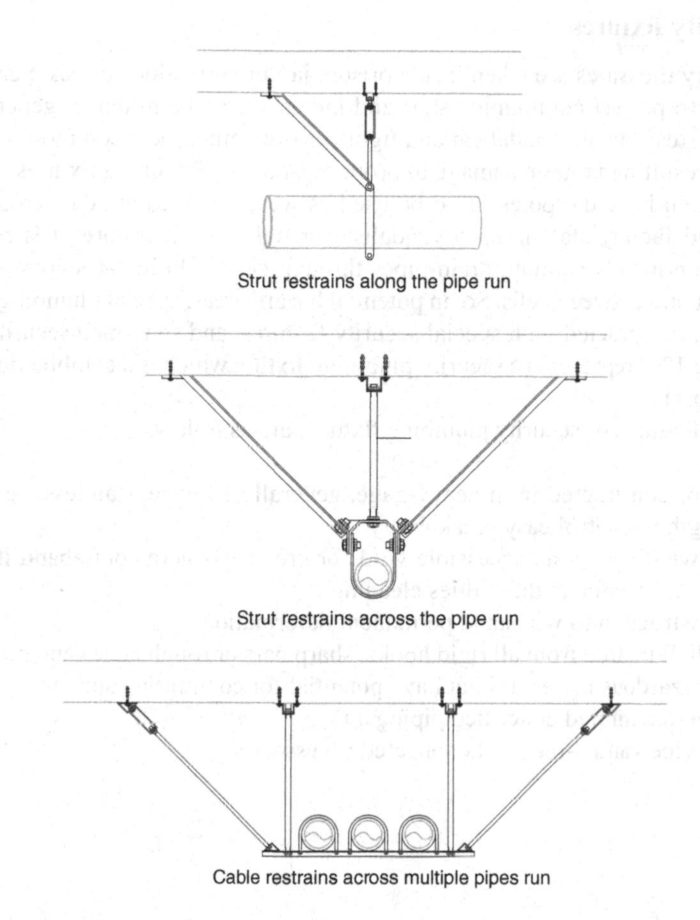

Figure 13.6 Seismic protection for suspended pipes.

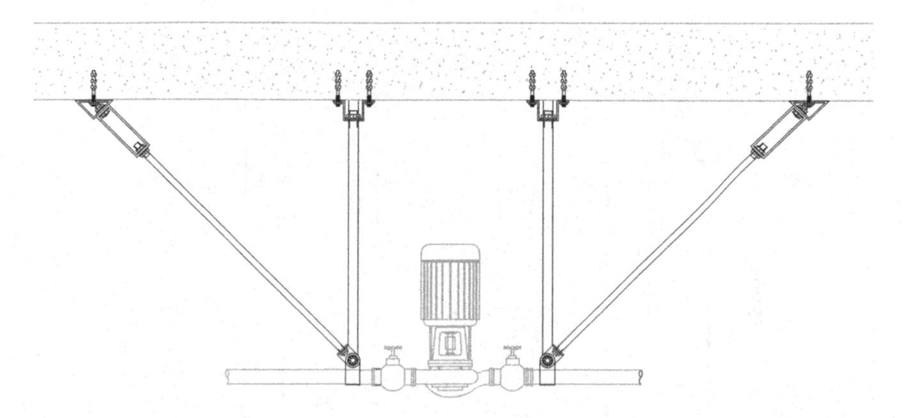

Figure 13.7 In-line pump hung independently of the piping

13.12 Security fixtures

A lot of security measures are taken inside prison, jail or correction centres. Security measures are taken to protect community, staff and inmates from harm that is generally caused by violent inmates. Inmate vandalism and fights or riots might be a common occurrence in these centres, resulting in huge damage to property and life. Plumbing fixtures that are vandalized or broken have the potential to be used as weapons, creating dangerous situations for inmates and facility staff in such vandalism or fights. Furthermore, it is reported that exposed and continually running drain pipes through cells help inmates communicate and transfer contraband between cells. So, in potential hazard areas, special plumbing fixtures are to be provided, constructed with special security features, and so termed security plumbing fixtures. Figure 13.8 represents a security plumbing fixture which is a combination of a basin and a water closet.

The special features of security plumbing fixtures are as follows.

1. Fixtures are constructed from heavy-gage, generally 14 gage, stainless steel sheets for high strength to control easy breakability
2. Seamless welding has no accessible voids or crevices where contraband items can be concealed and eliminate difficulties cleaning
3. Sturdy construction to withstand permanent deformation.
4. Fixtures shall be free from all rigid hooks, sharp bars or rough edges and projects which could be hazardous in general but have potential for committing suicide
5. Factory pre-assembled concealed piping and
6. Severe service vandal-resistant connected accessories.

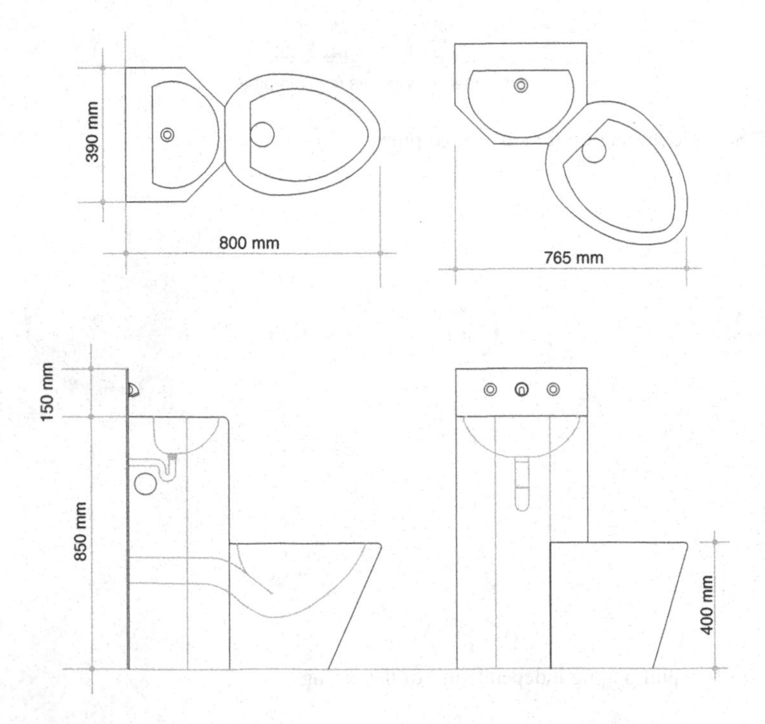

Figure 13.8 A security plumbing fixture combining a basin and a water closet

References

[1] Pricast (2009) "Cavitación en el bombeo de fluidos" Retrieved from file:///C:/Users/pm/Pictures/ESTexs%20Workshop/34521-Cavitacion-en-el-bombeo-de-fluidos.html, cited in Development of a Prototype Equipment for Cavitation Testing, Conference Paper, January 2017, Retrieved from https://www.researchgate.net/publication/324669223.

[2] Board of Building Standards (2018) "Ohio Plumbing Code" Chapter 4101:3–6. Water Supply Systems, Retrieved from http://codes.ohio.gov/oac/4101:3-6-01 on 29 December 2020.

[3] Wanamaker, Christopher (2019) "How to Calculate Water Evaporation Loss in a Swimming Pool" Retrieved from https://dengarden.com/swimming-pools/Determine-Evaporation-Rate-for-Swimming-Pool#:~:text=For%20an%20even%20easier%20and,gallons%20of%20water%20per%20day on 29 December 2020.

[4] Jain, Sudhir K. (UD) "Explanatory Examples on Indian Seismic Code IS 1893 (Part I)" Final Report: A – Earthquake Codes, p. 16, Retrieved from https://www.iitk.ac.in/nicee/IITK-GSDMA/EQ21.pdf on 21 June 2021.

[5] Quick Response Fire Supply (QRFS) Team (2020) "Guides for Fire Protection Equipment and Fire Safety Systems" Retrieved from www.qrfs.com/blog/329-seismic-bracing-for-ductwork-hvac-electrical-systems/ on 29 December 2020.

Appendices

Appendix A1

Dimensions and nominal mass of medium quality steel tubes as per IS 1239

PNC tube and pipe industries

Circular hollow sections | *Hot dip galvanized: BS 1461:1999*

British Standard Welded Steel Pipe

Thickness class	*Normal size*		*Outside diameter*		*Wall thickness*	*Calculated weight (plain end)*		*Hexagon packing*
			max.	*min.*				
	mm	*in.*	*mm*	*mm*	*mm*	*kg/m*	*kg/6m*	*pcs*
			D	D	t			
Class A (Light)	15	1/2"	21.4	21.0	2.0	0.947	5.68	91
	20	3/4"	26.9	26.4	2.3	1.38	8.28	91
	36	1"	33.8	33.2	2.6	1.98	11.88	91
	32	1–1/4"	42.5	41.9	2.6	2.54	15.24	44
	40	1–1/2"	48.4	47.8	2.9	3.23	19.38	44
	50	2"	60.2	59.6	2.9	4.08	24.48	44
	65	2–1/2"	76	75.2	3.2	5.71	34.26	19
	80	3"	88.7	87.9	3.2	6.72	40.32	19
	100	4"	113.9	113.0	3.6	9.75	58.50	10
Class B (Medium)	15	1/2"	21.7	21.1	2.6	1.21	7.26	91
	20	3/4"	27.2	26.6	2.6	1.56	9.36	91
	36	1"	34.2	33.4	3.2	2.41	14.46	91
	32	1–1/4"	42.9	42.1	3.2	3.10	18.60	44
	40	1–1/2"	48.8	48.0	3.2	3.57	21.42	44
	50	2"	60.8	59.8	3.6	5.03	30.18	44
	65	2–1/2"	76.6	75.4	3.6	6.43	38.58	19
	80	3"	89.5	88.1	4.0	8.37	50.22	19
	100	4"	114.9	113.3	4.5	12.2	73.20	10
Class C (Heavy)	15	1/2"	21.7	21.1	3.2	1.44	8.64	91
	20	3/4"	27.2	26.6	3.2	1.87	11.22	91
	36	1"	34.2	33.4	4.0	2.94	17.64	91
	32	1–1/4"	42.9	42.1	4.0	3.80	22.80	44
	40	1–1/2"	48.8	48.0	4.0	4.38	26.28	44
	50	2"	60.8	59.8	4.5	6.19	37.14	44
	65	2–1/2"	76.6	75.4	4.5	7.93	47.58	19
	80	3"	89.5	88.1	5.0	10.30	61.80	19
	100	4"	114.9	113.3	5.4	14.50	87.00	10

pnc.pipetech.doc/abc
Source: www.pnc.com.my/tube&pipeproducts2.htm

Appendix A2

uPVC pressure (water supply, irrigation and industrial use)

Nominal size (mm)	*Outside diameter (mm)*		*Wall thickness (mm)*									
			Class B		*Class C*		*Class D*		*Class E*		*Class O*	
	Min	*Max*	*Min*	*Max*	*Min*	*Max*	*Min*	*Max*	*Min*	*Max*	*Min*	*Max*
12	21.2	21.5										
19	26.6	26.9										
25	33.4	33.7										
32	42.1	42.4					2.2	2.7	2.7	3.2		
40	48.1	48.4					2.5	3.0	3.1	3.7	1.8	2.2
50	60.2	60.5			2.5	3.0	3.1	3.7	3.9	4.5	1.8	2.2
63	75.0	75.3			3.0	3.5	3.9	4.5	4.8	5.5	1.8	2.2
76	88.7	89.1	2.9	3.4	3.5	4.1	4.6	5.3	5.7	6.6	1.8	2.2
110	114.1	114.5	3.4	4.0	4.5	5.2	6.0	6.9	7.3	8.4	2.3	2.8
160	168.0	168.5	4.5	5.2	6.6	7.6	8.8	10.2	10.8	12.5	3.1	3.7
200	218.8	219.4	5.3	6.1	7.8	9.0	10.3	11.9	12.6	14.5	3.1	3.7

Pressure ratings: designated by the different classes at 20°C

Class	*B*	*C*	*D*	*E*	*O*
Bar	6	9	12	15	Non-pressure

Note:
2 percent of rated pressure should be reduced for each 1°C rise above 20°C.
Manufactured to: BS 3505/3506 Classes B, C, D & E, BS 3506, 1969 Class O.
Standard length: 5.8 and 6 meters.
Colour: dark grey (except Class O, which is grey).
Socket type: solvent weld/plain end.

Source: https://highstandardpipe.com/wp-content/uploads/2020/01/PVC-PIPE.pdf

Appendix A3

Taper pipe thread size; ASME B1.20.1 (NPT/API)

NPS	*Number of threads per 25.4 mm*	*Pitch of thread*	*Depth of thread*	*Truncation, max*	*Pitch diameter at plane of hand-tight engagement*
		P	*H*	*L*	*E*
		mm	*mm*	*mm*	*mm*
1/8"	27	0.941	0.753	0.091	9.489
1/4"	18	1.411	1.129	0.124	12.487
3/8"	18	1.411	1.129	0.124	15.926
1/2"	14	1.814	1.451	0.142	19.772
3/4"	14	1.814	1.451	0.142	25.117
1"	11.5	2.209	1.767	0.160	31.461
1–1/4"	11.5	2.209	1.767	0.160	40.218
1–1/2"	11.5	2.209	1.767	0.160	46.287
2"	11.5	2.209	1.767	0.160	58.325
2–1/2"	8	3.175	2.540	0.198	70.159
3"	8	3.175	2.540	0.198	86.068
3–1/2"	8	3.175	2.540	0.198	98.776
4"	8	3.175	2.540	0.198	111.433
5"	8	3.175	2.540	0.198	138.412
6"	8	3.175	2.540	0.198	165.252
8"	8	3.175	2.540	0.198	215.901
10"	8	3.175	2.540	0.198	296.772
12"	8	3.175	2.540	0.198	320.493
14"	8	3.175	2.540	0.198	352.365
16"	8	3.175	2.540	0.198	403.244
18"	8	3.175	2.540	0.198	454.025
20"	8	3.175	2.540	0.198	504.706
24"	8	3.175	2.540	0.198	606.066

Appendix A4

Hot water storage tank dimensions and capacities

Hot water storage capacity (litres)

Tank diameter (mm)	*Tank length (m)*									
	0.305	*0.61*	*0.914*	*1.22*	*1.52*	*1.83*	*2.13*	*2.44*	*2.74*	*3.05*
500	60.6	121	182	250	310	371	431	496	556	617
550	75.7	151	227	303	379	454	530	606	681	757
600	90.8	182	273	363	454	545	636	727	818	908
750	140	280	416	556	696	833	973	1113	1249	1389
900	201	401	602	802	1003	1200	1400	1601	1802	2002
1050	273	545	818	1090	1363	1635	1908	2180	2453	2725

Source: www.engineeringtoolbox.com/hot-water-storage-tank-capapacities-d_1676.html

Appendix A5

Estimating water demand against water supply fixture unit (WSFU)

Supply systems predominantly for flush tanks			*Supply systems predominantly for flush valves*		
Load	*Demand*		*Load*	*Demand*	
(WSFU)	*(gallons per minute)*	*(cubic feet per minute)*	*(WSFU)*	*(gallons per minute)*	*(cubic feet per minute)*
1	3.0	0.04104	–	–	–
2	5.0	0.0684	–	–	–
3	6.5	0.86892	–	–	–
4	8.0	1.06944	–	–	–
5	9.4	1.256592	5	15.0	2.0052
6	10.7	1.430376	6	17.4	2.326032
7	11.8	1.577424	7	19.8	2.646364
8	12.8	1.711104	8	22.2	2.967696
9	13.7	1.831416	9	24.6	3.288528
10	14.6	1.951728	10	27.0	3.60936
11	15.4	2.058672	11	27.8	3.716304
12	16.0	2.13888	12	28.6	3.823248
13	16.5	2.20572	13	29.4	3.930192
14	17.0	2.27256	14	30.2	4.037136
15	17.5	2.3394	15	31.0	4.14408
16	18.0	2.90624	16	31.8	4.241024
17	18.4	2.459712	17	32.6	4.357968
18	18.8	2.513184	18	33.4	4.464912
19	19.2	2.566656	19	34.2	4.571856
20	19.6	2.620128	20	35.0	4.6788
25	21.5	2.87412	25	38.0	5.07984
30	23.3	3.114744	30	42.0	5.61356
35	24.9	3.328632	35	44.0	5.88192
40	26.3	3.515784	40	46.0	6.14928
45	27.7	3.702936	45	48.0	6.41664
50	29.1	3.890088	50	50.0	6.684
60	32.0	4.27776	60	54.0	7.21872
70	35.0	4.6788	70	58.0	7.75344
80	38.0	5.07984	80	61.2	8.181216
90	41.0	5.48088	90	64.3	8.595624
100	43.5	5.81508	100	67.5	9.0234
120	48.0	6.41664	120	73.0	9.75864
140	52.5	7.0182	140	77.0	10.29336
160	57.0	7.61976	160	81.0	10.82808
180	61.0	8.15448	180	85.5	11.42964

Source: http://publicecodes.cyberregs.com/st/sc/b9v07/st_sc_st_b9v07_appe_par013.htm

Appendix A6

Manning's n for closed conduits flowing partly full (Chow, 1959)

Type of conduit and description	*Minimum*	*Normal*	*Maximum*
1 Brass, smooth	0.009	0.010	0.013
2 Steel			
Lockbar and welded	0.010	0.012	0.014
Riveted and spiral	0.013	0.016	0.017
3 Cast iron			
Coated	0.010	0.013	0.014
Uncoated	0.011	0.014	0.016
4 Wrought ron			
Black	0.012	0.014	0.015
Galvanized	0.013	0.016	0.017
5 Corrugated metal			
Subdrain	0.017	0.019	0.021
Stormdrain	0.021	0.024	0.030
6 Cement			
Neat surface	0.010	0.011	0.013
Mortar	0.011	0.013	0.015
7 Concrete			
Culvert, straight and free of debris	0.010	0.011	0.013
Culvert with bends, connections and some debris	0.011	0.013	0.014
Finished	0.011	0.012	0.014
Sewer with manholes, inlet etc. straight	0.013	0.015	0.017
Unfinished, steel form	0.012	0.013	0.014
Unfinished, smooth wood form	0.012	0.014	0.016
Unfinished, rough wood form	0.015	0.017	0.020
8 Wood			
Stave	0.010	0.012	0.014
Laminated, treated	0.015	0.017	0.020
9 Clay			
Common drainage tile	0.011	0.013	0.017
Vitrified sewer	0.011	0.014	0.017
Vitrified sewer with manholes, inlet etc.	0.013	0.015	0.017
Vitrified subdrain with open joint	0.014	0.016	0.018
10 Brickwork			
Glazed	0.011	0.013	0.015
Lined with cement mortar	0.012	0.015	0.017
Sanitary sewers coated with sewage slime with bends and connections	0.012	0.013	0.016
Paved invert, sewer, smooth bottom	0.016	0.019	0.020
Rubble masonry, cemented	0.018	0.025	0.030

Source: www.fsl.orst.edu/geowater/FX3/help/8_Hydraulic_Reference/Mannings_n_Tables.htm

Appendix A7

Hazen-Williams coefficients

Material	*Hazen-Williams coefficient (c)*	*Material*	*Hazen-Williams coefficient (c)*
ABS – Acrylonite butadiene styrene	130	Ductile iron pipe (DIP)	140
Aluminium	130–150	Ductile iron, cement lined	120
Asbestos cement	140	Fibre	140
Asphalt lining	130–140	Fibreglass pipe – FRP	150
Brass	130–140	Galvanized iron	120
Brick sewer	90–100	Lead	130–140
Cast iron – new unlined (CIP)	130	Metal pipes – very to extremely smooth	130–140
Cast iron 10 years old	107–113	Plastic	130–150
Cast iron 20 years old	89–100	Polyethylene, PE, PEH	140
Cast iron 30 years old	75–90	Polyvinyl chloride, PVC, CPVC	150
Cast iron 40 years old	64–83	Smooth pipes	140
Cast iron, asphalt coated	100	Steel new unlined	140–150
Cast iron, cement lined	140	Steel, corrugated	60
Cast iron, bituminous lined	140	Steel, welded and seamless	100
Cast iron, sea-coated	120	Steel, interior riveted, no projecting rivets	110
Cast iron, wrought plain	100	Steel, projecting girth and horizontal rivets	100
Cement lining	130–140	Steel, vitrified, spiral-riveted	90–110
Concrete	100–140	Steel, welded and seamless	100
Concrete lined, steel forms	140	Tin	130
Concrete lined, wooden forms	120	Vitrified clay	110
Concrete, old	100–110	Wrought iron, plain	100
Copper	130–140	Wooden or masonry pipe – smooth	120
Corrugated metal	60	Wood stave	110–120

Source: www.engineeringtoolbox.com/hazen-williams-coefficients-d_798.html

Appendix A8

Pump dimension with respect to size

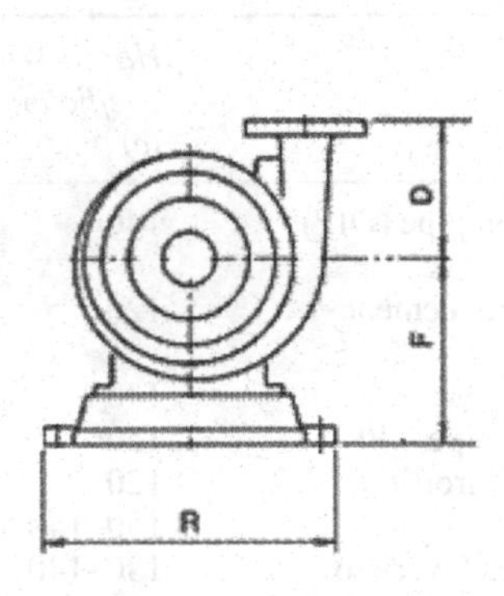

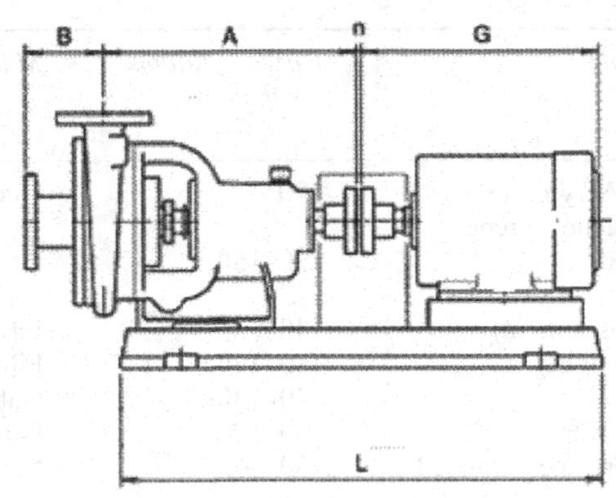

Dimensions (mm)													
Pump					Motor		Base plate					Total	
Bore suction/ discharge	A	B	D	F	Output power (kW)	Output power (BHP)	G	L	R	n		B+A+n+G Length	F+D Height
40	337	110	120	190	3.7	4.96	376	630	316	3		826	310
40	340	110	140	205	5.5	7.38	445	780	370	3		898	345
40	375	115	160	242	11	14.75	585	930	425	3		1078	402
50	340	115	140	205	7.5	10.06	445	780	370	3		903	345
50	375	115	170	230	5.5	7.38	445	780	370	3		938	400
50	433	120	200	250	7.5	10.06	483	880	396	3		1039	450
65	345	120	150	185	1.5	2.01	314	570	286	3		782	335
65	380	120	170	220	3.7	4.96	376	730	336	3		879	390
65	433	120	180	250	5.5	7.38	445	780	370	3		1001	430
80	380	125	170	242	18.5	24.81	630	930	425	3		1138	412
80	438	125	180	272	22	29.50	656	1030	470	3		1222	452
80	495	125	200	295	30	40.23	694	1030	470	3		1317	495
80	490	135	210	285	11	14.75	586	930	425	3		1214	495
100	438	135	180	272	30	40.23	694	1030	470	3		1270	452
100	438	135	200	250	7.5	10.06	483	880	396	3		1060	450
100	499	135	200	305	45	60.35	762	1130	571	4		1399	505
100	495	135	220	295	15	20.12	630	1030	470	3		1263	515
100	495	135	230	295	18.5	24.81	656	1030	470	3		1289	525
100	563	180	250	360	22	29.50	656	1380	571	3		1402	610

Source: www.aquadevice.com/english/03pump_uhn33.htm

Appendix A9

Conversion factors, mixed units

Length (L)

mile	*yard*	*ft*	*in*	*m*	*cm*
5.68×10^{-4}	1	3	36	0.9144	91.44
1.894×10^{-4}	0.333	1	12	0.3048	30.48
1.578×10^{-5}	0.028	0.083	1	0.0254	2.54
6.214×10^{-4}	1.094	3.281	39.37	1	100

Area (A)

mile²	*acre*	*yard²*	*ft²*	*in²*	*m²*
1	640	3.098×10^{6}	2.788×10^{7}	4.014×10^{9}	2.59×10^{6}
1.563×10^{-3}	1	4840	43,560	6.27×10^{6}	4047
3.228×10^{-7}	2.066×10^{-4}	1	9	1296	0.836
3.587×10^{-8}	2.3×10^{-5}	0.111	1	144	0.093
2.491×10^{-10}	1.59×10^{-7}	7.716×10^{-4}	6.944×10^{-3}	1	6.452×10^{-4}
3.861×10^{-7}	2.5×10^{-4}	1.196	10.764	1550	1

Volume (V)

acre-ft	*US val.*	*ft³*	*in.³*	*L*	*m³*	*cm³*
1	325,851	43,560	75.3×10^{-6}	1.23×10^{6}	1230	1.23×10^{9}
3.07×10^{-6}	1	0.134	231.6	3.875	3.875×10^{-3}	3875
2.3×10^{-5}	7.481	1	1728	28.317	0.028	28,317
1.33×10^{-8}	4.329×10^{-3}	5.787×10^{-4}	1	0.016	1.639×10^{-5}	16.39
8.1×10^{-7}	0.264	0.035	61.02	1	1×10^{-3}	1000
8.13×10^{-4}	264.2	35.31	6.10×10^{4}	1000	1	10^{6}

Time (T)

year	*month*	*day*	*hour*	*minute*	*second*
1	12	365	8760	525,600	3.1536×10^{7}

Source: Quasim, 2002

Velocity (L/T)				
ft/s	*ft/min*	*m/s*	*m/min*	*cm/s*
1	60	0.3048	18.29	30.48
0.017	1	5.08×10^{-3}	0.3048	0.5080
3.281	196.8	1	60	100
0.055	3.28	0.017	1	1.70
0.032	1.969	0.01	0.588	1

Discharge (L^3/T)					
mgd	*gpm*	*ft^3/s*	*ft^3/min*	*L/s*	*m^3/d*
1	694.4	1.547	92.82	43.75	3.78×10^3
1.44×10^{-3}	1	2.228×10^{-3}	0.134	0.063	5.45
0.646	448.9	1	60	28.32	2447
0.011	7.481	0.017	1	0.472	40.78
0.023	15.85	0.035	2.119	1	86.41
2.64×10^{-4}	0.183	4.09×10^{-4}	0.025	0.012	1

Mass (M)					
ton	*lb_m*	*grain*	*ounce (oz)*	*kg*	*g*
1	2000	1.4×10^7	32,000	907.2	907,185
0.0005	1	7000	16	0.454	454
7.14×10^{-8}	1.429×10^{-4}	1	2.29×10^{-3}	6.48×10^{-5}	0.065
3.125×10^{-5}	0.0625	437.6	1	0.028	28.35
1.10×10^{-3}	2.205	1.54×10^4	35.27	1	1000
1.10×10^{-6}	2.20×10^{-3}	15.43	0.035	10^{-3}	1

Temperature (T)			
°F	*°C*	*°K*	*°R*
°F	**5 / 9 (°F – 32)**	**5 / 9°F + 255.38**	**°F + 459.69**
9/5°C + 32	°C	°C + 273.16	9/5°C + 491.69
9/5°K – 459.69	°K – 273.16	°K	9/5°K
°R – 459.69	5 / 9°R – 273.16	5 / 9°R	°R

Density (M/L^3)				
lb/ft^3	*lb/gal (U.S)*	*kg/m^3*	*kg/L*	*g/cm^3*
1	0.1337	16.019	0.01602	0.01602
7.48	1	119.8	0.1198	0.1198
0.0624	8.345×10^{-3}	1	0.001	0.001
62.43	8.345	1000	1	1

Pressure (F/L²)

lb/in²	*ft water*	*in Hg*	*atm*	*mm Hg*	*kg/cm²*	*N/m²*
1	2.307	2.036	0.068	51.71	0.0703	6895
0.4335	1	0.8825	0.0295	22.41	0.0305	2989
0.4912	1.133	1	0.033	25.40	0.035	3386
14.70	33.93	29.92	1	760	1.033	1.013×10^5
0.019	0.045	0.039	1.30×10^{-3}	1	1.36×10^{-3}	133.3
14.23	32.78	28.96	0.968	744.7	1	98,070
1.45×10^{-4}	3.35×10^{-4}	2.96×10^{-4}	9.87×10^{-6}	7.50×10^{-3}	1.02×10^{-5}	1

Viscosity

Dynamic absolute viscosity (μ)

cp	*Ibf.ft²*	*Ib_m/ft.s*	*g/cms*	*N.s/m²*	*kg/m.s*	*dp*
1	2.09×10^{-5}	6.72×10^{-4}	0.01	1×10^{-3}	1×10^{-3}	1×10^{-3}
4.78×10^4	1	32.15	478.5	47.85	47.85	47.85
1488	0.031	1	14.88	1.488	1.488	1.488
100	2.09×10^{-3}	0.672	1	0.10	0.10	0.10
1000	0.021	0.672	10	1	1	1

Kinemate viscosity (V)

centistoke	*ft²/s*	*cm²/s*	*m²/s*	*myriastoke*
1	1.076×10^{-5}	0.01	1.0×10^{-6}	1.0×10^{-6}
9.29×10^4	1	929.4	0.093	0.093
100	1.076×10^{-3}	1	1.0×10^{-4}	1.0×10^{-4}
10^6	10.76	10^4	1	1

Force (F)

Ib_f	*N*	*dyne*
1	4.448	4.448×10^5
0.225	1	10^5
2.25×10^{-6}	10^{-5}	1

Energy (E)

kW.h	*hp.h*	*Btu*	*J*	*kj*	*calories*
1	1.341	3412	3.6×10^6	3600	8.6×10^5
0.7457	1	2545	2.684×10^6	2685	6.4×10^5
2.930×10^{-6}	3.929×10^{-4}	1	1055	1.055	252
2.778×10^{-7}	3.72×10^{-7}	9.48×10^{-4}	1	0.001	0.239
2.778×10^{-4}	3.72×10^{-4}	0.948	1000	1	239
1.16×10^{-6}	1.56×10^{-6}	3.97×10^{-3}	4.186	4.18×10^{-3}	1

Power (P)					
kW	*Btu/min*	*hp*	*ft.b/s*	*kg.m/s*	*cal/min*
1	56.89	1.341	737.6	102	14,330
0.018	1	0.024	12.97	1.793	252
0.746	42.44	1	550	76.09	10,690
1.35×10^{-3}	0.077	1.82×10^{-3}	1	0.138	19.43
9.76×10^{-3}	0.558	0.013	7.233	1	137.6
6.98×10^{-5}	3.97×10^{-3}	9.355×10^{-5}	0.0514	7.12×10^{-3}	1

Source: Quasim, 2002

Index

Note: Page numbers in *italic* indicate a figure and page numbers in **bold** indicate a table on the corresponding page.

Printed in the United States
by Baker & Taylor Publisher Services